Scientific Fundamentals of Robotics 1

M. Vukobratović
V. Potkonjak

Dynamics of Manipulation Robots

Theory and Application

With 149 Figures

Springer-Verlag Berlin Heidelberg NewYork 1982

D. Sc., Ph. D. MIOMIR VUKOBRATOVIC, corr. member of
Serbian Academy of Sciences and Arts
Institute »Mihailo Pupin«, Beograd,
Volgina 15, POB 15, Yugoslavia

Ph. D. VELJKO POTKONJAK, assistent professor,
Electrical Engineering Faculty of Belgrade University

ISBN 3-540-11628-1 Springer-Verlag Berlin Heidelberg NewYork
ISBN 0-387-11628-1 Springer-Verlag NewYork Heidelberg Berlin

Library of Congress Cataloging in Publication Data
Vukobratović, Miomir.
Dynamics of manipulation robots.
(Scientific fundamentals of robotics ; 1)
(Communications and control engineering series)
Bibliography: p.
Includes index.
1. Automata. 2. Manipulators (Mechanism)
I. Potkonjak, Veljko, 1931–. II. Title.
III. Series. IV. Series: Communications and control engineering series.
TJ211.V838 629.8'92 82-5602 AACR2

Offsetprinting: fotokop wilhelm weihert KG, Darmstadt. – Bookbinding: Konrad Triltsch, Würzburg.
2061/3020-54321

About the Series:
»Scientific Fundamentals of Robotics«

The age of robotics is the present age. The study of robotics requires different kinds of knowledge multidisciplinary in nature, which go together to make robotics a specific scientific discipline. In particular, manipulator and robot systems possess several specific qualities in both a mechanical and a control sense. In the mechanical sense, a feature specific to manipulation robots is that all the degrees of freedom are "active", i.e., powered by their own actuators, in contrast to conventional mechanisms in which motion is produced primarily by the so-called kinematic degrees of freedom. Another specific quality of such mechanisms is their variable structure, ranging from open to closed configurations, from one to some other kind of boundary conditions. A further feature specific of spatial mechanisms is redundancy reflected in an excess of the degrees of freedom for producing certain functional movements of robots and manipulators.

From a control viewpoint, robot and manipulator systems represent redundant, multivariable, essentially nonlinear automatic control systems. A manipulation robot is also an example of a dynamically coupled system, and the control task itself is a dynamic task.

The basic motivation for establishing the conception of this series has consisted in an intention to clearly define the role of dynamics and dynamic control of this class of system. The associates who have been engaged in the work on this series have primarily based their contributions on the development of mathematical models of dynamics of these mechanisms. They have thus created a solid background for systematic studies of robot and manipulator dynamics as well as for the synthesis of optimal characteristics of these mechanisms from the point of view of their dynamic performances. Having in mind the characteristics of robotic systems, the results concerning the problems of control of manipulation robots represent one of the central contributions of this series. In trying to bridge, or at least reduce, the gap existing between theoretical robotics and its practical application, considerable efforts have been made towards synthesizing such algorithms as would be suitable for implementation and, at the same time, base them on sufficiently accurate models of system dynamics.

The main idea underlying the conception of the series will be realized: to begin with books which should provide a broad education for engineers and "create" specialists in robotics and reach texts which open up various possibilities for the practical design of manipulation mechanisms and the synthesis of control algorithms based on dynamic models, by applying today's microelectronics and computer technologies.

Those who have initiated the publication of this series believe they will thus create a sound background for systematic work in the research and application of robotics in a wider sense.

Belgrade, Yugoslavia, February 1982 M.Vukobratović

Preface

This monograph represents the first book of the series entitled "SCIENTIFIC FUNDAMENTALS OF ROBOTICS". The aim of this monograph is to approach the dynamics of active mechanisms from the standpoint of its application to the synthesis of complex motion and computer-aided design of manipulation mechanisms with some optimal performances. The rapid development of a new class of mechanisms, which may be referred to as active mechanisms, contributed to their application in various environments (from underwater to cosmic). Because of some specific features, these mechanisms require very careful description, both in a mechanical sense (kinematic and dynamic) and in the synthesis of algorithms for precise tracking of the above motion under insufficiently defined operating conditions. Having also in mind the need for a very fast (even real-time) calculation of system dynamics and for eliminating, in principle, the errors made when forming mathematical models "by hand" this monograph will primarily present methods for automatic formulation of dynamic equations of motion of active spatial mechanisms. Apart from these computer-oriented methods, mention will be made of all those methods which have preceded the computer-oriented procedures, predominantly developed for different problems of rigid body dynamics.

If we wish to systematically establish the origins of the scientific discipline, which could be called robot dynamics, we must recall some groups and individuals, who, by solving actual problems in the synthesis and control of artificial motion, have contributed to a gradual formation of this discipline. Thus, Vukobratović and Juričić started (in 1968) research into the synthesis of artificial biped gaits and the dynamics of anthropomorphic mechanisms. Practically at the same time R.Mc Ghee and A.A.Frank (USA) studied (in 1970) the dynamics of the four-legged machine ("The Californian horse"). Slightly later, D.E.Okhotsimski (USSR, in 1971) and associates started an extensive and original investigation of the dynamics and synthesis of control algorithms with artificial intelligence elements of multi-legged locomotion systems. A few years later (in 1974) V.V.Beletskii (USSR) and his associates, as well as V.B.Larin (USSR), systematically continued the activity in the biped locomotion field. While Beletskii practically adopted the semi-inverse method by Vukobratović and later enlarged it to the completely inverse procedure of anthropomorphic gait synthesis, Larin, by adopting only the global control strategy of the Belgrade school, tried to apply the method of gait synthesis optimization to relatively simple anthropomorphic models.

In parallel with these investigations, the study of the dynamics and control of active spatial mechanisms was developed for application with manipulation robots. The first person to be mentioned here is the founder of modern mechanism theory I.I.Artobolevskii (USSR). Later, Yu.A.Stepanenko (USSR) was among the first to start working on algorithms for computer formulation of dynamic equations of open kinematic chains. The vital contribution by E.P.Popov (USSR) and his associates N.A.Lakota, A.F.Vereschagin, V.S.Kuleshov, A.S.Yuschenko and V.S.Medvedov should also be mentioned, as well as V.S.Yastrebov, and F.M.Kulakov in the field of the dynamics and control of manipulation robots.

R. Paul (USA) and his associates J.Luh, M.Walker and others should also be specially mentioned here for his work on the basis of the interactive procedure of forming mathematical models of manipulation systems. Based on mostly analytical forms of the equations some contributions to the manipulator dynamics analysis has been made by M.Renaud (France). J. Hollerbach and W.Silver (USA) have studied the manipulator dynamics by accepting recurrence relations suitable for computer forming of the mathematical models of manipulation systems.

Almost simultaneously with the discovery in the USSR of new results in the field of computer methods for constructing the mathematical models of spatial mechanisms, a systematic activity was initiated at the "Mihailo Pupin" Institute in Beograd (Yugoslavia) in the field of automatic (computer) forming of dynamic models of anthropomorphic systems. These algorithms developed later into general procedures for computer formulation of mathematical models of arbitrarily complex spatial mechanisms. Thus, apart from the previously elaborated method of forming mathematical models on the basis of general theorems of mechanics, the authors of this monograph elaborated general procedures for describing the dynamics of kinematic chains using Lagrange´s second-order equations and the Gibbs-Appel equations. Such development of the procedures for forming mathematical models of active mechanism motion while solving actual tasks of the synthesis and control of various movements applied in robotics has resulted in a specific unconventional system dynamics related to the forming of functional motion. Hence, such a dynamics, connected with the synthesis of functional motion in robotics, can be called functional dynamics. As the functional dynamics also assumes the calculation of driving forces, it may be concluded that in such problems the notion of "pure" dynamics overlaps with dynamic control, notably if one has the two-stage concept of suboptimal control synthesis in mind where the first stage is represented by the functional (nominal) dynamics synthesized under unperturbed conditions. The monograph contains five chapters. In Chapter 1 the various cases of active spatial mechanism are introduced and explained, their classification is presented, and we introduce the general postulates in the study of the dynamics of functional motion with manipulation and robots in general. The analytical methods of forming the dynamic equations of active, spatial, joint-connected rigid bodies are briefly stated.

Chapter 2 presents the computer methods for forming mathematical models of the dynamics of active spatial mechanisms on the basis of the general theorems method (the first chronologically), then those based on Lagrange´s second-order equations and those based on Appel´s equations and Gibb´s function of acceleration. This Chapter represents the central part of the monograph and for the first time gives a complete insight into the existing methods for automatically forming mathematical models of the dynamics of open kinematic chains of arbitrary spatial configuration. Chapter 3 presents applications of the methods in Chapter 2 to the synthesis of functional movements in the case of typical manipulation tasks. Various functional blocks are elaborated which solve the problems of the various ways of setting the manipulation tasks when applying manipulation mechanisms with different numbers of degrees of freedom. In the same chapter illustrative examples of the synthesis of functional movements of various conifiguration manipulation robots are given. Chapters 4 and 5 present the application of the dynamic models in the tasks of manipulation mechanism synthesis. While in Chapter 4 the results from the field of elastic manipulator dynamics are given, Chapter 5 presents a computer-oriented method for the design of manipulation robots based on their desired dynamic performances and optimal characteristics, taking into account various criteria and imposed constraints.

In this monograph the material from the field of forming the complete mathematical models of spatial active mechanisms, used mainly with manipulation robots, is presented for the first time. It should be emphasized that the same models can also be applied, along with setting specific conditions, in the case of locomotion systems synthesis. This monograph should provide in the first place a basis for serious study of robot dynamics by researchers engaged in applied robotics, as well as postgraduate students.

Finally, the authors have the pleasent duty of acknowledging, on this occasion, the activities of prof. D.Juričić, who, with the first author of this monograph, published thirteen years ago practically the first paper in applied robot dynamics, when the idea for the two-stage control synthesis of robots*) was also presented for the first time. The authors also express their gratitude to Dr D.Hristić, who contributed essentially to the work and development of the ideas in the synthesis of locomotion and manipulation mechanisms.

Apart from the results presented in the book, the results of Dr D.Stokić in the dynamics of the assembly process should be mentioned. These have contributed to the formation of an efficient algorithm for the dynamic control of the last phase of the above manipulation task. We also wish to take this opportunity of mentioning Miss Dr V.Cvetković, who obtained useful results in the field of calculating the manipulator dynamics in real time and forming the approximative models of manipulation systems. Mention should also be made of N.Kirćanski, M.Sc., Mrs M.Kirćanski, M.Sc., I.Nikolić, M.Sc. and B.Borovac, M.Sc. who, by their initial results in the field of approximative and exact dynamics, and the procedures of computer linearization of the models and analysis of the dynamic influence of manipulation robots actuators, contributed to extensive research in the field of robot dynamics and active spatial mechanisms. We should also mention Dr D.Surla who obtained new results in the dynamics of biped gait. The results mentioned will find also their place in the following books of this series.

The authors are grateful to Dr D.Hristić and Miss G.Aleksić for their help in preparing English version of this book. Our thanks also go to Dr T.Flannagan for improving the translation. Finally, our special appreciation goes to Miss V.Ćosić for her careful and excellent typing of the whole text.

Belgrade, Yugoslavia, February 1982 The Authors

---

*) Vukobratović M., Juričić D., "Contribution to the Synthesis of Biped Gait" IEEE Trans. on Biomed. Engn., BME, Vol. 16, Jan. 1969.

Contents

Chapter 1
General Remarks about Robot and Manipulator Dynamics

1.1. Introduction

During the last few years a new technical discipline has arisen in applied mechanics and the control of technical systems. The field in question may be called ROBOTICS. There are numerous indications of it being well-established. We mention only a few. On the one hand, for example, we have witnessed the rapid development of manipulation systems of different generations intended for a wide range of industrial applications. On the other hand, robot and manipulator theory has developed in many research centers without exercising any significant influence on the designers and manufacturers of actual manipulation devices or robots in general. Thus there has existed a certain discrepency between the existing practice and the often too academically oriented theory.

Recently, due to the needs for the best possible industrial manipulators and the best possible control algorithms, research has been more directed towards profitable application of theoretical results in the greatest possible measure to the development of better manipulators.

The development of industrial manipulators and robots in general, which in the course of time, has found wider and wider applications has demanded the development of new theoretical methods. When the expression "active mechanisms"[*)] is used in robotics, what is meant is robot mechanisms. Wider application of such mechanisms has uncovered a vacuum in theoretical methods.

The first work in the dynamics of spatial mechanisms was published by N.G. Bruyevich[**)] as far back as 1937. His paper applied the kineto-

---

*) Mechanisms, possessing in principle separate drive for each degree of freedom.

**) Reports VVA Zhukovskiy, Vol. 36, 37, 1937.

static method to the dynamic analysis of spatial lever mechanisms. Since electronic computers were not then available and since the analytical calculations are extremely involved for the greater number of mechanism members, the paper was of theoretical significance. For, many years afterwards the dynamics of spatial mechanisms was not investigated since, amongst other things, efficient methods of kinematic analysis had not been developed.

Later much more attention was paid to the study of spatial mechanisms, in particular of their dynamics. Several methods arose, in which attempts were made to find analytical procedures for deriving the mechanism dynamic equations. The methods were directed towards deriving the models "by hand", although some of them can be programmed for computer work. A common characteristic of several of these methods is that they were originally developed for various mechanisms in fields other than robotics. Later their authors strived to apply the same methods to the analysis of robot mechanisms.

A great advance in the field of the dynamics of active mechanisms was the appearance of the so-called automatic methods for the formation and solution of mathematical models of active mechanisms (the methods are described in detail in Ch. 2 of this monograph). The main feature of these methods is that the major part of forming and solving the mathematical model (of the dynamic equations) is done by computer. Of course, that was possible only when modern computers came to be used in scientific applications. Several things were responsible for the appearance of these automatic methods. First, formulating the dynamic equations "by hand" is very difficult even for a mechanism with a few degrees of freedom (d.o.f.). The great likelihood of making errors in the course of such a lengthy task should not be left unmentioned. Even the model obtained is so clumsy, that it is rather useless for practical applications. The solution of such a model "by hand" is impossible. Its programming for computer solution is in any case very complex. Finally, it should be stressed that it is frequently necessary to analyze a greater number of various configurations of robots, which, in the manner described is impossible. Hence the idea naturally emerged of transferring to the computer the whole procedure from the formulation of the model itself to its solution. Thus, the task of the researcher would only be to prepare the input data about the mechanism configuration, because the automatic methods work for arbitrary configurations. As computer output, solutions are obtained for the direct or inverse problem of dynamics*), depending on the task set. In order that these methods could demonstrate their full efficiency, the way of viewing mechanism kinematics and dynamics had to be changed somewhat, i.e. one should be primarily concerned with deriving the recursive kinematic and dynamic relations. Such recursions, written by hand, do not have a compact model form but they are indispensable to an efficient numerical computer calculation.

Let us now consider in more detail the question of functional dynamics. It will be seen that the term completely suits the dynamics of robots and manipulators. As a rule, the methods for the dynamic analysis of active mechanisms use generalized coordinates. From a purely theoretical standpoint, the dynamics problem is solved if, for known driving forces and torques, the corresponding motion, expressed in generalized coordinates, is obtained or vice versa. However, in practice such a solution is insufficient, because one needs to consider the so-called functional robot motion. This is a motion, satisfying cartain practical demands. Let us consider the case of the industrial manipulator. The manipulation task, i.e. its functional motion, can be prescribed in several ways. For instance, the law governing the manipulator tip motion can be given as can that of the gripper orientation in the space. Thus, we are only interested in such functional motions (from the set of all possible mechanism motions). It is therefore necessary to obtain the drives producing these functional motions. Hence we often speak about functional dynamics.

Let us now analyze the connection between the active mechanism dynamics and the control of such mechanisms. When regarding a functional movement (for instance a manipulation task) it can be seen that the essential problem lies in determining the driving torques and forces of the actuators which will produce the desired mechanism (or manipulator) motion. Now it is clear that the considerations of functional dynamics are closely connected with control and conversely; so the term dynamic control can be meaningfully introduced. Anyhow, the connection between dynamics and control can be regarded in several ways. First, dynamic analysis of the functional movement, notably the simulation algorithms, makes possible the calculation of nominal dynamics. On that basis, con-

---

*) Direct problem is in obtaining the driving forces and torques which will realize the prescribed mechanism (robot) motion, and the inverse problem is in obtaining the moiton when the drives are known.

trol synthesis is carried out for unperturbed working states*). This applies mainly to industrial robotics where there are firmly defined tasks which are performed in known, usually invariant working conditions. The question of control is different if a certain degree of uncertainty exists; e.g. manipulation in underwater explorations and the like. In this case the approach is different. However, about the connection between dynamics and control, more will be said later in this introduction.

Here we pose a basic question: what are the purposes and aims of studying active mechanism dynamics? One is connected with control, which has already been mentioned (and will be discussed more full later), and a second is the development of procedures for optimal design and a certain automation of the process of designing industrial manipulators. This second aim only emerged recently and more and more attention is being paid to it. Let us say a few words about this problem which features prominently in this monograph.

In practice up to now, in the design of the mechanical part of the robot the choice of the kinematic scheme, the choice of different parameters (dimensions, masses etc.), as well the choice of the actuator units, was a subject of free speculation, frequently based on experience but lacking any system or method. Hence, with the existing manipulators, many parameters, and often the motors, were overpowered. Such a device is not at all optimal from the point of view of energy consumption or operating speed. The need therefore arose to develop certain criteria and procedures for a systematic choice of manipulator configuration. The main aim of this monograph is to develope such criteria and procedures.

We now discuss one more aspect of the automatic formulation of the robot mathematical model. This is the question of real time. Contemporary computers are at the frontiers of formulating the mathematical model, i.e. computing the dynamics of the manipulator in real time. But the question of purposefulness of attaining real-time computation is posed.

---

*) If the two-stage control concept of robots and manipulators is adopted [1, 2, 3, 4], then of the first, so-called stage of unperturbed regimes the very calculations of the dynamic nominal regimes (the programmed trajectories) is performed.

If to this question we wanted to give a sufficiently simple and, at the same time, sufficiently exact answer, it could be said that automatic (computer) formation of mechanism differential motion equations would be a sufficient result, and the attainment of real-time computation is more of academic significance. This is surely the case when the application of the dynamics to optimal design is in question because in this case computing time is not of prime importance. Let us see how things are when the dynamics is used in connection with control. As already stated the questions of dynamics and control cannot be altogether separated. It was shown that it is more correct to speak about functional dynamics, or dynamic control, i.e., a control based on detailed knowledge of the system dynamic characteristics. Above all, this is a control which, by knowing the system completely and its energy requirements with the scope of the task defined, is based on the driving and control components having no unnecessary power reserve. Only now can the second question be posed, not about the justification of dynamic calculation, but about the justification of the synthesis of control algorithms in real time. Surely in most real applications of active mechanisms this feature is not necessary. For all manipulation systems predetermined to work permanently or during certain time periods, under the same working conditions, the same environments, or on the same programmed tasks, the control algorithms do not change during the process. This refers mainly to industrial robotics.

However, things are somewhat different for systems working under unsufficiently defined working conditions, i.e., in environments with a degree of uncertainty. We mention only one class of task, referring to manipulators for underwater applications. Calculation of the dynamic and control parameters in this case should be understood as being necessary for calculating the programmed kinematics (depending on the object in question) within the limits of the kinematic and geometrical capabilities of the manipulator, i.e., of its mechanism. And depending on other variable conditions such as weight (and within the capability limits of the actuators), it is also necessary to calculate the required driving forces and, in that connection, to select the corresponding gains in order to ensure and satisfy good tracking quality of the trajectories. Reasons for partial calculation of the dynamics of manipulation systems in real time become more convincing if for instance we consider the case of the assembly tasks of mechanical elements in various working environments, including underwater and cosmic space.

Redundancy is another frequent specification of active mechanisms in robotics. In this case the mechanism has more d.o.f. than is needed for performing the task in question. On the one hand, this permits a greater mechanism flexibility in task performance, while on the other hand, it complicates the control system by introducing optimizing procedures for solving the problem of the system redundancy. This surplus of d.o.f. can also be used to satisfy special additional requests.

1.2. Classification of Active Mechanisms in Robotics and Some of Their Specifications

From the point of view of mechanism theory, active mechanism in robotics are complex kinematic chains of variable structure, having a great number of members, some of which can be of variable length, with controlled degrees of freedom.

From the point of view of control theory they are complex, nonlinear, multivariable dynamic systems. Active mechanism can be divided according to the number of kinematic chains into:

- simple (consisting of a single kinematic chain)
- complex (comprising a number of simple shains)

According to their form, simple kinematic chains may be open or closed. Complex chains may be classified as:

- branched (comprising only simple open chains)
- combined (comprising both open and closed chains).

Depending on the kinematic constraints imposed on their end members, active mechanisms may be divided into:

- free or open
- connected or closed (connected by kinematic pairs to the fixed base)

Members of active mechanisms are interconnected by means of kinematic pairs. There is no difference between kinematic pairs of active and "classical" spatial mechanisms. Execution of kinematic pairs of both mechanisms classes are practically identical, except for the differ-

class of task	nr. of links	nr. of d.o.f.	TYPES OF PAIRS: I			II			III		
I	1	5	nr. of movem.	rot.	lin.						
			allowed	3	2						
			restricted 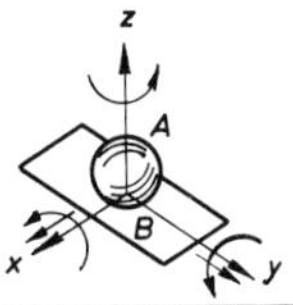	0	1						
II	2	4	nr. of movem.	rot.	lin.	nr. of movem.	rot.	lin.			
			allowed	3	1	allowed	2	2			
			restricted 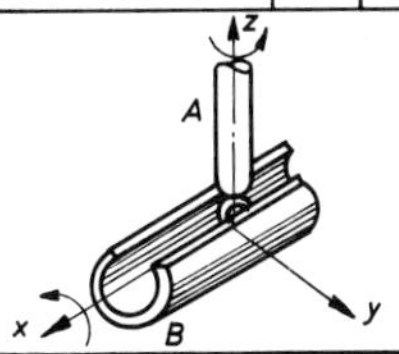	0	2	restricted 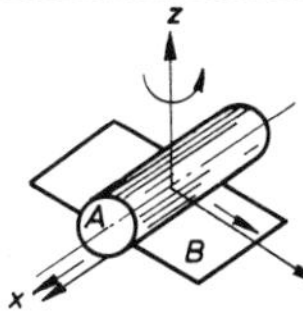	1	1			
III	3	3	nr. of movem.	rot.	lin.	nr. of movem.	rot.	lin.	nr. of movem.	rot.	lin.
			allowed	3	0	allowed	2	1	allowed	1	2
			restricted 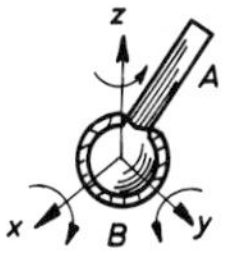	0	3	restricted 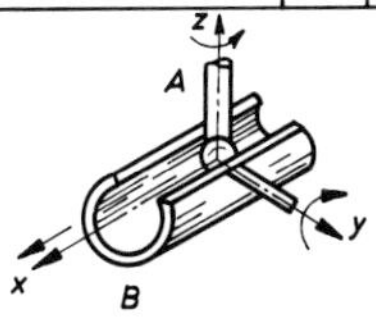	1	2	restricted 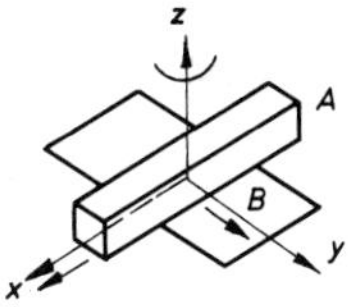	2	1
IV	4	2	nr. of movem.	rot.	lin.	nr. of movem.	rot.	lin.			
			allowed	2	0	allowed	1	1			
			restricted 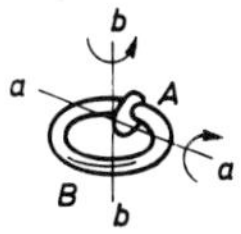	1	3	restricted 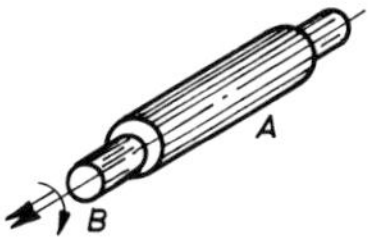	2	2			
V	5	1	nr. of movem.	rot.	lin.	nr. of movem.	rot.	lin.	nr. of movem.		
			allowed	1	0	allowed	0	1	allowed	0	0
			restricted 	2	3	restricted 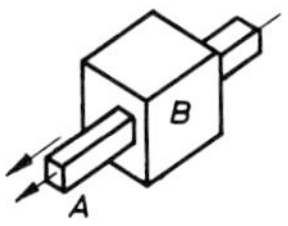	3	2	restricted 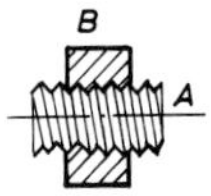	2	2

Fig. 1. Table of kinematic pairs

ences created by actuators mounted in the mechanism joints.

Kinematic pairs of various classes are presented in the table, Fig. 1, [5].

The class of a kinematic pair is determined by the number of constraining conditions on the connections concerning free relative motion of members. In the table kinematic pairs are arranged into five classes according to the number of the member relative motion, d.o.f. Kinematic pairs in the fifth class have one d.o.f. and the pairs in the first class have five d.o.f. of relative motion. Besides being partitioned into classes, kinematic pairs are divided into types, depending on the number of relative rotations within the scope of the total number of d.o.f. in the joint. Pairs of the first type allow the maximal number (3) of relative rotations, pairs of the second type two rotations, and pairs of the third type only one relative rotational motion. Besides the pairs, in which relative motions of members are mutually independent, there are pairs with interconnected motion. The simplest example is the screw-nut kinematic pair, in which the linear and rotational motions are linearly dependent; so this is a fifth class pair.

In the theory of machines and mechanisms, kinematic chains are calssified as simple or complex, complex chains being formed by several simple ones. Simple kinematic chains can be open or closed. In a closed chain, each member enters into two kinematic pairs, while in an open chain, the last member enters into one kinematic pair only. With complex kinematic chains, the individual members enter into three or more kinematic pairs. Here the notion and properties of open and closed kinematic chains should be examined more closely. In the literature on active mechanisms, neither the notion nor the conditions of the closed (open) state of the open and closed chain is discussed. The kinematic chain is closed when its terminal members are connected by means of kinematic pairs to one (or more) member(s), which can be: fixed (support), a member of another kinematic chain or a member of the initial chain. It should also be emphasized that in the course of working the active mechanism chains (either of manipulators or locomotion machines) change their configuration once or several times from open to closed or vice versa. The kinematic chain of the manipulator during its motion through the working space is open but during execution of the operation itself (e.g. insertion or screwing in) it becomes closed. The mechanism of the locomotion biped is open during the swing phase of the step (when one foot is not on the ground) but in the double support phase becomes closed (Fig. 2a and 2b). However, during the single support phase the anthropomorphic mechanism can also possess two configurations. The "foot" can rotate around its edges (Fig. 3a, b). The corresponding kinematic schemes are given in Fig. 3c, d. As can be seen, when the foot is supported alternately on one and then the other foot edge, the position of hinge "O" changes abruptly.

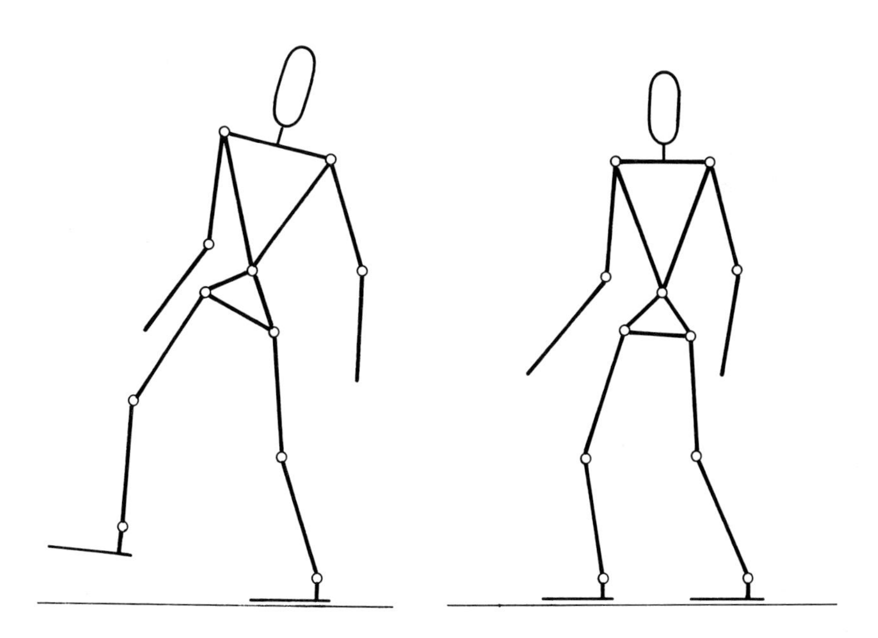

Fig. 2. Anthropomorphic locomotion mechanism

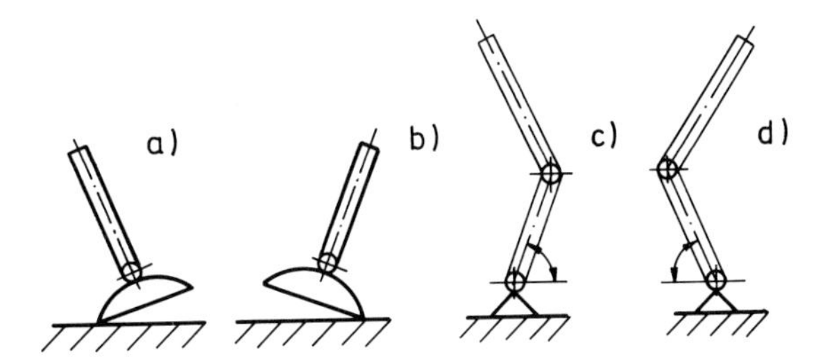

Fig. 3. Schematic of the uncontrollable d.o.f. of foot

In addition, joint O, because of constant changes in its position, cannot be equipped by a corresponding drive (actuator). On the other hand, the change of the coordinate q_o is exceptionally important because with greater values of q_o the system becomes statically unstable (it overturns). This feature of the anthropomorphic mechanism creates a special control problem because the uncontrollable d.o.f. must be controlled by means of the other d.o.f. Such an anthropomorphic mechanism also has the corresponding kinematic constraints at each joint, for the sake of imitating the human state. In addition, the mechanism of the locomotion biped (differing from the manipulator) is connected to the support surface by means of the frictional force only. Thus, the mechanism of an exoskeleton demonstrates a variable structure, the presence of an uncontrollable d.o.f., kinematic constraints and an essential influence of the frictional force. The manipulator kinematic scheme is somewhat different. Above all, problems of global system stability do not exist, unlike problems arising from the frictional force (except in some special working operations). Kinematic chains of manipulators are mainly simple. The kinematic chains of some industrial manipulator designs (telescopic manipulators) possess linear kinematic pairs. Such a connection between members permits an increase in the working space (the reach) of the manipulator because one member is inside the other. By means of corresponding drives one has been able to create linear motion of these members. It follows that some manipulators have, besides a variable structure, members of variable length (Fig. 4).

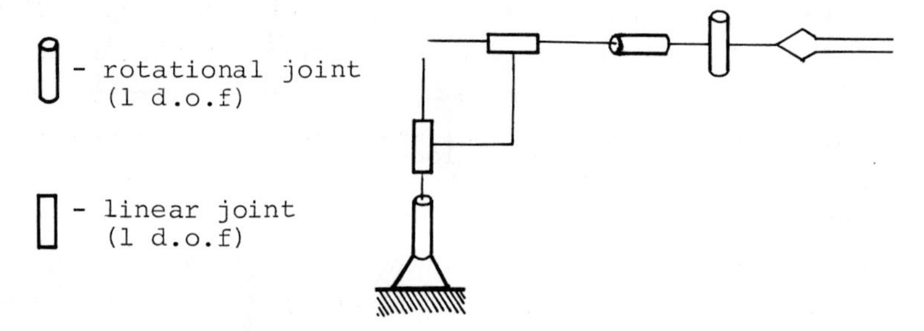

Fig. 4. One mechanism of "telescopis" manipulator

The variable structure of active mechanism chains presents an essential difference from the classical spatial mechanisms, the structure of which does not change during work. A second difference is in the number of d.o.f. With spatial mechanisms with driving member, number of d.o.f. is rarely greater than two and by synchronizing motion of the working members, the execution of the working operation is achieved in advance. The number of d.o.f. of active mechanisms is noticeably greater (up to ten and even more), so drives in the joints are indispensable. Only by the action of the torques and forces of these motors (during working operation) is the desired motion achieved. On the one hand this allows for an exceptional adaptibility of these mechanisms to the working environment and various tasks (which in the case of automation type mechanisms is impossible). On the other hand, it imposes exceptional difficulties in realizing control because some of the mechanism d.o.f. appear as redundant.

As already stated, robots represent active mechanisms of variable structure. For instance, one manipulator can, during its work, change the group which it would belong according to the given classification. We illustrate this change of structure by considering an example of an industrial manipulator in the course of inserting a cylindrical working object into a hole (Figs. 5 to 7). At first (Fig. 5a) the manipulator has an open kinematic scheme as in Fig. 5b (simple open chain). In the phase of transferring the working object (Fig. 6a) the kinematic chain does not change (Fig. 6b) but the last member (now the gripper and object together) changes its dimensions and mass, which cause the dynamics to change too. Finally, in the phase of object insertion (Fig. 7a), the kinematic scheme of the manipulator changes too and it becomes a simple closed kinematic chain (Fig. 7b).

Mechanisms of legged locomotion machines are, as a rule, complex kinematic chains. Fig. 8 shows an arbitrary, complex kinematic chain comprising four simple chains, the first three (formed by the members 1 - 6) being closed and the fourth (formed by the members 7 and 8) being open. The kinematic chains connected to the support are basic chains, while the chains connected to them, but not by means of the support, are satellite chains (satellites). With this said, the procedure of separating one complex chain into a number of simple ones is practically defined. The notion of an independent kinematic chain, introduced in Ref. [5] may be defined at this point. A kinematic chain is said to be independent if its motion with respect to the support is independent of the satellite chains, namely, if its last members are connected to the support. This means that a basic chain is independent. This definition is slightly different from that given in [6], but, as a result, only one autonomous chain is obtained in any complex connected mechanism, all the remaining chains being satellites with respect to the autonomous one. Taking the model of a human presented in Fig. 2. as an example, it is possible to isolate the chains of "legs", "body" and "arms"; here, the chain consisting of "legs" is independent since the chains of "body" and "arms" impose no kinematic constraints on it. For the mechanism presented in Fig. 8, if the motion of member 1 is

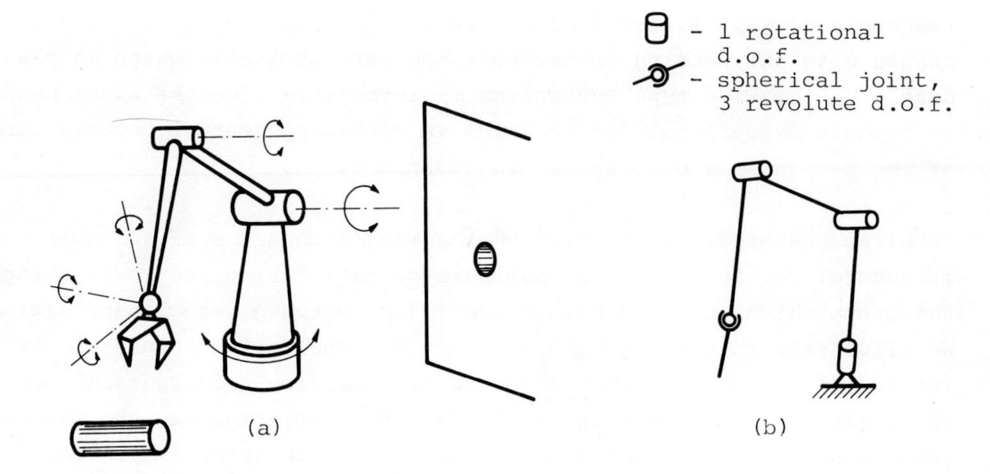

Fig. 5. Manipulator before grasping the object

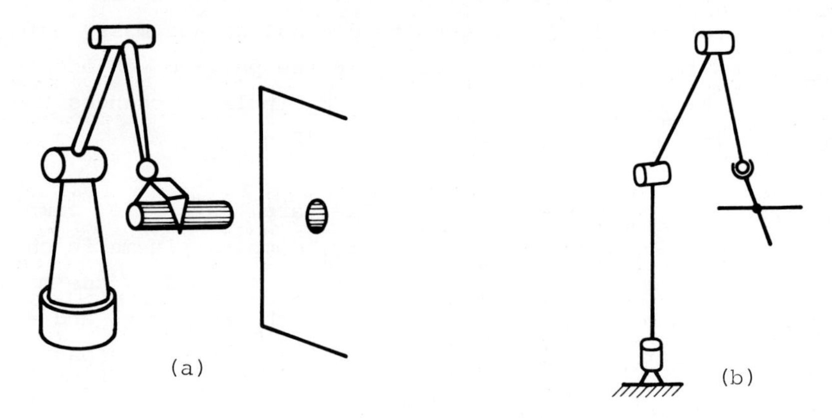

Fig. 6. Phase of working object transfer

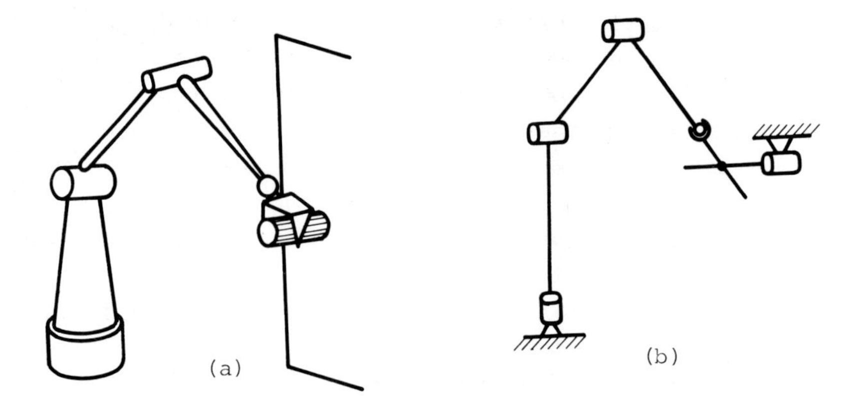

Fig. 7. Phase of object insertion into the hole

known, the chain I is independent, all the remaining ones being satellites. The sequence to be followed in performing kinematic and dynamic analyses has thus been determined; namely, the basic independent chains

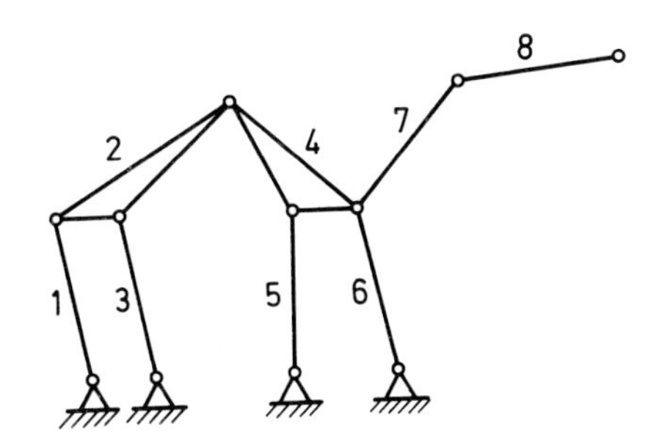

Fig. 8. Complex kinematic chain

should be analyzed first, and the satellite (guided) chains second. Figures 9 and 10 illustrate the kinematic chains of multilegged walking machines without presenting the foot-to-ground connection realized by kinematic pairs. If the global kinematic constraints imposed on the mechanism and the relative motions of some members are known, it is possible to define the autonomous chain and the satellites. Without considering the requirements concerning the gait type and foot-to-ground contact, it may be said that, for the mechanism presented in Fig. 9, it is necessary to know the motion of at least one of "the legs" so as to determine the guiding and the guided mechanism part; while for the mechanism of Fig. 10, such a division requires the kinematics of any two legs to be known.

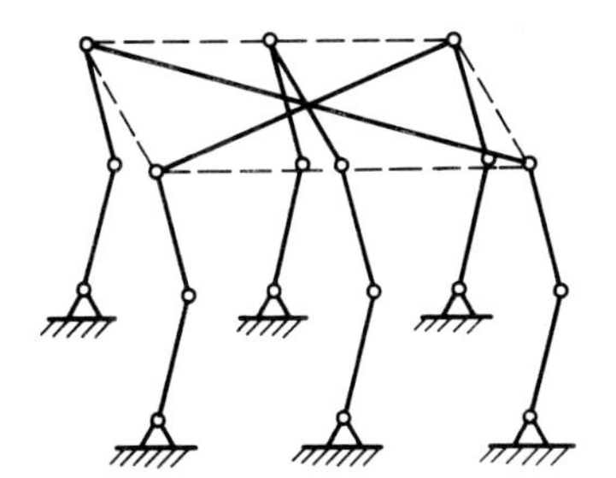

Fig. 9. Mechanism of a six-legged locomotion machine

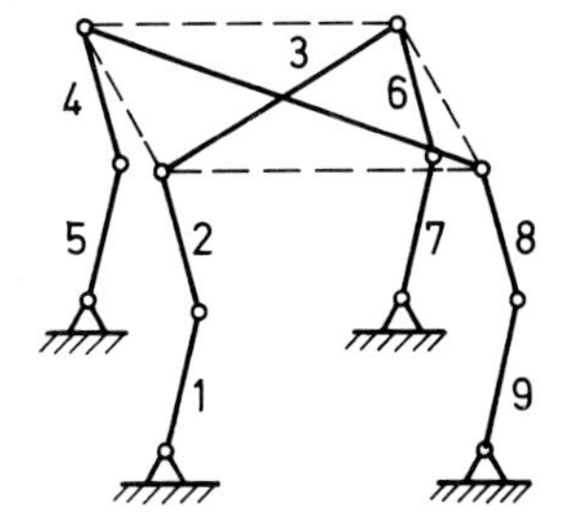

Fig. 10. Mechanism of a four-legged locomotion machine

The notion of the structure of a topological branch may be encountered in the literature [8, 9, 12]. What is understood by this notion is a branch-like (tree-like) system of rigid bodies connected by different kinematic pairs (Fig. 11).

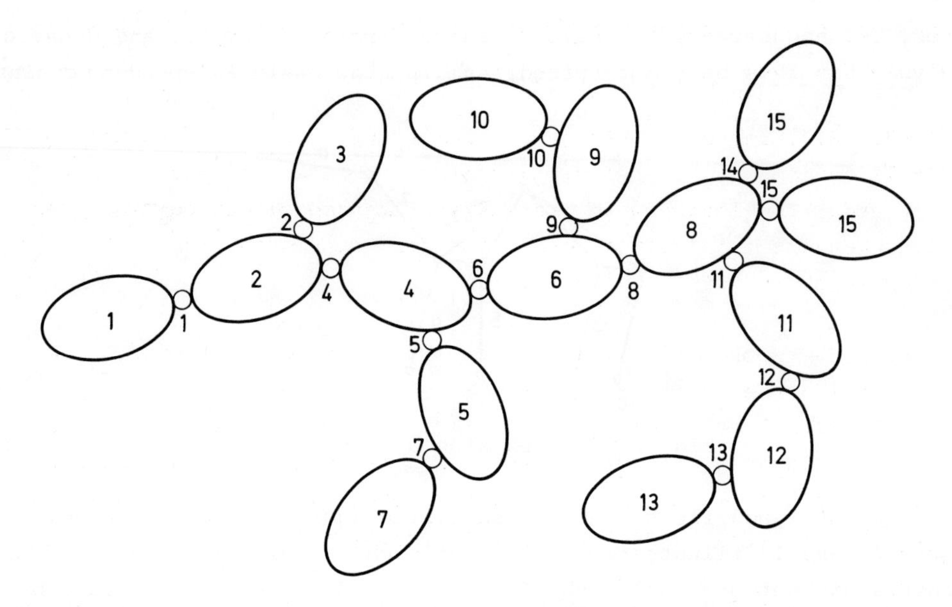

Fig. 11. Mechanism of free topological branch structure

Since it is our intention to present in this monograph only those methods for formulating mathematical models of active spatial mechanism dynamics which are computer-oriented, we will not discuss in this chapter the results of the dynamics of rigid connected bodies. These results were mainly based on a study of satellite dynamics and were available to the dynamics of spatial mechanisms, as applied in robotics. Hence the results are mentioned more with the intention of indicating the bibliographical sources than for the sake of a more detailed insight into the dynamics of spatial mechanisms; and also to draw attention to those researchers, who have precided the more modern approach in the range of new abilities of modern comptuer system.

1.3. Previous Results

The mathematical models developed in the preceding period of analytical methods can be devided into two main groups:

- Methods based on general theorems of dynamics,
- Methods based on second-order Lagrange equations.

Methods based on Newton-Euler equations. The motion equations of active mechanism can be written for every body taking into account the equations of the connections between the bodies and of the kinematical connections.

In 1963., H.J.Fletcher, L.Rongved and E.Y.Yu [7] studied the motion of a satellite composed of two rigid bodies, connected by a universal joint, under the load due to gravitation. The model can be simply expanded when driving torques act at the joints. Elimination of the connection force in this case is trivial.

In 1965., W.W.Hooker and G.Margulies [8], inspired by the preceding work, studied the general case, where n+1 bodies are connected by means of joints with 1 or 2 rotational d.o.f. Although by this method was a significant advance, it still possesses some unsuitable features such as that although the method uses matrix formalism, it is not able to obtain the matrices as functions of the system state.

R.E.Roberson and J.Wittenburg [9], in their approach, have defined the system of bodies as a graph, whereby the elaborated and known graph properties are used. The system is restricted to the form of a topological tree, two bodies being connected by a joint, so that the system consists of n bodies and n-1 joints. Isomorphism between the system of rigid bodies and the graph is estabilished in such way, that the mass centers form a set of graph knots $X = \{x_1, x_2,\ldots,x_n\}$, and the set $U = \{u_1, u_2,\ldots,u_n\}$ of the graph branches is defined as a set of joints connecting the bodies. The mapping $\Gamma(x)\to X$ is defined by directing the graph in Fig. 12.

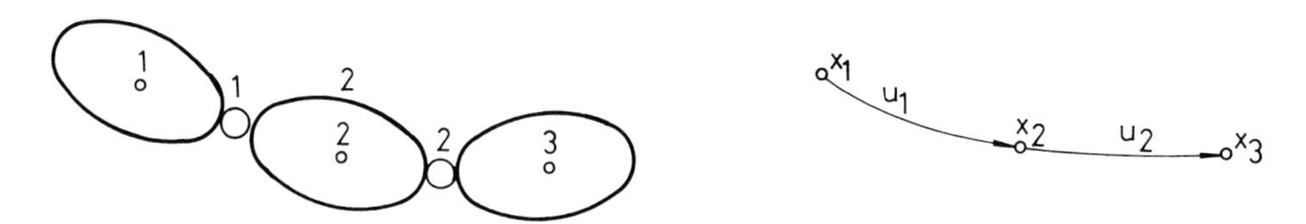

Fig. 12. Example of a system of bodies and the corresponding graph

This method remains interesting especially for its topological representation of the system.

The method of W.W.Hooker [10] provides a new possibility of eliminating constraints but presents a serious problem in deriving the motion equations.

The method of P.W.Likins [11], proposed in 1971, is a more direct application of the preceding method. Moreover, without reducing the problem´s generality, it simplifies the numerical designation of the bodies in the kinematical chain, and of the joints and unit vectors of the rotation axis. This method is based on the following two remarks:

Remark 1: The studies [9, 10] have demonstrated the significance of determining one reference body and some particular sequence of bodies in such way that their numbers grow during the description of a topological branch.

Remark 2: In practice, joints with several d.o.f. require in general the introduction of interconnected segments. This is particularly necessary when control of each individual d.o.f. is desired. Hence, in the majority of cases it is natural to separate a joint with several d.o.f. into several joints with one d.o.f. This is always possible by adding segments of zero mass.

P.W.Likins supposes that this separation has already been done. This approach results in a simplification of the equations proposed by W.W. Hooker [10] and enables Likins to express rotational system motion by matrix equations. This approach is suitable for computer simulation of joint-connected bodies in the form of a topological branch having relative rotations only.

Until now we have only discussed methods for constructing equations of systems in the form of topological branches and with relative rotations. If relative linear motions do not appear with anthropomorphic robots, such motions can be very interesting for industrial robots.

In the same way, the case, e.g., of studying gait in the double-support phase, demands the development of methods catering for closed chains in the mechanism. For this reason several people have recently been studying mechanisms of a more general structure. We will shortly illustrate a few of these methods.

J.Wittenburg has generalized the method [9] to systems having the structure of a topological branch, whose joints j permit rotational motion with r_j d.o.f. and linear motion with t_j d.o.f. [12], (r_j and t_j = 1, 2 or 3). The method is very general. It does not, however, include fully closed chains, so it remains too theoretical and does not offer an explicit matrix procedure for obtaining the equations.

F.W.Ossenberg-Franzes gave in 1973 a more general method for formulating equations (relative rotation and translation with closed chains), based on Newton's and Euler's equations [13]. First one obtains motion equations for n+1 bodies with 6(n+1) coordinates. This number is then reduced by introducing constraints. The method does not offer an explicit matrix procedure for obtaining equations.

All these methods, except the method of J.Wittenburg, were directed to the analysis of flying object dynamics. The control of orbital craft and satellites is very similar to the dynamics of active spatial mechanisms because it is necessary to determine the driving forces and torques in the joints between the bodies, these being based on previously determined inertial forces and moments and external forces and moments, for which a kinematic analysis of the system is necessary.

The requirements of and the dynamic analysis of active mechanisms hence do not differ from the global requirements of those methods.

From the point of view of the generality of the structural models considered, although most of the methods considered models of complex open kinematic chains with rotational joints, some methods considered translatory kinematic pairs and closed chains, so in that aspect they are completely acceptable. In all methods the dimensions of all bodies were finite, which is important for active mechanisms in which some members cannot be considered as canes or cylinders.

The basic deficiency of the above methods (except the method of P.W. Likins) lies in the fact, that they do not permit the formulation of recursive kinematic and dynamic equations, but they are analytic ("by hand"). However, with active mechanisms iterative calculations of driving forces and torques is necessary at each time instant and each mechanism joint. This requires the use of a digital computer for any more complex configuration. This in turn requires the dynamic equations to be recursive, which requires a consecutive and mutually connected enumeration of bodies. The preceding procedures are based on direct enumeration (except the method of P.W.Likins).

From that point of view, the procedure of P.W.Likins is completely adaptable to programming and the application of digital computers to the analysis of complex open chains with rotational kinematic pairs of all classes. It is a different question how "fast" this procedure would be in terms of computing time, i.e., how suitable it would be for control in real computer time compared with other algorithmic procedures. (Judging by the number of the kinematic and dynamic operations needed, it would not be too inferior). The sole deficiency of this procedure lies in the fact that it uses the values of the augmented body tensors of inertia, which is superfluous to computer calculations.

It should be emphasized that by changing the enumeration or formulating the recursive relations it is also possible to adjust the other methods for programming, but this changes the whole kinematic and dynamic calculations in these methods.

Using the notions of an augmented body and its mass center (the body barycenter) allows one to introduce the value of the tensor of inertia for the body barycenter. The dynamic equations can then be written in a somewhat condensed form, but their derivation remains analytical.

Finally, it should be noted that these methods have made a useful contribution to active mechanisms dynamics because the elimination of some of their deficiences has given rise to some new procedures in dynamics of robots and manipulators.

<u>Methods based on Lagrange´s equations.</u> Methods using Newton´s and Euler´s equations are in principle complex, because of the complexity of eliminating the constraints by forces and moments. Moreover, they do not directly show the algebraic values of forces and moments due to the action of actuators (motors), springs and dampers. On the other hand, Lagrange´s equations provide the possibility of directly regarding the equations as functions of the system control inputs. However, the inherent unsuitability of applying of Lagrange´s equation lies in the need to calculate the partial derivatives of Lagrange´s function, and hence of the kinetic energy.

The Lagrangean L of a system as is known, is the difference between the kinematic and potential energy: $L = E_k - E_p$. Defining the system in terms of their generalized forces Q_d^1, $Q_d^2, \ldots, HQ_d^n$, which are not derivatives of some potential function (friction forces, external forces and moments), the Lagrange equations of the system are written in the familiar form:

$$\frac{d}{dt}\frac{\partial L}{\partial \dot{q}_i} - \frac{\partial L}{\partial q_i} = Q_d^i, \qquad i = 1,2,\ldots,n$$

In 1968., J.J.Uicker proposed a method based on Lagrange´s equations, for the study of the dynamical behaviour of joint-connected systems of arbitrary structure (with an arbitrary number of closed chains) [14]. The author supposed that the joints permit either rotation or translation, so this method can be regarded as general. The equations are obtained for the case when the system is under the action of a force field, the forces being time-dependent and when, in addition to the generalized spring and friction forces act in each joint. The potential is the result of forces due to gravity and the elasticity of springs.

The method is sufficiently general to enable motion equations to be simulated on a computer. The method has been used by several authors to treat particular cases.

M.E.Kahn applied the method by J.J.Uicker to analytically obtain the motion equations of a three-segmented body, connected by means of rotation axes [15, 16]. Although two or three rotations are made around parallel axes, the equations obtained are too complex and only by a final arrangement of certain members is a simplification of the equations obtainable.

A.K.Bejczy and R.A.Lewis attempted to apply Uicker´s method to analytically obtain the motion equations of a telescopic manipulator with three revolute d.o.f.

A.K.Bejczy [17] obtained, in analytical form, the expressions for the manipulator kinetic and potential energy.

The expressions obtained are complex and the author proposes to simplify them by decoupling the motions of the manipulator and the gripper. R.A.Lewis [18] proposes a simplified calculation of some Lagrange equations coefficients.

Independently from these efforts, G.V.Koronev [19] proposed a method for formulating the equations of mechanically joint-connected system, having a topological branching, structure the joints of which belong to certain classes defined by the author. This method treats the connections only in a later stage of calculation when the equations of the free system have been obtained in Lagrange´s form. The author starts by writing the quadratic form of the kinetic energy of the body system as a function of the derivatives of 6 coordinates, defining body position and orientation. Thus, these equations represent the mathematical model of the complete free system as a function of 6n coordinates. These 6 coordinates are valid for the real system provided that the generalized forces due to external forces and moments are augmented by generalized forces due to connections. In addition, coupling of the various kinds is taken into account by the equations in a set of 6n preceding equations. As the number of generalized coordinates is 6n - h, it is possible to express these coordinates as functions of the generalized 6n - h coordinates. Introducing the 6n coordinates values and their first and second derivatives as functions of the corresponding values of generalized coordinates in the 6n preceding equations leads, after eliminating the generalized connection forces, to 6n - h Lagrange equations for the real system. This method is of theoretical interest only and does not provide any possibility of obtaining a matrix form of the system of equations.

M.Renaud [20, 21] derives the motion equations for the system of n+1 bodies, forming the structure of a tree in relation to the referent body O with the coordinate system O_o. The permissible relative motions of the adjacent bodies are rotation and translation (Fig. 13) $|\tau_i$ is the sign of translation.

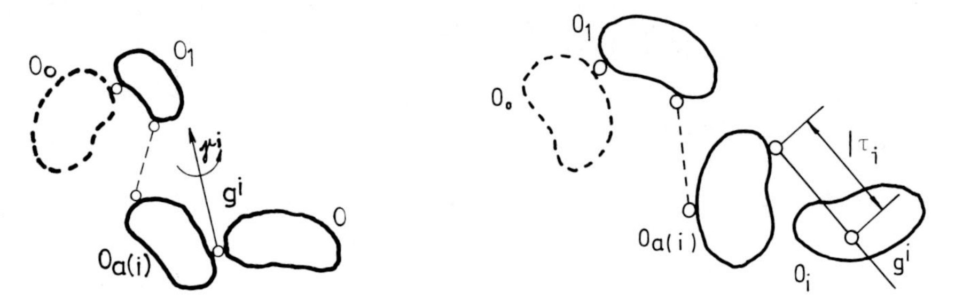

Fig. 13. Definition of relative body motion

With each body i ($i\varepsilon s$) is associated the coordinate system $Q_i(x_i, y_i, z_i)$, chosen in such a way that $z_i = g_i$, where g_i is the unit vector defining relative rotation or translation of the body relative to the preceding body. Each coordinate system O_i ($i\varepsilon s$) is defined in relation to $O_{a(i)}$ and is given by the set α_i, β_i, γ_i, the turning angles of one system relative to the other (Fig. 14).

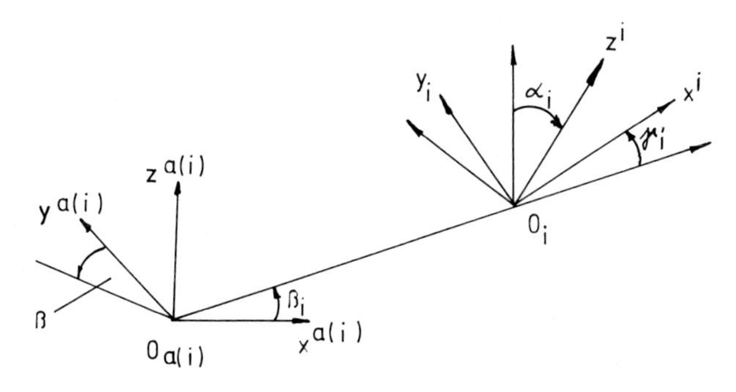

Fig. 14. Connected coordinate system

Values are defined according to the graph of the chain structure:

$$[e] = [e_{ij}]; \qquad i, j \varepsilon s$$

$$e_{ij} = \begin{cases} 1 & \text{if body j is directly connected to body i, and is not between bodies 0 and 1,} \\ 0 & \text{otherwise} \end{cases}$$

In addition, one defines the matrix:

$$[\varepsilon] = [\varepsilon_{ij}]; \qquad i, j \varepsilon s$$

$$\varepsilon_{ij} = \begin{cases} 1 & \text{if body i is on the chain, connecting bodies 0 and j} \\ 0 & \text{otherwise} \end{cases}$$

Index a(i) refers to the body connected directly to body i and is between bodies 0 and i. The coefficient σ_i is defined by:

$$\sigma_i = \begin{cases} 0 & \text{if body i rotates around the axis on body a(i),} \\ 1 & \text{if body i slides along a straight line fixed to body a (i)} \end{cases}$$

The generalized coordinates are defined as $q_i = \bar{\sigma}_i \gamma_i + \sigma_i \tau_i$, where $\sigma_i q_i = \sigma_i \tau_i$, $\bar{\sigma}_i q_i = \bar{\sigma}_i \gamma_i$.

Renaud uses one of the usual forms of Lagrange´s equations:

$$\frac{d}{dt}\left(\frac{\partial E_K}{\partial \dot{q}_\ell}\right) - \frac{\partial E_K}{\partial q_\ell} = Q^\ell,$$

where E_K is the kinetic system energy, while the potential system energy is given by $U = U_p + U_{art}$, where U_p is the potential energy due to the earth´s gravity and U_{art} is potential energy given to the system.

By means of this method, all stages of formulating the differential motion equations of a kinematical chain have been systematically treated sufficiently well. As the for suitability of the method for practical applications, nothing more can reliably be said.

The method by J.J. Uicker [14] is geared to analytical calculation or to recursive calculation of the motion equations of some arbitrary mechanical joint-connected system. However, this method has led to an important but unnecessary complication of the equations [15, 16]. To overcome this inconveniance M.Renaud has proposed a method for the study of mechanical, joint-connected bodies in the form of a topological branch performing rotations around axes only. This method is based on the tensor calculus [20] and is explained in much more detail with the aid of the matrix calculus in [21, 22]. An example of the calculation of kinetical energy of a chain-form mechanism of 4 bodies is given in ref. [21, 22].

As a last procedure we mention one closed form solution by Lagrange´s method, suitable for kinematic chains with rotational joints [24].

As stated already, a direct application of Lagrange´s equations to the formulation of dynamic models is unsuitable for automatic model construction mainly because of undesirable numerical differentiation. In this procedure this can be avoided and the coefficients of Lagrange´s equations are expressed explicitly.

The methods using Lagrange´s equations are an advance over those using Newton-Euler equations because they do not require elimination of the forces and moments of constraints. This advantage is partly due to the form of Lagrange´s second order equations, on the right hand side of which are the generalized forces (forces or moments) in terms of the respective generalized coordinates. By the choice of the relative displacements in the joints in terms of the generalized coordinates, the generalized forces are represented by the reaction forces and moments between the bodies. The basic deficiency in these methods is the need to calculate the partial derivatives of the kinetic and potential energy with respect to the generalized coordinates and velocities.

In a structural sense these methods are sufficiently general. They assume rotational and translatory joints between members. By means of these all other kinematic pairs can be described. Uicker's method concerns complex closed chains and the method of M.Renaud, apart from the topological branch structure, provides the possibility of describing closed kinematic chains. The method by G.Korenev is based on other suppositions and although it is completely general from the structural point of view, it does not lead to a practical derivation of the equations. The enumeration in Uicker's method is consecutive although it is given in a somewhat different form: (The author has found a consecutive enumeration of the joints, starting independently from the base in each closed chain; but then the i-th body is always between the i-th and (i+1)-th joint). Accordingly, there is a direct connection between the body enumeration and the enumeration of joints. A shortcoming of this method is that all kinematic relations are based on the matrix equation of the vector closure of the kinematic closed chain contour. In order to apply this method to open kinematic chains it is necessary to make certain modifications to the kinematic analysis. Finally, it should be noted that this method, like that of P.Likins, can, with modification, be fully adapted to computer calculation since it can be reduced to an algorithmic level.

The enumeration of members and joints in the method by M.Renaud is direct. This is why the procedure for formulating equations is analytical.

For further practical calculations, according to [21, 22], it is possible to mechanize the equations obtained using a computer, because the members, the derivatives of which must be found, are trigonometric functions of the internal angles. In 1977. J.Zabala [23] proposed an algorithm based on M.Renaud's method for automatically formulating the motion equations.

References

[1] Vukobratović M., "How to Control the Artificial Anthropomorphic Systems", Trans. on System, Man and Cybernetics, Sept. 1973.

[2] Vukobratović M., Stokić D., "Simplified Control Procedure for Strongly Coupled Nonlinear Large-Scale Mechanical Systems", (in Russian), Avtomatika i Telemekhanika, No. 11, 1978.

[3] Vukobratović M., Stokić D., "One Engineering Concept of Dynamical Control of Manipulators", Trans. of ASME, Journal of Dynamic Systems, Measurement and Control, Special issue, June, 1981.

[4] Vukobratović M., Stokić D., "Contribution to the Decoupled Control of Large-Scale Mechanical Systems", Automatica, Jan. 1980.

[5] Artobolevskii I.I., Theory of Mechanisms and Machines, (in Russian), Moscow, 1953.

[6] Vukobratović M., Legged Locomotion Robots and Anthropomorphic Mechanisms, Monograph, Mihailo Pupin Institute, 1975, P.O.B. 15, Beograd, Yugoslavia.

[7] Fletcher H.J., Rongved L. et Yu E.Y., "Dynamics Analysis of a Two-Body Gravitationally Oriented Satellite", The Bell System Technical Journal, pp. 2239 - 2266, September 1963

[8] Hooker W.W. et Margulies G., "The Dynamical Attitude Equations for a n-Body Satellite", The Journal of the Astronautical Sciences, Vol. XII, No. 4, pp. 123 - 128, Winter 1965.

[9] Roberson R.E. et Wittenburg J., "A Dynamical Formalism for an Arbitrary Number of Interconnected Rigid Bodies, with Reference to the Problem of Satellite Attitude Control", 3rd IFAC-CONGRESS, Session 46, Paper 64 D, 24th June 1966.

[10] Hooker W.W., "A Set of r Dynamical Attitude Equations for an Arbitrary n-Body Sattelite Haivng r Rotational Degrees of Freedom", AIAA Journal, Vol. 8, No 7, July 1970.

[11] Likins P.W., Passive and Semi-Active Attitude Stabilizations--Flexible Spacecraft AGARD-LS - 45 - 71 - Panel 3b, October 1971.

[12] Wittenburg J., Automatic Construction of Non Linear Equations of Motion for Systems with Many Degrees of Freedom, Euromech 38 Colloquium, Louvain La Neuve - Belgium - 3 - 5 September 1973.

[13] Ossenberg - Franzes, Equations of Motion of Multiple Body Systems with Translational and Rotational Degrees of Freedom. Euromech 38 Colloquium, Louvain La Neuve, Belgium, 3-5 September 1973.

[14] Uicker J.J., "Dynamic Behaviour of Spatial Linkages", Transactions of the ASME, No 68 Mech - 5 - pp. 1-15 - 1969.

[15] Kahn M.E., The Near Minimum Time Control of Open Loop Articulated Kinematic Chains, Stanford University Ph. D. Engineering Mechanical, 1970.

[16] Kahn M.E. et Roth B., "The Near Minimum Time Control of Open

Articulated Kinematic Chains", Transactions of the ASME. Journal of Dynamic Systems, Measurement and Control - pp. 164 - 172 - September 1971.

[17] Bejczy A.K., Robot Arm Dynamics and Control, JET Propulsion Laboratory NASA Technical Memorandum 33 - 669 - February 15, 1974.

[18] Lewis R.A., Autonomous Manipulation on a Robot: Summary of Manipulator Software Functions. JET Propulsion Laboratory, NASA Tehnical Memorandum 33 - 679 - March 15, 1975.

[19] Koronev G.V., "Movements of a Man to Reach a Specified Goal", Automatika and Telemekhanika No. 6, pp. 131 - 142 - June 1972.

[20] Liageois A. et Renaud M., Modele Mathématique des Systems Mécaniques Articulés en vue de Commande Automatique de leurs Mouvements. C.R.A.S.T. 278 - Serie B-29 Avril 1974.

[21] Renaud M., Energie Cinétique d'un Bras Mécanique Articulé, Note Interne LASS - SMA 74.I.09, Mai 1974.

[22] Liageois A., Renaud M. et al, "Synthese de la Commande Dynamique d'un Teléopérateur Redundant", Revue RAIRO Jaune J.2 - pp 98 - 103 Mai 1975.

[23] Zabala J., "Automatic Computer Construction of Equations of Motion of Articulated Mechanisms", Proc. I Yugoslav Symp. on Industrial Robots, Beograd, Yugoslavia, 1977.

[24] Mahil S., "Contribution to the Construction of Mathematical Models for Active Mechanisms via Lagrange's Equations", Proc. VII Symp. on Industrial Robots, Tokyo, 1977.

Chapter 2
Computer-Aided Methods for Setting and Solving Mathematical Models of Active Mechanisms in Robotics

2.1. Introduction

In this chapter the computer-aided methods*) for setting and solving mathematical models of active mechanisms will be explained. By the term mathematical model, we mean a set of dynamic equations i.e. differential equations of motion. This concept will be expanded later and a more precise definition will be given.

To write down the differential equations of motion by hand is a very complicated task. Although, in principle, it is always possible, it hardly makes sense to do so when more complex mechanisms are involved. In addition, a problem to be kept in mind is the always present risk of making numerous errors when handling such complex task. This has given rise to the idea of using a digital computer for both forming and solving mathematical models. As a result, computer-aided methods of forming and solving mathematical models have been developed.

It appears necessary to develop computer methods of mathematical modelling for at least two reasons. One of them is that it is impossible to immediately choose the most convenient configuration when designing robots. The term configuration should be interpreted as the structure (i.e. kinematical scheme) and parameters (i.e. dimensions, masses etc.). Thus, it is necessary to analyze a number of different robot configurations and choose the one most appropriate to the future purpose of the device. Knowing how complex a task it is to write a mathematical model by hand, the need for an algorithm that would enable a computer to perform the task seems quite logical.

The other reason is the need in some applications for real-time (on--line) control of robots. The development of such computer methods which perform real-time calculations of robot dynamics is a direct contribution to the synthesis of control algorithms for practical purposes.

---

*) Computer oriented methods, computer methods or automatic methods.

Bearing in mind the reasons underlying the development of computer--aided methods, we may now formulate the requirements which have to be satisfied by a certain method, if the method is to be regarded as "computer-aided":

1. The input data for the algorithm are: robot configuration (kinematical scheme and parameters), information on the problem being solved, and the initial state.

Using such input data, the computer itself forms and solves the mathematical model i.e. robot dynamics. The algorithm therefore operates for an arbitrarily given configuration and dynamical problem. It should be said that, in principle, two problems of dynamics may be solved, a direct and an inverse one. A combination may also occur. The direct problem consists in calculating the mechanism joints driving torques and forces needed to realize the prescribed motion. In this case, motion is the input information. The inverse problem consists in calculating the motion for the given driving forces and torques. The forces and torques now represent the input information.

2. The algorithm includes no numerical differentiation.

This requirement follows directly from the fact that numerical differentiation of an expression is an undesired task, even for modern computers. That is why such tasks are to be avoided in the algorithm.

The algorithms meeting these requirements allow simple dynamical analysis of different robot configurations (by changing the input data only). If used in high speed computers, these algorithms are also suitable for the synthesis of control algorithms for real-time operation.

The first methods satisfying the requirements imposed appeared independently of each other [1, 2]. The first approach was developed in connection with the dynamical analysis of manipulators [1], while the second resulted from the efforts toward the synthesis of artificial gait [2]. Other computer-aided methods of forming and solving the mathematical model of active mechanisms have also been developed and they will all be treated in this chapter.

2.2. The Basic Ideas of Computer-Aided Formation and Solution of a Mathematical Model

In this paragraph the basic ideas of computer-aided formation and solution of mathematical model of active mechanisms will be explained. These ideas are common to all computer-aided methods (c.-a. methods in the subsequent text).

Let us consider a general case: a mechanism consists of m arbitrary rigid bodies (mechanism members or segments) which are connected by arbitrary joints (arbitrary kinematical pairs). Let the mechanism have n degrees of freedom (d.o.f. in the subsequent text).

There are several approaches to the description of mechanical system dynamics. There are different forms of dynamical equation system and on the other hand the dynamics may be described via the formulation of certain principles (variational principles of mechanics). Many of these approaches where used as the basis for developing c.-a. methods; so different methods have recently appeared. This is why a general formulation of mathematical models will be given here. This general formulation encompasses all different approaches and is especially suitable when the c.-a. methods are in question.

The aim of each c.-a. method is to derive functions f and g, such that:

$$\ddot{u} = f(u, \dot{u}, P, \text{mechanism configuration}) \tag{2.2.1}$$

and

$$P = g(u, \dot{u}, \ddot{u}, \text{mechanism configuration}) \tag{2.2.2}$$

In the expressions (2.2.1), (2.2.2) the vector u represents the set of variables determining the position of the mechanism. If the mechanism is described via generalized coordinates $q_1,\ldots,q_n$, then

$$u = q = [q_1 \cdots q_n]^T \tag{2.2.3}$$

and the elements of the vector u are independent. The vector u may also be of dimension 6m if the position of each mechanism segment is determined by 6 variables (for instance 3 Cartesian coordinates of the center of gravity and 3 Euler´s angles). In this case

$$u = [x_1 y_1 z_1 \theta_1 \psi_1 \varphi_1 \; \cdots \; x_m y_m z_m \theta_m \psi_m \varphi_m]^T \qquad (2.2.4)$$

and the elements of vector u are not independent. This is due to kinematic pairs constraints (joints). The dependance has to be built into the functions f and g.

P represents the vector of driving forces and torques (called the drives in the sequel) acting in mechanism joints.

The functions f and g are not some explicitly prescribed or explicitly derived functions. They represent large computation algorithms. f represents the algorithm for computing $\ddot{u}$ for the known u, $\dot{u}$, P and the configuration, and g represents the algorithm for computing P for the known u, $\dot{u}$, $\ddot{u}$ and the configuration.

The realization of the algorithms f and g is specific to and characteristic of each method and it depends on the mechanical approach. It should be said that the algorithms f and g, although mutually inverse, are sometimes realized in rather different ways.

So, we form the mathematical model for one time-instant by carrying out the algorithms f or g; f for the inverse problem and g if a direct problem is involved.

We now consider solving the mathematical model for a finite time interval T. The c.-a. methods operate with discrete time. Let us divide the interval T into small subintervals Δt_k by introducing the sequence of time-instants $t_o, t_1, \ldots, t_{end}$. The subintervals Δt may be, but need not be, the same length.

First consider the inverse dynamic problem. It involves solving the motion u(t) for given drives P(t). In this book the notation u(t), P(t) and the like represent no explicit time functions but discrete time dependences, given by sequences of time points. As we said earlier, the initial state of the mechanism i.e. $u(t_o)$, $\dot{u}(t_o)$ is given. The mathematical model (2.2.1) should be formed for this initial time-instant t_o. The algorithm f is performed and $\ddot{u}(t_o) = f(u(t_o), \dot{u}(t_o), P(t_o),$ configuration) computed. The value obtained for acceleration $\ddot{u}(t_o)$ is now considered constant over a small subinterval Δt_1. So, by simple integration

$$u(t_o+\Delta t_1) = \frac{1}{2}\ddot{u}(t_o)\Delta t_1^2 + \dot{u}(t_o)\Delta t_1 + u(t_o)$$

$$\dot{u}(t_o+\Delta t_1) = \ddot{u}(t_o)\Delta t_1 + \dot{u}(t_o) \tag{2.2.5}$$

we obtain the mechanism state for the next time-instant $t_1 = t_o + \Delta t_1$ i.e. $u(t_1)$, $\dot{u}(t_1)$. The whole procedure is now repeated for time-instant t_1. Hence, a time-recursive procedure is obtained. The output is the mechanism motion $u(t)$, $\dot{u}(t)$.

If the use of some standard integration methods is desirable, then eq. (2.2.1) is written in the canonical form

$$\begin{aligned} \dot{u} &= v \\ \dot{v} &= f(u,\ v,\ P) \end{aligned} \tag{2.2.6}$$

Now consider a direct dynamical problem. This is a much easier task. By given motion we mean the known u, $\dot{u}$, $\ddot{u}$ in a sequence of time-instants. But the input is only $\ddot{u}(t)$ because u, $\dot{u}$ for each time-instant can be computed by simple integration (2.2.5) from the previous one. The procedure is performed in this way: for each time-instant the model (2.2.2) is formed i.e. the algorithm g is carried out and the drives P computed. The output is $P(t)$.

There is another property common to most c.-a. methods, namely, the method of treating kinematic pairs (i.e. joints). Most methods consider the 5-th class kinematic pairs i.e. the joints permitting one relative rotation or one relative translation of two connected segments. If a compound joint is in question, then it is dissembled into a sequence of 5-th class joints with small parameter segments between them. An explanation is needed. With manipulation mechanisms, joints with one rotational or linear d.o.f. are most frequent. Hence most methods have been derived in such way that they consider joints of that type. More complex joints with two or more d.o.f., which sometimes appear in robot mechanisms, are, as a rule, so designed that to each d.o.f. in the joint there corresponds an exactly determined rotation or translation axis (hardware axes). Such joints can be simply separated into a series of simple joints with one d.o.f. each, according to the real joint axes. It is necessary to say this because such separation is not possible for all types of complex joints. For instance, a spherical joint with three rotational d.o.f. cannot be equivalently separated into a series like this. The presentation will be mainly restricted to joints which can be separated, and for the other the separation holds only aproximatively. In spite of the possibility of separation, some methods are derived in such may that they can directly consider the complex joints most frequently encountered. These are joints with 2 d.o.f., permitting one rotation and one translation. This has been done with computer time economy in mind. Some methods have also been derived which directly consider complex joints with three rotational d.o.f., including the spherical joints.

A survey of the known c.-a. methods will be given later in the book. As we said earlier the realization of the algorithms f and g is a characteristic of each method. The main difference between methods is in the mechanical approach. Hence, all c.-a. methods may be divided into three groups:

- methods based on general theorems of dynamics and Newton-Euler equations,
- methods based on second-order Lagrange equations
- methods based on Appel´s equations and the Gauss principle.

Within the individual groups, the methods differ significantly in the manner of mathematical interpretation and derivation, in the generality of the kinematical scheme in question and in the types of joint they are operating with. In the sequel, methods will be presented according to their group. In the course of presenting each method, the original notation will not be respected. In all methods, it will usually be unified.

Methods Based on General Theorems of Dynamics and Newton-Euler Equations

2.3. The Method of General Theorems

This is the first method for c.-a. formation a mathematical model of active mechanisms. It has been derived independently by Yu. Stepanenko [1] and M.Vukobratović [2]. The method is based on kinetostatics i.e. D´Alambert´s principle and is often called the kinetostatical method [3 - 6]. Modification of the method has been found [7 - 10] and the method presented here will be in that modified form. Some particulars of the original version will be pointed out at the end of this paragraph.

Basic ideas. Let us dicuss the basic ideas in this method for deriving the algorithms f and g.

A mechanism with n d.o.f. is considered so we introduce the n-dimensional generalized coordinates vector:

$$q = [q_1 \cdots q_n]^T \tag{2.3.1}$$

The dynamics of such a mechanical system can be described by a differential equation system in matrix form:

$$W\ddot{q} = P + U \tag{2.3.2}$$

where P is the n-dimensional vector of driving forces and torques in the mechanism joints. The $n\times n$ matrix W depends on generalized coordinates q and the $n\times 1$ matrix U depends on the mechanism state q, $\dot{q}$. The algorithm for computing W and U is derived from general theorems of dynamics: the theorem about moment of momentum and the theorem about the center of gravity (c.o.g. in the sequel) motion.

Now $u = q$ and the function f represents solving the system (2.3.2) for the unknown vector $\ddot{q}$. This can be done by using suitable numerical procedures or directly by matrix inversion:

$$\ddot{q} = f(q, \dot{q}, P, \text{konfiguration}) = W^{-1}(P+U) \tag{2.3.3}$$

The function g is obtained directly from (2.3.2):

$$P = g(q, \dot{q}, \ddot{q}, \text{configuration}) = W\ddot{q}-U \tag{2.3.4}$$

Mechanism configuration. This method considers the mechanism of open chain type consisting of n arbitrary rigid bodies (Fig. 2.1). Also, there is no branching in the mechanism.

The joints connecting the mechanism segments have one d.o.f. each. That d.o.f. may be rotational or linear. A rotational joint S_i (Fig. 2.2) allows a relative rotation around an axis determined by a unit vector $\vec{e}_i$. A linear joint S_j allows a relative translation along an axis determined by a unit vector $\vec{e}_j$.

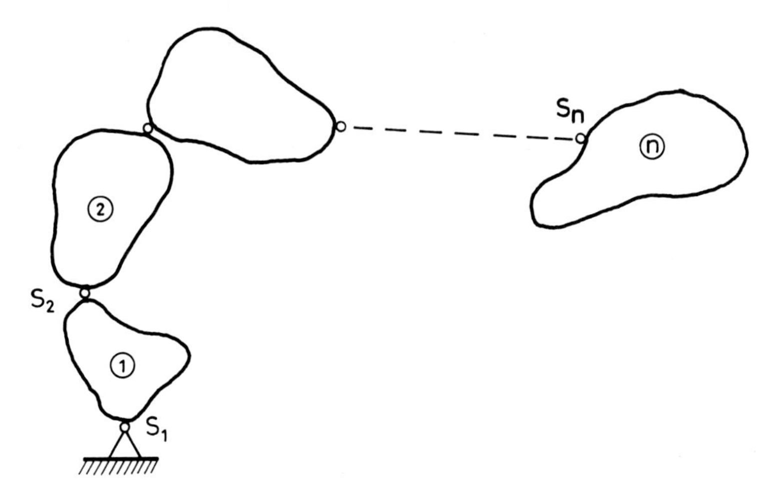

Fig. 2.1. Open kinematic chain without branching

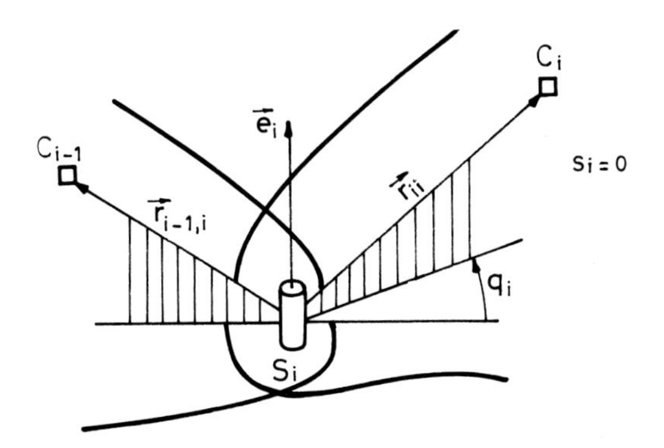

Fig. 2.2. Rotational joint

C_i, C_j and quadrats are used to mark the centers of gravity (c.o.g) of each segment in the figures in the text. s_i, s_j are indicators determing the type of joints:

$$s_k = \begin{cases} 0, & \text{if } S_k \text{ is a rotational joint} \\ 1, & \text{if } S_k \text{ is a linear joint.} \end{cases}$$

The prescription of the configuration will be descussed later.

Driving forces and torques. There is a driving motor in each mechanism joint. So, there is a driving torque $\vec{P}_i$ acting in the revolute joint S_i:

$$\vec{P}_i = P_i^M \vec{e}_i \tag{2.3.5a}$$

and a driving force $\vec{P}_j$ acting in the linear joint S_j:

$$\vec{P}_j = P_j^F \vec{e}_j \qquad (2.3.5b)$$

Now, the vector of the drives is

$$P = [P_1 \cdots P_n]^T \qquad (2.3.5c)$$

In the expression (2.3.5c) the upper indexes M, F are omitted because the indicators s_k are used to determinine the type of each joint and each drive.

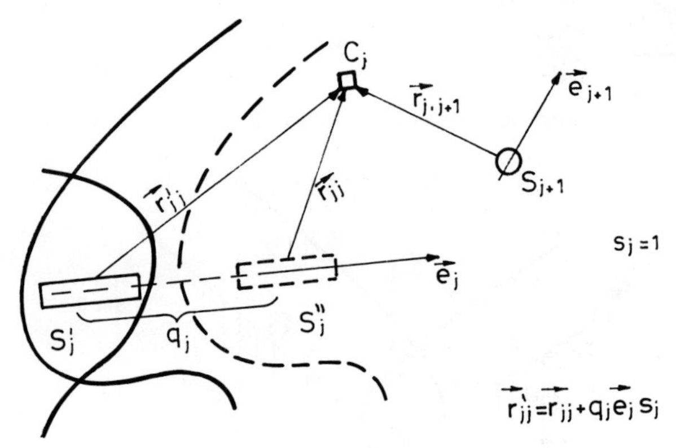

Fig. 2.3. Linear joint

Generalized coordinates. A set of n generalized coordinates $q_1, \ldots, q_n$ is used to determine the mechanism position. Each coordinate corresponds to one d.o.f., i.e., to one joint.

For a rotational joint S_i the corresponding generalized coordinate is defined as an angle of rotation in the joint around the exis $\vec{e}_i$. That angle may be regarded as the angle between the projection of the vectors $-\vec{r}_{i-1,i}$ and $\vec{r}_{ii}$ onto the plane perpendicular to the joint axis $\vec{e}_i$ (Fig. 2.2).

A particular case occurs when $\vec{r}_{ii} || \vec{e}_i$ or $\vec{r}_{i-1,i} || \vec{e}_i$. Then, the angle of rotation may not be considered in the previous way. If $\vec{r}_{i-1,i} || \vec{e}_i$ we call it the "specificity" of (i-1)-th segment on the upper end. Then we introduce a unit vector $\vec{r}^{\,*}_{i-1,i}$ perpendicular to $\vec{e}_i$ ($\vec{r}^{\,*}_{i-1,i} \perp \vec{e}_i$) (Fig. 2.4a). Further, the vector $\vec{r}^{\,*}_{i-1,i}$ is used instead of $\vec{r}_{i-1,i}$ for determining the generalized coordinate q_i. If $\vec{r}_{ii} || \vec{e}_i$ we call it the "specifity" of i-th segment on the down end. Then we introduce a unit vector $\vec{r}_{ii}^{*}$ perpendicular to $\vec{e}_i$ ($\vec{r}_{ii}^{*} \perp \vec{e}_i$) (Fig. 2.4b) and use it instead of $\vec{r}_{ii}$.

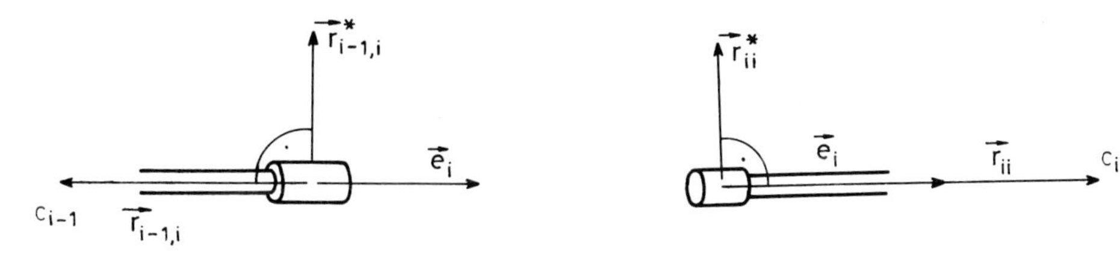

Fig. 2.4. (a) "specificity" of (i-1)-th segment on the upper end (b) "specificity" of i-th segment on the down end

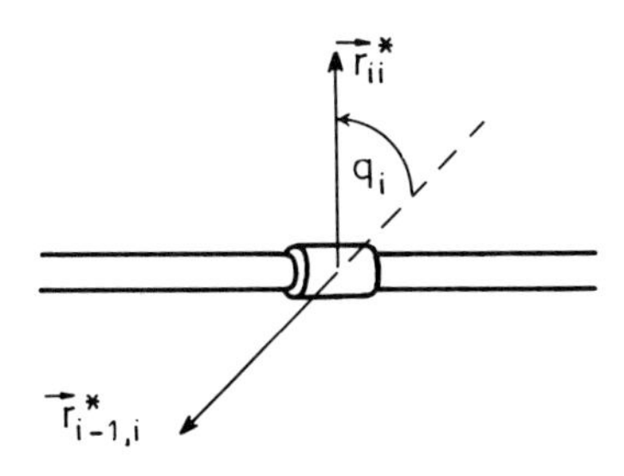

Fig. 2.5. Definition of the generalized coordinate in the case of "specificity"

The definition of generalized coordinates in the case of "specificity" is shown in Fig. 2.5.

The existence of "specificity" have to be given to the algorithm via special indicators.

If S_j is a linear joint, then the corresponding generalized cordinate q_j is defined as a relative linear displacement along the joint axis $\vec{e}_j$ i.e. $q_j = |\overrightarrow{S_j^{\prime} S_j^{\prime\prime}}|$ (Fig. 2.3).

Computation of transition matrices. Let us introduce the coordinate systems. First, there is an external non-moving Cartesian coordinate system Oxyz. A vertical z-axis is suitable but is not obligatory. Further, for each segment "i", a body-fixed (b.-f. in the sequel) Cartesian coordinate system $O_i x_i y_i z_i$ is defined. The origin O_i of such a system coincides with the c.o.g. C_i of the segment and the axes are oriented along the inertial principal axes.

We introduce the notation $\vec{a}_i$ to designate a vector corresponding to the i-th segment or i-th joint and which is expressed via three projections onto the axes of the external coordinate system. $\tilde{\vec{a}}_i$ designates the same vector but expressed by projections onto axes of i-th body-fixed system. $\underset{\sim}{\vec{a}}_i$ denotes the same vector but expressed with respect to the (i-1)-th b.-f. system.

Now, the transition matrix from the i-th b.-f. system to the external system (matrix A_i) is defined as follows:

$$\vec{a}_i = A_i \tilde{\vec{a}}_i \tag{2.3.6a}$$

There is also a transition matrix $A_{i-1,i}$ from the i-th to the (i-1)-th b.-f. system:

$$\underset{\sim}{\vec{a}}_i = A_{i-1,i} \tilde{\vec{a}}_i \tag{2.3.6b}$$

or inversely:

$$\tilde{\vec{a}}_i = A^{-1}_{i-1,i} \underset{\sim}{\vec{a}}_i = A_{i,i-1} \underset{\sim}{\vec{a}}_i \tag{2.3.6c}$$

A few things should be pointed out. The vectors $\vec{r}_{ii}$ and $\vec{r}_{i,i+1}$ (Figs. 2.2, 2.3) which determine the position of joints relative to the segment c.o.g. are proper to each segment. So they are constant vectors if expressed by projections onto the axes of i-th b.-f. system. That is, $\tilde{\vec{r}}_{ii}$ and $\tilde{\vec{r}}_{i,i+1}$ are constants. Further, the axis $\vec{e}_i$ of the joint S_i has constant position with respect to the i-th and (i-1)-th system. So the axis vector is constant if expressed via projections onto the i-th or (i-1)-th b.-f. system. That is, $\tilde{\vec{e}}_i$ and $\underset{\sim}{\vec{e}}_i$ are constants. Such vectors $\tilde{\vec{r}}_{ii}$, $\tilde{\vec{r}}_{i,i+1}$, $\tilde{\vec{e}}_i$, $\underset{\sim}{\vec{e}}_i$ which determine the geometry of the i-th segment and the i-th joint have to be prescribed for each segment and joint.

The computation of transition matrices is recursive. In each iteration the next segment is added to the chain and the corresponding transition matrix computed recursively. So A_i is computed, when adding the i-th segment, assuming that A_{i-1} is already computed.

When A_{i-1} is known, $\vec{r}_{i-1,i} = A_{i-1}\tilde{\vec{r}}_{i-1,i}$ and $\vec{e}_i = A_{i-1}\underset{\sim}{\vec{e}}_i$ can be easily computed.

Let us now suppose that S_i is a rotational joint. The following vectors should be computed:

$$\vec{a}_i = \frac{-\vec{e}_i \times (\vec{r}_{i-1,i} \times \vec{e}_i)}{|\vec{e}_i \times (\vec{r}_{i-1,i} \times \vec{e}_i)|} \quad ; \quad \vec{\tilde{a}}_i = \frac{\vec{\tilde{e}}_i \times (\vec{\tilde{r}}_{ii} \times \vec{\tilde{e}}_i)}{|\vec{\tilde{e}}_i \times (\vec{\tilde{r}}_{ii} \times \vec{\tilde{e}}_i)|} \qquad (2.3.7)$$

(a) (b)

The vectors $\vec{a}_i$ and $\vec{\tilde{a}}_i$ are perpendicular to $\vec{e}_i$ and $\vec{\tilde{e}}_i$ respectively. $\vec{a}_i$ is the unit vector of the axis "a" and (2.3.7b) holds for $q_i = 0$ (Fig. 2.6).

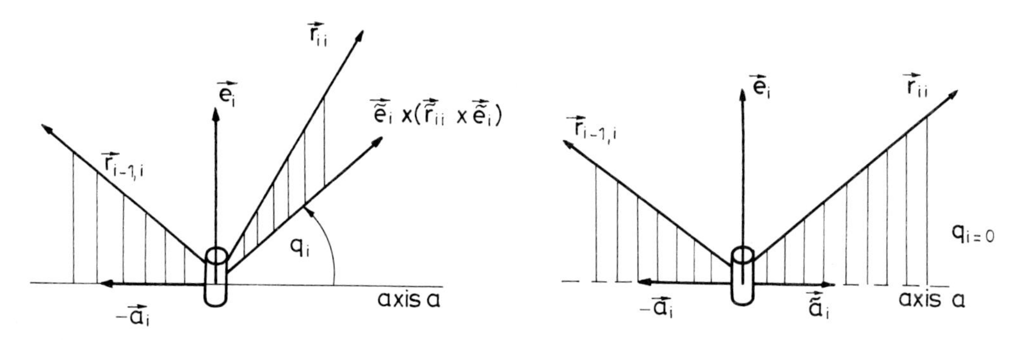

Fig. 2.6. Determination of the transition matrix

Introducing $\vec{b}_i = \vec{e}_i \times \vec{a}_i$, the three linearly independent vectors $\{\vec{e}_i, \vec{a}_i, \vec{b}_i\}$ are obtained. Introducing $\vec{\tilde{b}}_i = \vec{\tilde{e}}_i \times \vec{\tilde{a}}_i$, we also obtain the three linearly independent vectors $\{\vec{\tilde{e}}_i, \vec{\tilde{a}}_i, \vec{\tilde{b}}_i\}$.

Let A_i^o be the transition matrix corresponding to $q_i = 0$. Then (2.3.7b) holds and so

$$\vec{e}_i = A_i^o \vec{\tilde{e}}_i, \qquad \vec{a}_i = A_i^o \vec{\tilde{a}}_i, \qquad \vec{b}_i = A_i^o \vec{\tilde{b}}_i \qquad (2.3.8)$$

Now matrix notation will be introduced. Let e_i be the 3×1 matrix corresponding to the vector $\vec{e}_i$. Analogous matrix notation will be used for all other vectors in the text. Now, expressions (2.3.8) can be written together in matrix form

$$[e_i \ a_i \ b_i] = A_i^o [\tilde{e}_i \ \tilde{a}_i \ \tilde{b}_i] \qquad (2.3.9)$$

It follows that

$$A_i^o = [e_i \ a_i \ b_i][\tilde{e}_i \ \tilde{a}_i \ \tilde{b}_i]^{-1} \qquad (2.3.10)$$

By computing the matrix A_i^o, the process of "assembling" the joint is completed. The columns of the matrix A_i^o represent the unit vectors of b.-f. system axes for $q_i = 0$. So turning should be performed. The so-called finite turnings formula (i.e. Rodrigues´ formula) is used for turning each unit vector around the axis $\vec{e}_i$ for the angle q_i.

Let v_{i1}, v_{i2}, v_{i3} denote the columns of the matrix A_i^o i.e. the unit vectors of b.-f. system:

$$A_i^o = [v_{i1}\ v_{i2}\ v_{i3}] \tag{2.3.11}$$

Now, by turning:

$$\vec{V}_{ij} = \vec{v}_{ij}\cos q_i + (1-\cos q_i)(\vec{e}_i \cdot \vec{v}_{ij}) \cdot \vec{e}_i + \vec{e}_i \times \vec{v}_{ij}\sin q_i \qquad j=1,2,3 \tag{2.3.12}$$

where V_{ij} is the j-th column (i.e. the unit vector) after turning. So, the transition matrix

$$A_i = [V_{i1}\ V_{i2}\ V_{i3}] \tag{2.3.13}$$

is obtained.

In the case of "specificity" of the (i-1)-th segment on the upper end, the vector $\vec{\tilde{r}}^*_{i-1,i}$ is used instead of $\vec{\tilde{r}}_{i-1,i}$. If there is a "specificity" of the i-th segment on the down end, the vector $\vec{\tilde{r}}^*_{ii}$ is used instead of $\vec{\tilde{r}}_{ii}$.

If S_i is a linear joint, the transition matrix is computed in a different way. For simplicity, a unit vector $\vec{u}_i$ is defined. The vector is constant with respect to the i-th and (i-1)-th segment and is not parallel to $\vec{e}_i$. It allows us to define the three linearly independent vectors:

$$\vec{e}_i = A_{i-1}\vec{\underset{\sim}{e}}_i; \qquad \vec{a}_i = A_{i-1}\vec{\underset{\sim}{u}}_i; \qquad \vec{b}_i = \vec{e}_i \times \vec{a}_i$$

and another three linearly independent vectors

$$\vec{\tilde{e}}_i, \qquad \vec{\tilde{a}}_i = \vec{\tilde{u}}_i, \qquad \vec{\tilde{b}}_i = \vec{\tilde{e}}_i \times \vec{\tilde{a}}_i$$

As there is no "turning" the transition matrix may be obtained directly:

$$A_i = [e_i\ a_i\ b_i][\tilde{e}_i\ \tilde{a}_i\ \tilde{b}_i]^{-1}$$

Input data for the algorithm. Let us return to the definition and prescription of the mechanism configuration. The term "configuration" we mean the structure and the parameters. By the structure we mean the number of segments and the number and type of joints. By the parameters we mean all the information about the segments (dimensions, inertial properties etc.). So here is a list of the input data defining the configuration:

n = number of d.o.f.(=number of segments = number of joints),

s_i, $i=1,\dots,n$, determine the types of joints,

$\tilde{\vec{e}}_i$, $\vec{e}_{\sim i}$, $i=1,\dots,n$, determine the orientation of joint axes relative to the connected segments,

$\tilde{\vec{u}}_j$, $\vec{u}_{\sim j}$, if S_j is a linear joint,

$\tilde{\vec{r}}_{ii}$, $\tilde{\vec{r}}_{i,i+1}$, $i=1,\dots,n$ and $\vec{r}_{o1}$, determine the position of joints relative to segment c.o.g.,

$\tilde{\vec{r}}^{\,*}_{ii}$ or $\tilde{\vec{r}}^{\,*}_{i,i+1}$, in the case of "specificity" of i-th segment on the down or upper end,

m_i, $\tilde{J}_i$, $i=1,\dots,n$, m_i is the mass of i-th segment and J_i is the inertia tensor of the same segment with respect to the corresponding b.-f. system.

The initial state of the mechanism must also be prescribed. So the initial generalized coordinates and the initial generalized velocities are also input data:

$q_i(t_o)$, $i=1,\dots,n$ (i.e. $q(t_o) = [q_1(t_o) \cdots q_n(t_o)]^T$)

$\dot{q}_i(t_o)$, $i=1,\dots,n$ (i.e. $\dot{q}(t_o)$)

Depending on the type of dynamical problem there are also the input data:

$\ddot{q}(t_k)$, $k = 0,1,\dots,k_{end}$,

if a direct problem of dynamics is to be solved, or

$P(t_k)$, $k = 0,1,\dots,k_{end}$,

in the case of the inverse problem of dynamics. If a mixed problem is considered, some accelerations and some drives are prescribed.

Kinematical relations. Let us consider the kinematical chain containing the rotational and linear joints. Let $\vec{v}_i$ be the velocity and $\vec{w}_i$ the acceleration of i-th segment c.o.g. Further, let $\vec{\omega}_i$ be the angular velocity and $\vec{\varepsilon}_i$ the angular acceleration of the same segment. Then for the velocities,

$$\vec{\omega}_i = \vec{\omega}_{i-1} + \dot{q}_i\vec{e}_i(1-s_i) \tag{2.3.14}$$

$$\vec{v}_i = \vec{v}_{i-1} - \vec{\omega}_{i-1} \times \vec{r}_{i-1,i} + \vec{\omega}_i \times \vec{r}'_{ii} + \dot{q}_i\vec{e}_i s_i \tag{2.3.15}$$

$$\tilde{\vec{r}}'_{ii} = \tilde{\vec{r}}_{ii} + q_i\tilde{\vec{e}}_i s_i$$

and for the accelerations,

$$\vec{\varepsilon}_i = \vec{\varepsilon}_{i-1} + (\ddot{q}_i\vec{e}_i + \dot{q}_i\vec{\omega}_{i-1} \times \vec{e}_i)(1-s_i) \tag{2.3.16}$$

$$\vec{w}_i = \vec{w}_{i-1} - \vec{\varepsilon}_{i-1} \times \vec{r}_{i-1,i} - \vec{\omega}_{i-1} \times (\vec{\omega}_{i-1} \times \vec{r}_{i-1,i}) + $$
$$+ \vec{\varepsilon}_i \times \vec{r}'_{ii} + \vec{\omega}_i \times (\vec{\omega}_i \times \vec{r}'_{ii}) + (\ddot{q}_i\vec{e}_i + 2\vec{\omega}_{i-1} \times \vec{e}_i\dot{q}_i)s_i \tag{2.3.17}$$

with the boundary conditions

$$\vec{v}_o = 0, \ \vec{\omega}_o = 0, \ \vec{\varepsilon}_o = 0, \ \vec{w}_o = 0 \tag{2.3.18}$$

<u>Forming the equation system</u>. Forming the equation system i.e. computation of the matrices W and U (for the system (2.3.2)) is performed on the basis of the general theorems of mechanics: the theorem about moment of momentum and the theorem about c.o.g. motion.

Let the mechanism have ℓ rotational and m linear joints ($\ell + m = n$).

First, let us consider a rotational joint $S_k(s_k=0)$, and further, let us fictitiously interrupt the chain in the joint S_k. Now consider the part of the mechanism from S_k up to the free end. The rest of the mechanism is replaced by a reaction force $\vec{F}_{Rk}$ and a reaction moment $\vec{M}_{Rk}$. $\vec{M}_{Rk}$ is perpendicular to the joint rotation axis $\vec{e}_k$ i.e. $\vec{M}_{Rk} \perp \vec{e}_k$ (Fig. 2.7).

Now let us apply the theorem about moment of momentum to the part of the mechanism considered. All moments are consider relative to the point S_k. It follows that

$$\sum_{i=k}^{n} \vec{r}_i^{(k)} \times m_i\vec{w}_i + \sum_{i=k}^{n} \vec{M}_i = \vec{P}_k + \vec{M}_{Rk} + \sum_{i=k}^{n} \vec{r}_i^{(k)} \times m_i\vec{g} \tag{2.3.19}$$

i.e.

$$\sum_{i=k}^{n} (\vec{r}_i^{(k)} \times m_i\vec{w}_i - \vec{r}_i^{(k)} \times m_i\vec{g}) + \sum_{i=k}^{n} \vec{M}_i = \vec{P}_k + \vec{M}_{Rk} \tag{2.3.20}$$

where the gravity acceleration vector is $\vec{g} = \{0, 0, -9.81\}$, and $\vec{r}_i^{(k)} = \overrightarrow{S_kC_i}$ (Fig. 2.7). $\vec{M}_i$ represents the change in momentum moment of the i-th segment relative to its c.o.g. So, $\vec{M}_i$ is determined by Euler´s equations

$$\vec{M}_i = A_i\vec{\tilde{M}}_i \tag{2.3.21}$$

$$\vec{\tilde{M}}_i = \tilde{J}_i\vec{\tilde{\varepsilon}}_i - (\tilde{J}_i\vec{\tilde{\omega}}_i) \times \vec{\tilde{\omega}}_i \tag{2.3.22}$$

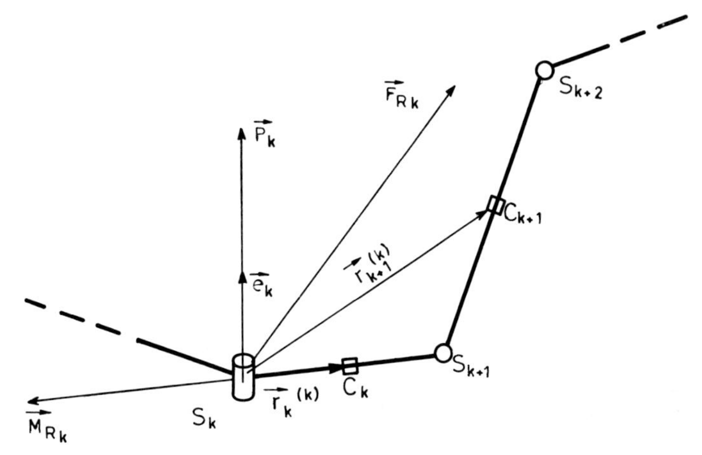

Fig. 2.7. Fictitious interruption of the chain and reactions in the rotational joint S_k

From scalar multiplication of (2.3.20) by $\vec{e}_k$ it follows that

$$\sum_{i=k}^{n} \underbrace{((\vec{r}_i^{(k)} \times m_i\vec{w}_i)\vec{e}_k - (\vec{r}_i^{(k)} \times m_i\vec{g})\vec{e}_k)}_{B_{ik}} + \sum_{i=k}^{n} \vec{M}_i\vec{e}_k = P_k^M \tag{2.3.23}$$

where

$$\vec{r}_i^{(k)} = \overrightarrow{S_kC_i} = \sum_{p=k}^{i-1} (\vec{r}\,'_{pp} - \vec{r}_{p,p+1}) + \vec{r}\,'_{ii} \tag{2.3.24}$$

If there are ℓ rotational joints, then in the way described, we obtain ℓ scalar equations of the form (2.3.23).

Now consider a linear joint S_k. Let us interrupt the chain fictitiously in the joint S_k and substitute the rest of the mechanism (from S_k to the base) with a reaction force $\vec{F}_{Rk}$ and a reaction moment $\vec{M}_{Rk}$ (Fig. 2.8). Then $\vec{F}_{Rk} \perp \vec{e}_k$.

If we apply the theorem about c.o.g. motion to the part of the mechanism considered (from S_k up to the free end), it follows that

$$\sum_{i=k}^{n} m_i \vec{w}_i = \vec{P}_k + \sum_{i=k}^{n} m_i \vec{g} + \vec{F}_{Rk} \tag{2.3.25}$$

After transformation and scalar multiplication by $\vec{e}_k$ the following form is obtained:

$$\sum_{i=k}^{n} \underbrace{(m_i \vec{w}_i \vec{e}_k - m_i \vec{g}\vec{e}_k)}_{C_{ik}} = P_k^F \tag{2.3.26}$$

If there are m linear joints we obtain m scalar equations of the form (2.3.26).

So, by using both the theorem about moment of momentum and the theorem about c.o.g. motion, $\ell + m = n$ scalar equations are obtained.

It should be mentioned that only one external force, gravity, was considered. If there are other external forces they should be added to the gravity force. If necessary, the external forces acting upon a segment are reduced with respect to the c.o.g. of the segment, and then they appear in the equations (2.3.22), (2.3.23) and (2.3.26).

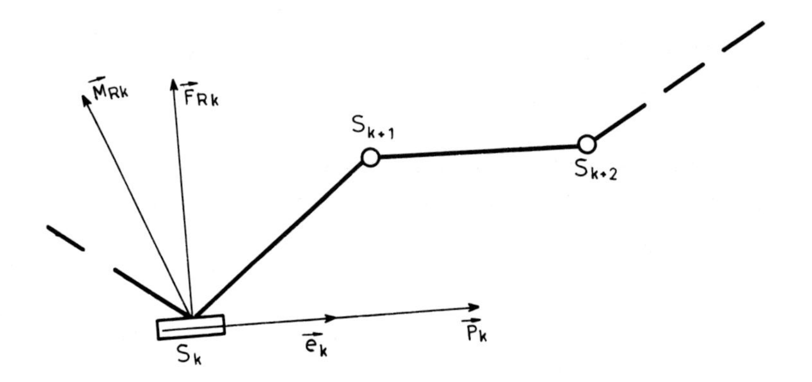

Fig. 2.8. Fictitious interruption and reactions in the linear joint S_k

Now matrix notation is introduced. We show how the n scalar equations (ℓ+m=n eqs.) may be written in matrix form, (2.3.2) which is necessary for the computational algorithm.

Formation of the system (2.3.2) i.e. computation of the matrices W, U is performed recursively. In each iteration the next segment is added to the mechanism chain.

Let Ω be the 3×n matrix the columns of which represent the coeficients of the generalized accelerations in the expression for segment c.o.g. acceleration, and let Θ be the 3×1 matrix containing the free member of the same expression. In the i-th iteration the i-th segment is considered and the matrices Ω an Θ refer to its acceleration $\vec{w}_i$. So

$$w_i = \Omega\ddot{q} + \Theta \qquad (2.3.27)$$

The columns of matrices are designated

$$\Omega = [\beta_1^i \cdots \beta_i^i \; 0 \cdots 0] \qquad (2.3.28)$$

$$\Theta = [\delta^i]$$

Further, let Γ be the 3×n matrix, containing the coefficients of the generalized accelerations in the expression for segment angular acceleration, and let Φ be the 3×1 matrix containing the free member of the expression. In the i-th iteration the matrices refer to $\vec{\varepsilon}_i$ i.e.

$$\varepsilon_i = \Gamma\ddot{q} + \Phi \qquad (2.3.29)$$

The columns of the matrices are designated

$$\Gamma = [\alpha_1^i \cdots \alpha_i^i \; 0 \cdots 0]$$

$$\Phi = [\gamma^i] \qquad (2.3.30)$$

In each iteration the next segment is added to the chain. The modifications and supplementations of the matrices Ω, Θ, Γ and Φ are performed in order to make them correspond to the new segment. These are performed on the basis of recursive expressions for velocities and accelerations (2.3.14) - (2.3.17). From these expressions follow the formulae for modifications and supplementations of the matrices in the i-th iteration:

$$\left.\begin{aligned} \vec{\alpha}_j^i &= \vec{\alpha}_j^{i-1}, \quad j = 1,\dots,i-1 \\ \vec{\alpha}_i^i &= (1-s_i)\vec{e}_i \end{aligned}\right\} \tag{2.3.31}$$

$$\left.\begin{aligned} \vec{\gamma}^i &= \vec{\gamma}^{i-1} + \vec{f} \\ \vec{f} &= \dot{q}_i(\vec{\omega}_{i-1}\times\vec{e}_i)(1-s_i) \end{aligned}\right\} \tag{2.3.32}$$

$$\left.\begin{aligned} \vec{\beta}_j^i &= \vec{\beta}_j^{i-1} - \vec{\alpha}_j^{i-1} \times \vec{r}_{i-1,i} + \vec{\alpha}_j^i \times \vec{r}\,'_{ii}, \quad j = 1,\dots,i-1 \\ \vec{\beta}_i^i &= \vec{e}_i s_i + \vec{\alpha}_i^i \times \vec{r}\,'_{ii} \end{aligned}\right\} \tag{2.3.33}$$

$$\left.\begin{aligned} \vec{\delta}^i &= \vec{\delta}^{i-1} - \vec{\gamma}^{i-1} \times \vec{r}_{i-1,i} + \vec{\gamma}^i \times \vec{r}\,'_{ii} + \vec{h} \\ \vec{h} &= -\vec{\omega}_{i-1} \times (\vec{\omega}_{i-1}\times\vec{r}_{i-1,i}) + \vec{\omega}_i\times(\vec{\omega}_i\times\vec{r}\,'_{ii}) + 2\vec{\omega}_{i-1}\times\vec{e}_i s_i \dot{q}_i \end{aligned}\right\} \tag{2.3.34}$$

Let us consider the expression B_{ij} which appears in (2.3.23)

$$B_{ij} = \vec{e}_j(\vec{r}_i^{(j)} \times m_i\vec{w}_i) - \vec{e}_j(\vec{r}_i^{(j)} \times m_i\vec{g}) \tag{2.3.35}$$

If we substitute (2.3.27) into (2.3.35) it follows that

$$B_{ij} = \underbrace{e_j^T\, \underline{\underline{r}}_i^{(j)} m_i \Omega}_{b^T}\cdot\ddot{q} + \underbrace{e_j^T\, \underline{\underline{r}}_i^{(j)} m_i \Theta - e_j^T\, \underline{\underline{r}}_i^{(j)} m_i g}_{v_1} \tag{2.3.36}$$

where $\underline{\underline{r}}_i^{(j)}$ designates the matrix

$$\underline{\underline{r}}_i^{(j)} = \begin{bmatrix} 0 & -r_{i3}^{(j)} & r_{i2}^{(j)} \\ r_{i3}^{(j)} & 0 & -r_{i1}^{(j)} \\ -r_{i2}^{(j)} & r_{i1}^{(j)} & 0 \end{bmatrix} \tag{2.3.37}$$

which corresponds to the vector $\vec{r}_i^{(j)} = \{r_{i1}^{(j)}, r_{i2}^{(j)}, r_{i3}^{(j)}\}$ and is used to form the vector product by matrix calculus.

Further, let us consider the expression $\vec{e}_j\vec{M}_i$ which also appears in (2.3.23). By using (2.3.21) and (2.3.22) the expression may be written in the form

$$\vec{e}_j\vec{M}_i = \vec{e}_j(A_i\vec{\tilde{M}}_i) = \vec{e}_j[A_i(\tilde{J}_i\vec{\tilde{\varepsilon}}_i - (\tilde{J}_i\vec{\tilde{\omega}}_i) \times \vec{\tilde{\omega}}_i)] =$$
$$= \vec{e}_j[A_i\tilde{J}_iA_i^{-1}\vec{\varepsilon}_i - A_i(\tilde{J}_i\vec{\tilde{\omega}}_i) \times \vec{\tilde{\omega}}_i] \tag{2.3.38}$$

Substituting (2.3.29) into (2.3.38) it follows that

$$\vec{e}_j\vec{M}_i = \underbrace{e_j^T A_i\tilde{J}_iA_i^{-1}\Gamma}_{c^T}\cdot\ddot{q} + \underbrace{e_j^T A_i\tilde{J}_iA_i^{-1}\Phi - K}_{v_2} \tag{2.3.39}$$

$$K = [A_i(J_i\vec{\tilde{\omega}}_i) \times \vec{\tilde{\omega}}_i]\cdot\vec{e}_j$$

Further, let us perform a transformation of the expression

$$C_{ij} = \vec{e}_jm_i\vec{w}_i - \vec{e}_jm_i\vec{g} \tag{2.3.40}$$

which appears in (2.3.26), by introducing (2.3.27) into (2.3.40). It follows that

$$C_{ij} = \underbrace{e_j^Tm_i\Omega}_{d^T}\cdot\ddot{q} + \underbrace{e_j^Tm_i\Theta - e_j^Tm_ig}_{v} \tag{2.3.41}$$

Now, the algorithm for transforming the equation system from the form (2.3.23), (2.3.26), $k=1,\ldots,n$ into the matrix form (2.3.2), i.e., the algorithm for computation of the system matrices W and U, can be represented by a flow-chart given in Fig. 2.9.

Now, when the matrices W and U are computed, i.e., the system (2.3.2) formed, the inverse and the direct problem of dynamics are solved by using (2.3.3) and (2.3.4) according to the basic ideas of c.-a. methods of forming and solving a mathematical model.

If should be mentioned also that the method described permits simple computation of reaction forces and moments in the mechanism joints. Namely, after computing the generalized accelerations $\ddot{q}$, and then accelerations $\vec{\varepsilon}_i$, $\vec{w}_i$, $i=1,\ldots,n$ by means of (2.3.27) and (2.3.29), the reactions $\vec{F}_{Rk}$, $\vec{M}_{Rk}$ in the k-th joint may be obtained from (2.3.20) and (2.3.25).

<u>Some characteristics of the original version of the method (the kinetostatical approach)</u>. As it has already been said that the method of general theorems differs a little from the original version of the

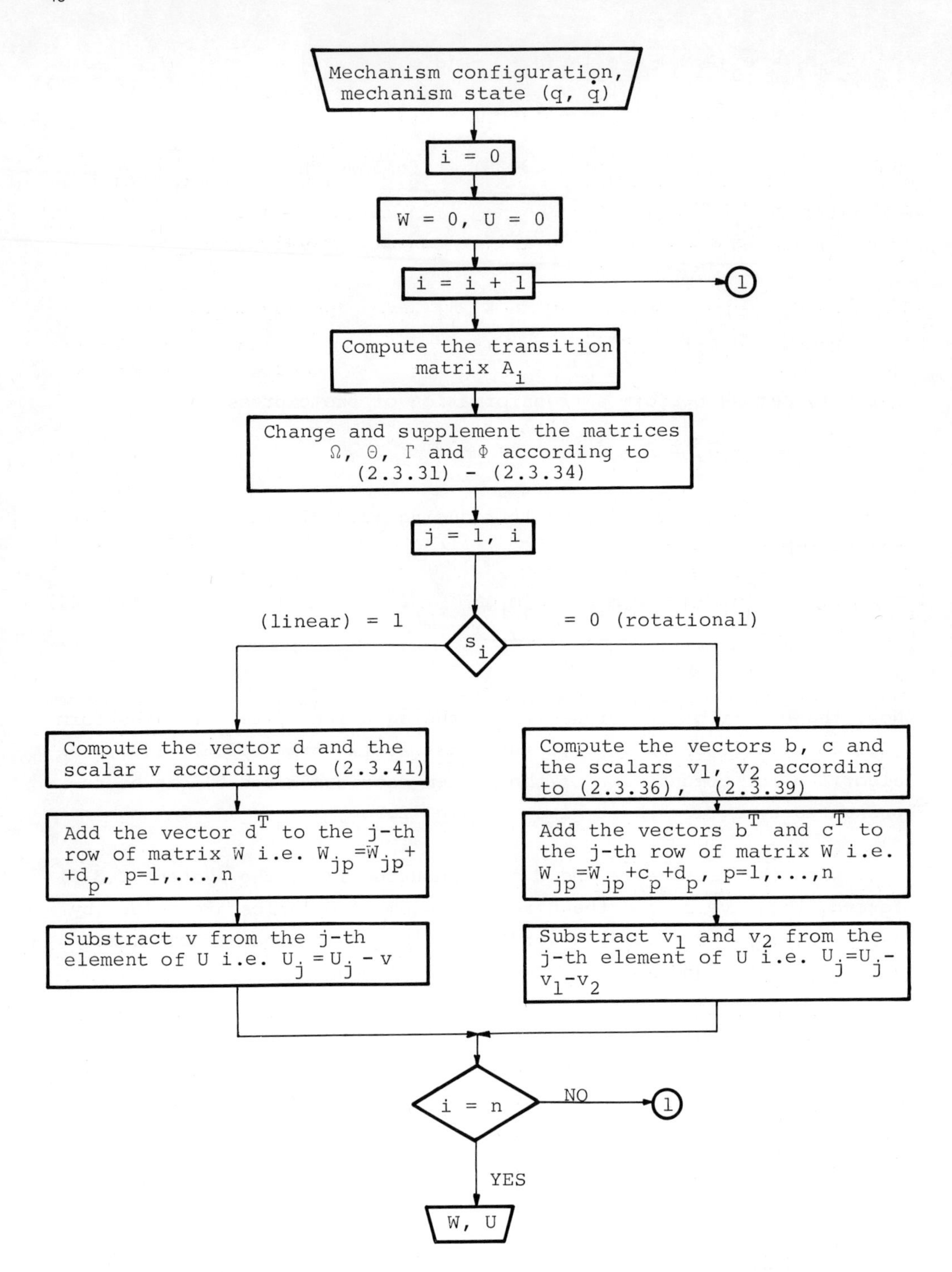

Fig. 2.9. Flow-chart of the general theorems method

method [3, 4, 6].

In the original version of the method, the kinetostatic approach i.e. the use of D´Alambert´s principle was used. The procedure of fictitious interruption in successive joints has also been applied. When the interruption in a joint is made then D´Alambert´s principle is applied to the mechanism part from the interruption up to the free end.

In order to use the kinetostatical approach, the inertial forces of each segment have to be reduced with respect to the segment c.o.g. So, the resultant force and the resultant couple (resultant moment) relative to the segment c.o.g. are introduced. The resultant inertial force of the i-th segment is expressed as

$$\vec{F}_{Ii} = -m_i \vec{w}_i \tag{2.3.42}$$

The resultant couple of the inertial forces, M_{Ii} of the i-th segment is obtained from Euler´s equations, i.e., the relations (2.3.21) and (2.3.22), by putting a minus sign before the right hand side of equation (2.3.22).

One characteristic of the original version of the method is the treatment of the cane segments. The cane segment is that segment the length of which is five or more times its diameter.

As cane segments are very often used in robot mechanisms, the method in such cases is programed to compute the resultant vector of inertial forces in simpler and faster ways.

A cane segment is characterized by its length 2ℓ, mass m, and two inertial moments I_N and I_L. I_N is the inertial moment with respect to the c.o.g. axis perpendicular to the cane. I_L is the moment with respect to the longitucinal cane axis.

Let us introduce the concept of equivalent angular acceleration $\vec{\tau}_i$. It is defined by the condition that the moment of inertial forces due to its action is equal to the moment of inertial forces due to angular velocity $\vec{\omega}_i$. For a cane segment,

$$\vec{\tau}_i = (\vec{\omega}_i \cdot \vec{L}_i)(\vec{L}_i \times \vec{\omega}_i), \tag{2.3.43}$$

where $\vec{L}_i$ is the unit vector of the longitudinal cane axis.

Now, the moment of inertial forces relative to the segment c.o.g. may be written in the form

$$\vec{M}_{Ii} = -J_i(\vec{\varepsilon}_i + \vec{\tau}_i), \tag{2.3.44}$$

where J_i is the tensor of inertia.

Let $\vec{M}_{Ii}$ and $\vec{\varepsilon}_i$ be separated into two components, one perpendicular to the longitudinal cane axis (index N) and the other parallel to it (index L):

$$\vec{M}_{iN} = (\vec{L}_i \times \vec{M}_i) \times \vec{L}_i, \qquad \vec{M}_{iL} = (\vec{M}_i \cdot \vec{L}_i) \cdot \vec{L}_i$$
$$\vec{\varepsilon}_{iN} = (\vec{L}_i \times \vec{\varepsilon}_i) \times \vec{L}_i, \qquad \vec{\varepsilon}_{iL} = (\vec{\varepsilon}_i \cdot \vec{L}_i) \cdot \vec{L}_i \tag{2.3.45}$$

Since $\vec{\tau}_i$ is perpendicular to the longitudinal cane axis the expression (2.3.44) reduces to

$$\vec{M}_{Ii} = -I_{iN}(\vec{\varepsilon}_{iN} + \vec{\tau}_i) - I_{iL}\vec{\varepsilon}_{iL} \tag{2.3.46}$$

The equation obtained determines the resultant couple of inertial forces which is expressed in the external coordinate system. By using such a simplification, the calculation speed may be increased two - three times in the case of cane segments.

Let us consider the fictitious interuption in a rotation joint S_k and apply the D´Alambert´s principle to the mechanism part from S_k up to the free end. So, the sum of moments of external and inertial forces relative to S_k as well as the driving torque and the reaction moment in the joint, equals zero. The vector equation is thus obtained. After scalar multiplication by $\vec{e}_k$ one obtains the scalar equation (2.3.23).

Now, let S_k be a linear joint. From D´Alambert´s principle it follows that the sum of external and inertial forces as well as the driving and the reaction forces, equals zero. After scalar multiplication by $\vec{e}_k$, one obtains the scalar equation (2.3.26).

If the described procedure is carried out for all joints and then (2.3.27) and (2.3.29) used, one obtains the matrix system (2.3.2). If the original notation is used, the matrix system is

$$H\ddot{q} + h = P \tag{2.3.47}$$

The case of a moving base has been also considered by using this method. If the acceleration of the base is $\vec{w}_o$ and the angular acceleration $\vec{\varepsilon}_o$, the system (2.3.47) transforms into

$$H\ddot{q} = H_1P + H_2w_o + H_3\varepsilon_o + \xi \qquad (2.3.48)$$

However, no details will be considered here.

2.4. Method of Block Matrices

This method is less computer oriented than the previous one. It is described in detail in [23]. The method represents the analytically derived mathematical model. But, by using the suitable block-matrices formalism, the model reduces to the compact matrix form suitable for solving on a computer. The derivation of the mathod is very long. So, only the basic dynamical and kinematical relations will be explained here, as well as the methodology of derivation.

Basic ideas. We consider a mechanism with n degrees of freedom (d.o.f.) and introduce the n-dimensional generalized coordinates vector. $q = [q_1 \cdots q_n]^T$. The dynamics of such a mechanical system can be described by differential equation system in matrix form:

$$W(q)\cdot\ddot{q} = P + U(q, \dot{q}) \qquad (2.4.1)$$

The dimensions of the matrices W, U, P are $n\times n$, $n\times 1$ and $n\times 1$ respectively. P represents the column vector of driving forces and torques in the mechanism joints. The matrix W depends on the generalized coordinates q, and U depends on q, $\dot{q}$. Of course, these matrices also depend on the mechanism configuration.

Now, the functions f and g defined by (2.2.1) and (2.2.2) may be derived in the form

$$\ddot{q} = f(q, \dot{q}, P, \text{configuration}) = W^{-1}(U+P) \qquad (2.4.2)$$

and

$$P = g(q, \dot{q}, \ddot{q}, \text{configuration}) = W\ddot{q} - U \qquad (2.4.3)$$

We derive the expressions for W(q) and U(q, $\dot{q}$).

The mechanical approach in this method is the same as in the original version of the previous method (kinetostatical approach). However, we give a short explanation of the approach.

Kinematical scheme of the mechanism considered: This is an open kinematical chain without branching (Fig. 2.10). The joints connecting the mechanism segments have one d.o.f. each. That d.o.f. may be rotational or linear (5-th class kinematic pairs). Hence, there are n joints, n d.o.f. and n segments. Let $\vec{e}_i$ be the unit vector of the rotation axis if S_i is a rotational joint and let it be a unit vector of the translation axis if S_i is a linear joint. The indicator s_i will be used to notate the type of the joint.

$$s_i = \begin{cases} 0, & \text{if the kinematical pair (i-1,i), i.e., the joint } S_{i-1}\text{, is rotational} \\ 1, & \text{if the kinematical pair (i-1,i), i.e., the joint } S_{i-1}\text{, is linear} \end{cases} \tag{2.4.4}$$

Generalized coordinates. A set of n generalized coordinates is used to determine the mechanism position. Each coordinate corresponds to one d.o.f. If a rotational joint S_{i-1} connects the i-th and the (i-1)-th segments, then the angle θ_i of relative rotation of the i-th segment with respect to the (i-1)-th segment around the axis $\vec{e}_{i-1}$, is chosen for the corresponding generalized coordinate. If S_{i-1} is a linear joint, then the corresponding generalized coordinate is defined as the relative linear displacement u_i along the joint axis $\vec{e}_{i-1}$.

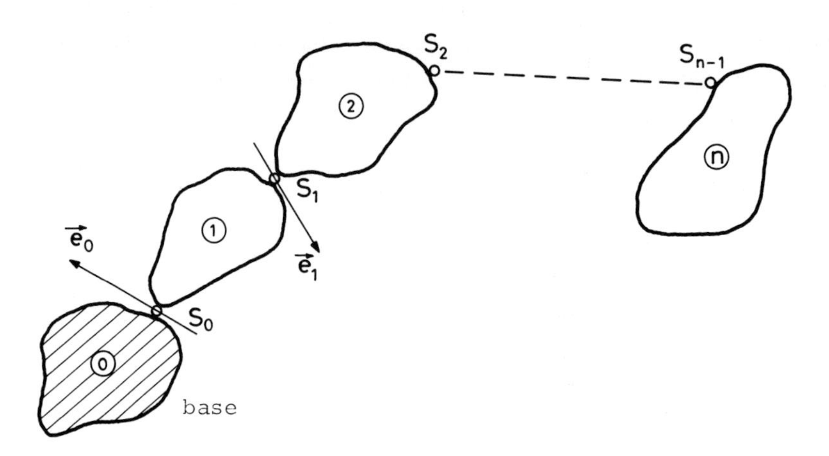

Fig. 2.10. Open kinematic chain without branching

Hence:

$$q_i = (1-s_i)\theta_i + s_i u_i, \tag{2.4.5}$$

where s_i is defined via (2.4.4).

Dynamic equations. The dynamic equations of the mechanism are derived on the basis of D´Alambert´s principle. We introduce the following notation.

- $\vec{G}_k$, the gravity force vector of the k-th segment;
- $\vec{F}_{Ek}$, the resultant of other external forces acting on the k-th segment;
- $\vec{M}_{Ek}$, the resultant external moment acting on the k-th segment (resultant couple relative to the segment c.o.g.);
- $\vec{F}_{Ik}$, the resultant of the inertial forces of the k-th segment;
- $\vec{M}_{Ik}$, the resultant moment of inertial forces of the k-th segment (resultant couple relative to the segment c.o.g.);
- $\vec{P}_i$, the vector of the drive in the joint S_{i-1};
- $\vec{P}_i = P_i^M \vec{e}_{i-1}$; if S_{i-1} is a rotation joint;
- $\vec{P}_i = P_i^F \vec{e}_{i-1}$; if S_{i-1} is a linear joint;
- P_i^F is a driving force and P_i^M is a driving torque;
- $\vec{\rho}_{i,j} = \overrightarrow{S_i C_j}$; C_j, the c.o.g. of the j-th segment.

Let the mechanism have ℓ rotational and m linear joints.

Let us interrupt the chain fictitiously in the joint S_{i-1} and consider the part of the mechanism from S_{i-1} up to the free end (Fig. 2.11). The rest of the mechanism is replaced by a reaction force $\vec{F}_{Ri}$ and a reaction moment $\vec{M}_{Ri}$.

Let S_{i-1} be a linear joint. Then $\vec{F}_{Ri} \perp \vec{e}_{i-1}$. If D´Alambert´s principle of the real and the inertial forces is applied, it follows that

$$\sum_{k=i}^{n} (\vec{G}_k+\vec{F}_{Ek}+\vec{F}_{Ik}) + \vec{F}_{Ri} + \vec{P}_i = 0 \tag{2.4.6}$$

Scalar multiplication of (2.4.6) by $\vec{e}_{i-1}$ gives

$$\sum_{k=i}^{n} (\vec{G}_k+\vec{F}_{Ek}+\vec{F}_{Ik})\vec{e}_{i-1} + P_i^F = 0 \tag{2.4.7}$$

m scalar equations of the form (2.4.7) are obtained by performing the interuption for all m linear joints.

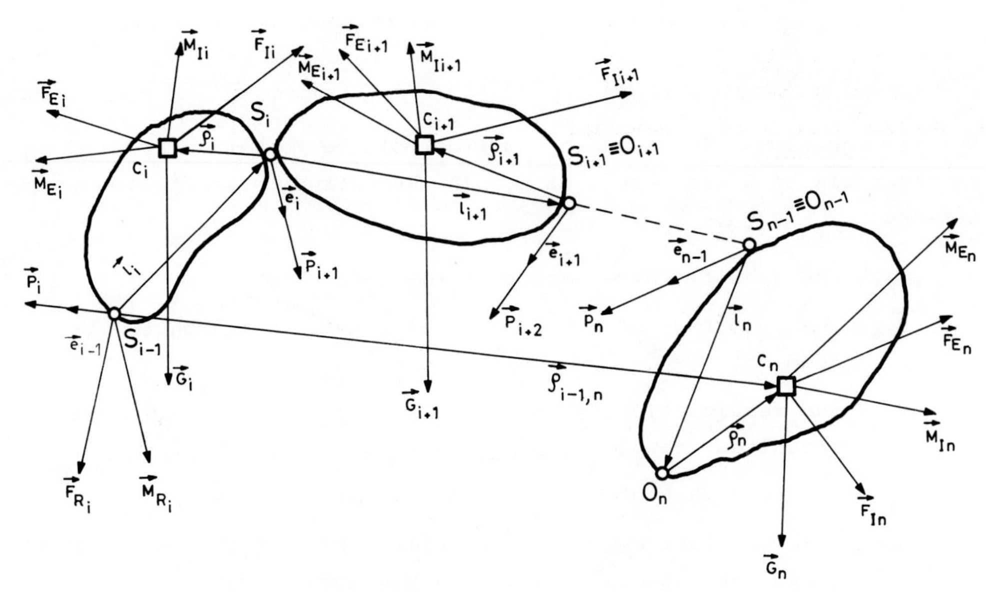

Fig. 2.11. Fictitious interruption of the chain in the joint S_{i-1}

Now let S_{i-1} be a rotational joint. Then $\vec{M}_{Ri} \perp \vec{e}_{i-1}$. If D´Alambert´s principle is applied to the real and inertial moment relative to S_{i-1}, then

$$\sum_{k=i}^{n} [\vec{M}_{Ek}+\vec{M}_{Ik}+\vec{\rho}_{i-1,k}\times(\vec{G}_k+\vec{F}_{Ek}+\vec{F}_{Ik})] + \vec{M}_{Ri} + \vec{P}_i = 0 \tag{2.4.8}$$

Scalar multiplication of (2.4.8) by $\vec{e}_{i-1}$ gives

$$\sum_{k=i}^{n} [\vec{M}_{Ek}+\vec{M}_{Ik}+\vec{\rho}_{i-1,k}\times(\vec{G}_k+\vec{F}_{Ek}+\vec{F}_{Ik})] \cdot \vec{e}_{i-1} + P_i^M = 0. \tag{2.4.9}$$

ℓ scalar equations of the form (2.4.9) are obtained by repeating the procedure for all ℓ rotational joints.

In all, $\ell + m = n$ scalar equations of the form (2.4.7) and (2.4.9) are obtained. Such a system can be transformed into the matrix form (2.4.1) by using block matrix formalism.

The coordinate system and transition matrices. Let us introduce the coordinate systems. First, there is an external non-moving Cartesian system Oxyz. A vertical z-axis is suitable, but is not obligatory. Further, for each segment "i", a body-fixed (b.-f.) Cartesian coordinate system $O_i x_i y_i z_i$ is defined. The origin O_i of such a system coincides with S_i. It should be pointed out that although the system origin O_i is in the joint S_i, the system is fixed with respect to the segment "i". The z_i-axis of the b.-f. system should be along the joint axis $\vec{e}_i$. The x_i-axis if perpendicular to z_{i-1} and z_i. y_i is perpendicular to x_i and z_i so the system is orthogonal (Fig. 2.12)

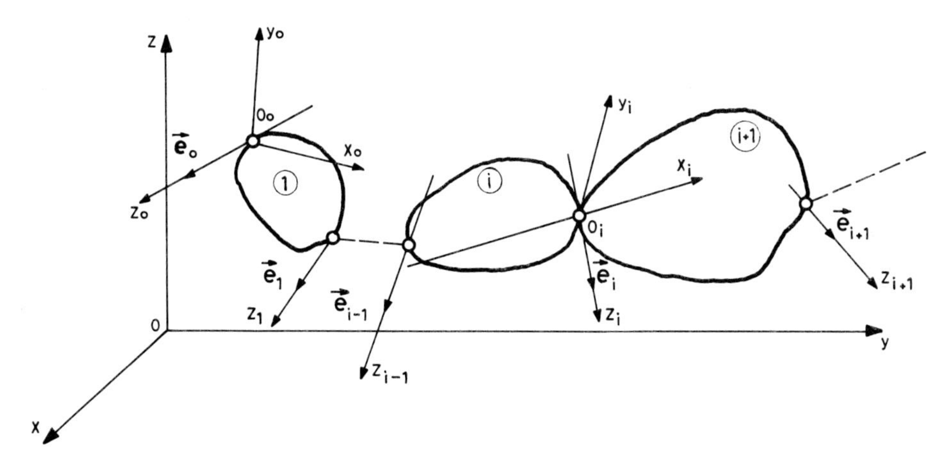

Fig. 2.12. Coordinate systems

We introduce the following notation: $\vec{a}_i^{(r)}$ denotes a vector which is characteristic of the i-th segment and is expressed via three projections onto the axes of the r-th b.-f. system, i.e., $\vec{a}_i^{(r)}=\{a_{i_{xr}}, a_{i_{yr}}, a_{i_{zr}}\}$. For instance $\vec{e}_i^{(i)}=\{0, 0, 1\}$. Further, let $\vec{a}_i$ notate the same vector but expressed in the external coordinate system Oxyz.

The following vectors should be introduced: $\vec{\ell}_i = \overrightarrow{O_{i-1}O_i}$, $\vec{\rho}_i = \vec{\rho}_{i,i} = \overrightarrow{O_iC_i}$ (Fig. 2.13). These vectors are constants if expressed in the corresponding b.-f. system i.e. $\vec{\ell}_i^{(i)}$ and $\vec{\rho}_i^{(i)}$ are constants and characteristics of the i-th segment.

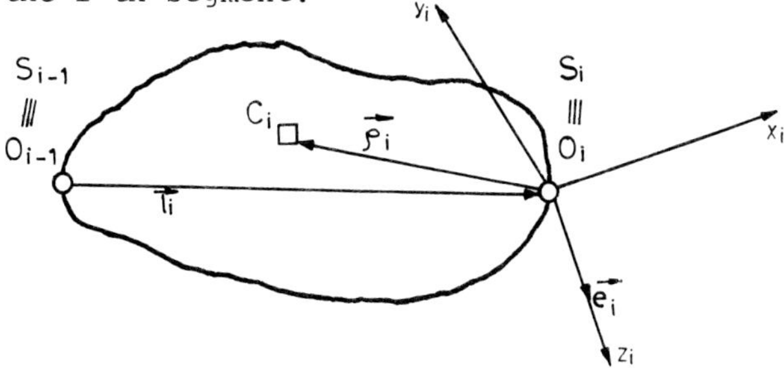

Fig. 2.13. Configuration of a segment and the corresponding b.-f. system

Let us introduce matrix notation and write a_i for the 3×1 matrix corresponding to a vector $\vec{a}_i$.

The transformation of the system $O_{i-1}x_{i-1}y_{i-1}z_{i-1}$ into $O_ix_iy_iz_i$ will now be considered. This transformation has four phases (Fig. 2.14):

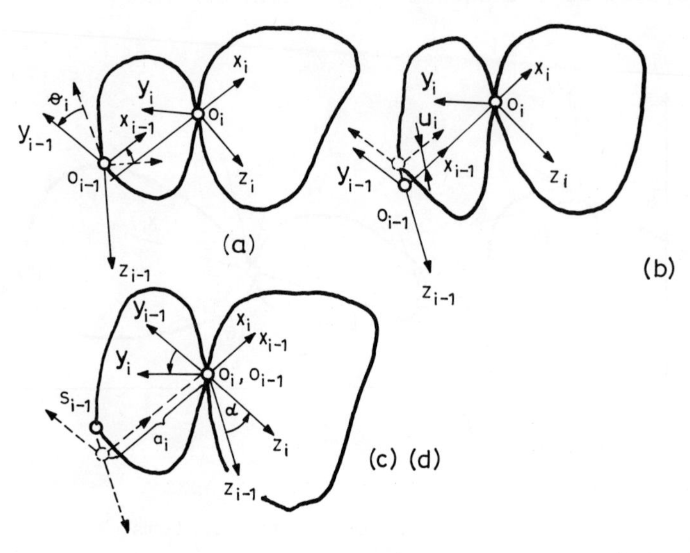

Fig. 2.14. Phase of transforming the (i-1)-th system into i-th one

(a) rotation around the z_{i-1}-axis until x_{i-1} becomes parallel with x_i. Rotation angle is θ_i;

(b) translation along z_{i-1}-axis until x_{i-1} coincides with x_i. Translation displacement is u_i;

(c) translation along x_i-axis until O_{i-1} coincides with O_i. Displacement is a_i;

(d) rotation around x_i-axis until all axes of the two coordinate systems coincide. Rotation angle is α_i.

The transition matrix $A'_{i,i-1}$ corresponds to phase (a), and the transition matrix $A''_{i,i-1}$ to the phase (d):

$$A'_{i,i-1} = \begin{bmatrix} \cos\theta_i & \sin\theta_i & 0 \\ -\sin\theta_i & \cos\theta_i & 0 \\ 0 & 0 & 1 \end{bmatrix} \qquad (2.4.10)$$

$$A''_{i,i-1} = \begin{bmatrix} 1 & 0 & 0 \\ 0 & \cos\alpha_i & \sin\alpha_i \\ 0 & -\sin\alpha_i & \cos\alpha_i \end{bmatrix} \qquad (2.4.11)$$

So, the whole transition matrix is

$$A_{i,i-1}=A''_{i,i-1}A'_{i,i-1} = \begin{bmatrix} \cos\theta_i & \sin\theta_i & 0 \\ -\sin\theta_i\cos\alpha_i & \cos\theta_i\cos\alpha_i & \sin\alpha_i \\ \sin\theta_i\sin\alpha_i & -\cos\theta_i\sin\alpha_i & \cos\alpha_i \end{bmatrix} \qquad (2.4.12)$$

The phases (b) and (c) together represent the translation of the coordinate system origin for the vector $\vec{\ell}_i$ which can be expressed as

$$\ell_i^{(i-1)} = [a_i\cos\theta_i \quad a_i\sin\theta_i \quad u_i]^T \qquad (2.4.13)$$

The inverse transformation i.e. the transformation of the i-th b.-f. system into (i-1)-th, has the transition matrix

$$A_{i-1,i} = A^{-1}_{i,i-1} = A^T_{i,i-1} \qquad (2.4.14)$$

due to the orthogonality of the systems.

Now, for some vector $\vec{a}_i$,

$$a_i^{(i-1)} = A_{i-1,i}a_i^{(i)} \qquad (2.4.15)$$

The transformation of coordinates of some point M from one system (i-th) into another (i-1)-th is defined by

$$\begin{bmatrix} x_{i-1} \\ y_{i-1} \\ z_{i-1} \end{bmatrix} = A_{i-1,i} \begin{bmatrix} x_i \\ y_i \\ z_i \end{bmatrix} + \begin{bmatrix} \ell_{ix_{i-1}} \\ \ell_{iy_{i-1}} \\ \ell_{iz_{i-1}} \end{bmatrix} \qquad (2.4.16a)$$

i.e.

$$r^{(i-1)} = A_{i-1,i}r^{(i)} + \ell_i^{(i-1)} \qquad (2.4.16b)$$

where $r^{(k)}$ is the position vector of the point M with respect to the origin O_k (i.e. $\vec{r}^{(k)} = \overrightarrow{O_kM}$), expressed by projections onto the axes $O_k x_k y_k z_k$. For the external coordinate system, $\vec{r} = \overrightarrow{OM}$.

We repeat that for rotational joint S_{i-1}, $\theta_i = q_i$ i.e. θ_i represents the corresponding generalized coordinate and u_i is a constant characteristic of the segment "i". If S_{i-1} is a linear joint then $u_i = q_i$ and θ_i is a constant.

The transition matrix from the i-th b.-f. system into the external one is

$$A_i = \prod_{j=0}^{i} A_{j-1,j} \tag{2.4.17}$$

and so it holds

$$a_i = A_i a_i^{(i)} \tag{2.4.18}$$

and

$$r = \begin{bmatrix} x \\ y \\ z \end{bmatrix} = \sum_{j=0}^{i} A_j \ell_j^{(j)} + A_i r^{(i)} \tag{2.4.19}$$

I should be pointed out that in the case of immobile, or inertial-moving base the external system is usually adopted to be connected to the base i.e. $Oxyz \equiv O_o x_o y_o z_o$. Then it holds $a_i \equiv a_i^{(o)}$ for each vector.

Kinematical relations. The relative angular velocity $\tilde{\omega}_i^{(i)}$ of the i-th segment with respect to the (i-1)-th one may be expessed in the form

$$\tilde{\omega}_i^{(i)} = (1-s_i)\nu_i \dot{q}_i \tag{2.4.20}$$

and the relative linear velocity $\tilde{v}_i^{(i)}$ in the form:

$$\tilde{v}_i^{(i)} = s_i \nu_i \dot{q}_i, \tag{2.4.21}$$

where

$$\nu_i = [0 \quad \sin\alpha_i \quad \cos\alpha_i]^T \tag{2.4.22}$$

These vectors are expressed in the corresponding b.-f. system.

After a long derivation [23], the expression for the absolute velocity of the i-th segment c.o.g. C_i (i.e. $v_{iC}^{(i)}$) is obtained as is the expession for the segment absolute angular velocity $\omega_i^{(i)}$:

$$\omega_i^{(i)} = \sum_{m=1}^{i} A_{im}\nu_m(1-s_m)\dot{q}_m \tag{2.4.23}$$

$$v_{iC}^{(i)} = -[\sum_{j=1}^{i} A_{ij}\lambda(\ell_j^{(j)})\sum_{m=1}^{j} A_{jm}\nu_m(1-s_m)\dot{q}_m + \lambda(\rho_i^{(i)})\sum_{m=1}^{i} A_{im}\nu_m(1-s_m)\dot{q}_m + \sum_{j=1}^{i} A_{ij}\nu_j s_j \dot{q}_j] \tag{2.4.24}$$

The vectors are expressed by projections onto the axis of the corresponding b.-f. system.

For some vector a_i, the notation $\lambda(a_i)$ designates the matrix

$$\lambda(a_i) = \begin{bmatrix} 0 & -a_{iz} & a_{iy} \\ a_{iz} & 0 & -a_{ix} \\ -a_{iy} & a_{ix} & 0 \end{bmatrix}$$

which is used to perform the vector product in matrix calculus.

The same velocities (2.4.23) and (2.4.24) may be expressed in the external coordinate system:

$$\omega_i = \sum_{m=1}^{i} A_m\nu_m(1-s_m)\dot{q}_m = \sum_{m=1}^{i} e_{m-1}(1-s_m)\dot{q}_m \tag{2.4.25}$$

$$v_{iC} = \sum_{m=1}^{i} [(1-s_m)\lambda(e_{m-1})\rho_{m-1,i} + s_m e_{m-1}]\dot{q}_m, \tag{2.4.26}$$

where

$$\rho_{m-1,i} = \sum_{j=m}^{i} \ell_j + \rho_i \tag{2.4.27}$$

One can derive the expressions for accelerations. If $w_{iC}^{(i)}$ is the acceleration of the i-th segment c.o.g. C_i, and $\varepsilon_i^{(i)}$ the segment angular acceleration, then

$$\varepsilon_i^{(i)} = \sum_{m=1}^{i} A_{im}[\lambda(\omega_m^{(m)})\nu_m(1-s_m)\dot{q}_m + \nu_m(1-s_m)\ddot{q}_m] \tag{2.4.28}$$

$$w_{iC}^{(i)} = [-\sum_{j=1}^{i} A_{ij}\lambda(\ell_j^{(j)})\sum_{m=1}^{j} A_{jm}\nu_m(1-s_m)\ddot{q}_m - \lambda(\rho_i^{(i)})\sum_{m=1}^{i} A_{im}\nu_m(1-s_m)\ddot{q}_m$$

$$+\sum_{m=1}^{i} A_{im}\nu_m s_m \ddot{q}_m] + [-\sum_{j=1}^{i} A_{ij}\lambda(\ell_j^{(j)}) \sum_{m=1}^{j} A_{jm}\lambda(\omega_m^{(m)})\cdot$$

$$\cdot\nu_m(1-s_m)\dot{q}_m - \sum_{j=1}^{i} A_{ij}\lambda(\omega_j^{(j)})\lambda(\ell_j^{(j)}) \sum_{m=1}^{j} A_{jm}\nu_m(1-s_m)\dot{q}_m -$$

$$-\lambda(\rho_i^{(i)}) \sum_{m=1}^{i} A_{im}\lambda(\omega_m^{(m)})\nu_m(1-s_m)\dot{q}_m -$$

$$-\lambda(\omega_i^{(i)})\lambda(\rho_i^{(i)}) \sum_{m=1}^{i} A_{im}\nu_m(1-s_m)\dot{q}_m +$$

$$+2\sum_{m=1}^{i} A_{im}\lambda(\omega_m^{(m)})\nu_m s_m \dot{q}_m] \qquad (2.4.29)$$

and if expressed in the external system,

$$\varepsilon_i = \sum_{m=1}^{i} (1-s_m)e_{m-1}\ddot{q}_m + \sum_{m=2}^{i} \sum_{k=1}^{m-1} (1-s_k)(1-s_m)\lambda(e_{k-1})e_{m-1}\dot{q}_k\dot{q}_m \qquad (2.4.30)$$

$$w_{iC} = \sum_{m=1}^{i} [\lambda(e_{m-1})(1-s_m)\rho_{m-1,i} + s_m e_{m-1}]\ddot{q}_m +$$

$$+ \sum_{m=1}^{i} \{ \sum_{k=1}^{m} \lambda(e_{k-1})\lambda(e_{m-1})\rho_{m-1,i}(1-s_m)(1-s_k)\dot{q}_k +$$

$$+ \sum_{k=m+1}^{i} \lambda(e_{m-1})\lambda(e_{k-1})\rho_{k-1,i}(1-s_m)(1-s_k)\dot{q}_k +$$

$$+ \sum_{k=m+1}^{i} \lambda(e_{m-1})e_{k-1}s_k(1-s_m)\dot{q}_k +$$

$$+ \sum_{k=1}^{m-1} \lambda(e_{k-1})e_{m-1}(1-s_k)s_m\dot{q}_k \}\dot{q}_m \qquad (2.4.31)$$

The derivation and the expressions obtained are so complex because of the kinematical approach via analytical expressions. In the previous method the complexity is avoided by using recursive expressions for velocities and accelerations. Such an approach is much more suitable for numerical computation.

Introducing block matrices. It is useful to use the block matrix formalism to obtain more compact forms of the equations.

Let $a_1,\ldots,a_n$ be a set of vectors, and let us define the block vectors a and a^o of dimensions 3n×1. The block vector a is defined as

$$a = [a_1^{(1)} \cdots a_n^{(n)}]^T \qquad (2.4.32a)$$

and it represents the vectors expressed in the corresponding b.-f. systems. The block vector a^o represents the vectors expressed in the external coordinate system

$$a^o = [a_1 \cdots a_n]^T \qquad (2.4.32b)$$

If the external system coincides with the base coordinate system, then

$$a^o = [a_1^{(o)} \cdots a_n^{(o)}]^T \qquad (2.4.33)$$

Further, let us introduce the 3n×3n block matrix

$$V = \begin{bmatrix} E_3 & 0 & \cdots & 0 \\ E_3 & E_3 & & 0 \\ \vdots & & & \vdots \\ E_3 & E_3 & \cdots & E_3 \end{bmatrix} \qquad (2.4.34)$$

where E_3 is the 3×3 unit matrix. Let us also introduce the block matrix

$$A = \begin{bmatrix} E_3 & 0 & 0 & \cdots & 0 \\ A_{21} & E_3 & 0 & & 0 \\ A_{31} & A_{32} & E_3 & & 0 \\ \vdots & & & & \\ A_{n1} & A_{n2} & A_{n3} & \cdots & E_3 \end{bmatrix} \qquad (2.4.35)$$

Now, the kinematic expressions (2.4.20) and (2.4.21) can be written in the form

$$\tilde{\omega} = \nu(E-s)\dot{q} \qquad (2.4.36)$$

$$\tilde{v} = \nu s\dot{q} \qquad (2.4.37)$$

where ν = diag $[\nu_1 \nu_2 \cdots \nu_n]$ is the 3n×n block matrix, s = diag $[s_1 s_2 \cdots s_n]$ is the n×n matrix, and E is the n×n unit matrix.

The other kinematical expressions can be written in more compact form if block matrix formalism is used. The expressions (2.4.23) and (2.4.24) then become

$$\omega = A\nu(E-s)\dot{q} \tag{2.4.38}$$

$$v = -(A\Lambda(\ell) + \Lambda(\rho))A\nu(E-s)\dot{q} + A\nu s\dot{q} \tag{2.4.39}$$

For a set of vectors $a_1,\ldots,a_n$, the notation $\lambda(a)$ designates

$$\Lambda(a) = \mathrm{diag}[\lambda(a_1)\lambda(a_2) \cdots \lambda(a_n)] \tag{2.4.40}$$

If the $6n\times 1$ block vector $\dot{x} = [v\ \omega]^T$ is introduced, then the expressions (2.4.38) and (2.4.39) can be written together:

$$\dot{x} = B\dot{q} \tag{2.4.41}$$

where the $6n\times n$ matrix B is:

$$B = \begin{bmatrix} B_1 \\ \\ B_2 \end{bmatrix}, \qquad \begin{aligned} B_1 &= -(A\Lambda(\ell) + \Lambda(\rho))A\nu(E-s) + A\nu s \\ B_2 &= A\nu(E-s) \end{aligned} \tag{2.4.42}$$

The velocities expressed in the external system, i.e., the expressions (2.4.25) and (2.4.26), may be written in the form:

$$\omega^o = Ve^o(E-s)\dot{q} = B_2^o\dot{q} \tag{2.4.43}$$

$$v^o = -\Lambda^*(\rho)e^o(E-s)\dot{q} + Vse^o\dot{q} = B_1^o\dot{q}, \tag{2.4.44}$$

where $e^o = \mathrm{diag}[e_o \cdots e_{n-1}]$ is a $3n\times n$ matrix, and

$$\Lambda^*(\rho) = \begin{bmatrix} \lambda(\rho_{o1}) & 0 & \cdots & 0 \\ \lambda(\rho_{o2}) & \lambda(\rho_{12}) & & 0 \\ \vdots & & & \vdots \\ \lambda(\rho_{on}) & \lambda(\rho_{1n}) & \cdots & \lambda(\rho_{n-1,n}) \end{bmatrix} \tag{2.4.45}$$

The expressions (2.4.43) and (2.4.44) may be united into the form

$$\dot{x}^o = B^o\dot{q}, \tag{2.4.46}$$

where

$$B^o = \begin{bmatrix} B_1^o \\ \\ B_2^o \end{bmatrix} \qquad (2.4.47)$$

The accelerations are given via (2.4.28) and (2.4.29) if expressed in b.-f. systems. Introducing the block matrices, a more compact form is obtained:

$$\varepsilon = A\Lambda(\omega)\nu(E-s)\dot{q} + A\nu(E-s)\ddot{q} \qquad (2.4.48)$$

$$\begin{aligned} w = & \left[-(A\Lambda(\ell) + \Lambda(\rho))A\nu(E-s) + A\nu s\right]\ddot{q} + \\ & + \left[-(A\Lambda(\ell) + \Lambda(\rho))A\Lambda(\omega)\nu(E-s) - (A\Lambda(\omega)\Lambda(\ell) + \right. \\ & \left. + \Lambda(\omega)\Lambda(\rho))A\nu(E-s) + 2A\Lambda(\omega)\nu s\right]\dot{q} \end{aligned} \qquad (2.4.49)$$

Introducing $\ddot{x} = [w\ \varepsilon]^T$ it follows that

$$\ddot{x} = B\ddot{q} + D\dot{q}, \qquad (2.4.50$$

where B is determined by (2.4.42), and

$$D = \begin{bmatrix} D_1 \\ \\ D_2 \end{bmatrix}; \quad \begin{aligned} D_1 = & (A\Lambda(\ell)+\Lambda(\rho))A\Lambda(\omega)\nu(E-s)-(A\Lambda(\omega)\Lambda(\ell)+ \\ & + \Lambda(\omega)\Lambda(\rho))A\nu(E-s) + 2A\Lambda(\omega)\nu s \\ D_2 = & A\Lambda(\omega)\nu(E-s) \end{aligned} \qquad (2.4.51)$$

In the external coordinate system, i.e., (2.4.30) and (2.4.31), it follows (for the angular accelerations), that

$$\varepsilon^o = B_2^o\ddot{q} + D_2^o\dot{q}, \qquad (2.4.52)$$

where B_2^o is determined by (2.4.43). $D_2^o = [D_1^2 \cdots D_n^2]^T$ is the $3n\times n$ block matrix, where

$$D_i^2 = \dot{Q} \times H_i^2 \text{ (dimension } 3\times n) \qquad (2.4.53)$$

$$\dot{Q} = [\dot{q}_1E_3\ \dot{q}_2E_3 \cdots \dot{q}_nE_3]$$

$$H_i^2 = \begin{bmatrix} 0 & 0 & \cdots & 0 & \cdots & 0 \\ \sigma_1\sigma_2\lambda(e_o)e_1 & 0 & \cdots & 0 & \cdots & 0 \\ \sigma_1\sigma_3\lambda(e_o)e_2 & \sigma_2\sigma_3\lambda(e_1)e_2 & \cdots & 0 & \cdots & 0 \\ \vdots & \vdots & & & & \vdots \\ \sigma_1\sigma_i\lambda(e_o)e_{i-1} & \sigma_2\sigma_i\lambda(e_1)e_{i-1} & \cdots & \sigma_{i-1}\sigma_i\lambda(e_{i-2})e_{i-1} & \cdots & 0 \\ 0 & 0 & \cdots & 0 & & 0 \\ \vdots & \vdots & & & & \vdots \\ 0 & 0 & \cdots & 0 & \cdots & 0 \end{bmatrix}$$

where $\sigma_i = (1-s_i)$,

It also follows (for c.o.g. accelerations) that

$$w^o = B_1^o\ddot{q} + D_1^o\dot{q}, \tag{2.4.54}$$

where B_1^o is determined by (2.4.44). $D_1^o = [D_1^1 \cdots D_n^1]^T$ is the $3n\times n$ block matrix, where

$$D_i^1 = \dot{Q}(D_{i1}^1 + D_{i2}^1) \text{ (dimension } 3\times n)$$

$$D_{i1}^1 = \begin{bmatrix} d_{11}^1 & \cdots & d_{1i}^1 & 0 & \cdots & 0 \\ \vdots & & \vdots & \vdots & & \vdots \\ d_{i-1,1}^1 & \cdots & d_{i-1,i}^1 & 0 & \cdots & 0 \\ 0 & \cdots & 0 & 0 & \cdots & 0 \\ \vdots & & \vdots & \vdots & & \vdots \\ 0 & \cdots & 0 & 0 & \cdots & 0 \end{bmatrix}$$

$$d_{kj}^1 = \lambda(e_{j-1})\lambda(e_k)\rho_{j-1,i}(1-s_k)(1-s_j)$$

$$D_{i2}^{1} = \begin{bmatrix} 0 & d_{12}^{2} & \cdots & d_{1i}^{2} & 0 & \cdots & 0 \\ d_{21}^{2} & 0 & \cdots & d_{2i}^{2} & 0 & \cdots & 0 \\ \vdots & \vdots & & \vdots & \vdots & & \vdots \\ d_{i1}^{2} & d_{i2}^{2} & \cdots & 0 & 0 & \cdots & 0 \\ 0 & 0 & \cdots & 0 & 0 & \cdots & 0 \\ \vdots & \vdots & & \vdots & \vdots & & \vdots \\ 0 & 0 & \cdots & 0 & 0 & \cdots & 0 \end{bmatrix}$$

$$d_{kj}^{2} = \lambda(e_{k-1})e_{j-1}(1-s_k)s_j$$

$$d_{jk}^{2} = d_{kj}^{2}$$

The expressions (2.4.52) and (2.4.54) may be united into the form

$$\ddot{x}^{o} = \begin{bmatrix} w^{o} \\ \varepsilon^{o} \end{bmatrix} = B^{o}\ddot{q} + D^{o}\dot{q} \qquad (2.4.55)$$

where

$$D^{o} = \begin{bmatrix} D_{1}^{o} \\ D_{2}^{o} \end{bmatrix}$$

Finally, let us introduce the notation

$$\Lambda(\omega) = \Omega \qquad \Lambda(\omega^{o}) = \Omega^{o}$$

$$\Lambda(\ell) = L \qquad \Lambda(\ell^{o}) = L^{o}$$

$$\Lambda(\rho) = R \qquad \Lambda(\rho^{o}) = R^{o}$$

Forming the equation system. The resultant inertial force of the i-th segment, in vector notation, is:

$$F_{Ii} = -m_i w_{iC} \qquad (2.4.56)$$

By introducing the block matrices and taking care about (2.4.54),

$$F_I^o = -mw^o = -m(B_1^o\ddot{q} + D_1^o\dot{q}), \tag{2.4.57}$$

where $m = \text{diag}[m_1E_3 \cdots m_nE_3]$ (dimension 3n×3n).

Let us now determine the resultant moment (resultant couple) of inertial forces (M_{Ii}) for each segment, relative to its c.o.g. C_i. Let us introduce the so-called center of gravity - fixed system (c.o.g.-f. system) $O_{Ci}x_{Ci}y_{Ci}z_{Ci}$. The origin O_{Ci} coincides with the c.o.g. C_i and the system axes are parallel to the axes of the b.-f. system $O_ix_iy_iz_i$. Let $J_i^{(iC)}$ be the tensor of inertia with respect to the c.o.g.-f. system, i.e.,

$$J_i^{(iC)} = \begin{bmatrix} I_{x_{Ci}x_{Ci}} & -I_{x_{Ci}y_{Ci}} & -I_{x_{Ci}z_{Ci}} \\ -I_{x_{Ci}y_{Ci}} & I_{y_{Ci}y_{Ci}} & -I_{y_{Ci}z_{Ci}} \\ -I_{x_{Ci}z_{Ci}} & -I_{y_{Ci}z_{Ci}} & I_{z_{Ci}z_{Ci}} \end{bmatrix} \tag{2.4.58}$$

The moment of momentum $\mathfrak{G}_i^{(iC)}$ in the c.o.g.-f. system is

$$\mathfrak{G}_i^{(iC)} = J_i^{(iC)}\omega_i^{(i)} \tag{2.4.59}$$

Expressed in the external system (in this case, the base system):

$$\mathfrak{G}_i^o = A_{oi}\,\mathfrak{G}_i^{(iC)}. \tag{2.4.60}$$

The resultant moment of inertial forces is now

$$M_{Ii} = -\frac{d}{dt}\mathfrak{G}_i^o = -\frac{d}{dt}[A_{oi}\,\mathfrak{G}_i^{(iC)}] \tag{2.4.61}$$

Then

$$M_{Ii} = -A_{oi}\lambda(\omega_i^{(i)})J_i^{(iC)}\omega_i^{(i)} - A_{oi}J_i^{(iC)}\varepsilon_i^{(i)} \tag{2.4.62}$$

By introducing the block matrices, relation (2.4.62) is written in the form

$$M_I^o = -A_o^T\Omega J\omega - A_o^TJ\varepsilon, \tag{2.4.63}$$

where

$$J = \text{diag}[J_1^{(1C)} \cdots J_n^{(nC)}]$$

$$A_o = \text{diag}[A_{10} \cdots A_{n0}]$$

Using the expressions previously derived for ω and ε, equation (2.4.63) becomes

$$M_I^o = -A_o^T J B_2 \ddot{q} - A_o^T (\Omega J B_2 + J D_2)\dot{q} \tag{2.4.64}$$

and expressed in the b.-f. coordinate systems

$$M_I = -JB_2\ddot{q} - (\Omega J B_2 + J D_2)\dot{q} \tag{2.4.65}$$

The block vector M_I^o can be expressed as a function of ω^o, ε^o, so that

$$M_I^o = -J^o B_2^o \ddot{q} - (\Omega^o J^o B_2^o + J^o D_2^o)\dot{q} \tag{2.4.66}$$

with

$$J^o = \mathrm{diag}\,[A_i J_i^{(iC)} A_i^T] = A_o J A_o^T$$

$$B_2^o = Ve^o(E-s), \quad \Omega^o = \Lambda(B_2^o \dot{q})$$

The expressions for the block vectors of the resultant inertial forces and the resultant inertial moments, i.e., expressions (2.4.57) and (2.4.66), can be written together:

$$\begin{bmatrix} F_I^o \\ M_I^o \end{bmatrix} = -\begin{bmatrix} m\,B_1^o \\ J^o B_2^o \end{bmatrix}\ddot{q} - \begin{bmatrix} m\,D_1^o \\ \Omega^o J^o B_2^o + J^o D_2^o \end{bmatrix}\dot{q} = -\mathcal{J}\,B^o\ddot{q} - C^o\dot{q}, \tag{2.4.67}$$

where

$$\mathcal{J} = \begin{bmatrix} m & 0 \\ 0 & J^o \end{bmatrix}, \qquad C^o = \begin{bmatrix} m\,D_1^o \\ \Omega^o J^o B_2^o + J^o D_2^o \end{bmatrix} \tag{2.4.68}$$

Let us further introduce the vector of the drives P of dimension $n\times 1$:

$$P = [P_1 \cdots P_n]^T \tag{2.4.69}$$

Let us now return to the dynamic equations (2.4.7) and (2.4.9). Using the block matrices, the equations can be written together in the form

$$[(G^o + F_E^o + F_I^o)^T \vdots (M_E^o + M_I^o)^T]B^o + P^T = 0 \tag{2.4.70}$$

or

$$B^{oT}[G^o + F_E^o + F_I^o \vdots M_E^o + M_I^o]^T + P = 0 \tag{2.4.71}$$

Substituting the expressions for the block vectors M_I^o and F_I^o, i.e. the expressions (2.4.57) and (2.4.66), into the equations (2.4.71), we get

$$B^{oT}[G^{o} + F_E^{o} \vdots M_E^{o}]^T - B^{oT}\mathcal{J}B^{o}\ddot{q} - B^{oT}C^{o}\dot{q} + P = 0. \tag{2.4.72}$$

The equation obtained can be written in the form

$$W(q)\ddot{q} = B'(q, \dot{q})\dot{q} + C'(q)M_E^{o} + D'(q)(G^{o}+F_E^{o}) + P, \tag{2.4.73}$$

where

$$W = B^{oT}\mathcal{J}B^{o}, \qquad B' = -B^{oT}C^{o},$$

$$C' = B_2^{oT}, \qquad D' = B_1^{oT}$$

and the parentheses demonstrate that the $n\times n$ matrix W, the $n\times 3n$ matrix C' and the $n\times 3n$ matrix D' depend on the generalized coordinates q, and that the matrix B' depends on the coordinates q and velocities $\dot{q}$. By introducing

$$U(q, \dot{q}) = B'\dot{q} + C'M_E^{o} + D'(G^{o}+F_E^{o}), \tag{2.4.74}$$

equation (2.4.73) acquires the form (2.4.1), i.e.,

$$W(q)\ddot{q} = U(q, \dot{q}) + P.$$

2.5. The Method of the Newton-Euler Equations

This method for c.-a. solution of the direct problem of dynamics is based on the Newton-Euler equations already used. It was proposed in [12] after longer experience with deriving the mathematical model analytically [11].

We are concerned with an algorithm for computer realization of the function g defined in (2.2.2), i.e.:

$$P = g(q, \dot{q}, \ddot{q}, \text{mechanism configuration}) \tag{2.5.1}$$

where P is the vector of the driving forces and torques in the joints.

The inverse problem, i.e. the function f, cannot be treated by this method.

In presenting this method, the original designations will be significantly modified for the sake of brevity.

The kinematical scheme of the considered mechanism is in the form of a kinematical chain without branching (Fig. 2.15) with rotational and linear joints with one degree of freedom each. There are in all n degrees of freedom.

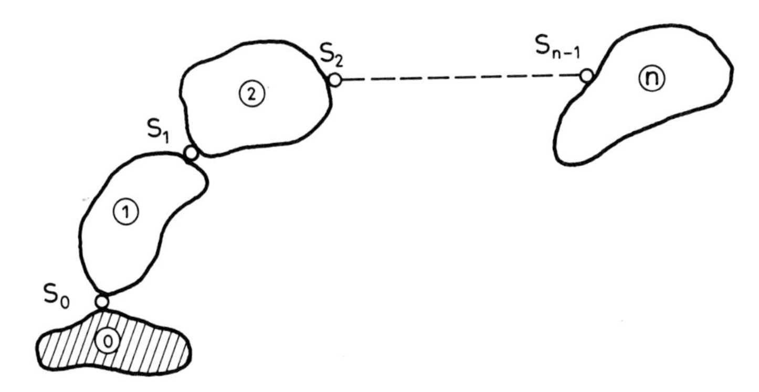

Fig. 2.15. Open kinematical chain

Generalized coordinates are chosen for each joint. If S_{i-1} is a rotational joint, the corresponding generalized coordinate q_i is defined as the angle of rotation θ_i around the rotation axis. If S_{i-1} is a linear joint, the corresponding generalized coordinate q_i is defined as the displacement u_i along the translation axis.

Coordinate systems and transition matrices. For each segment a body-fixed (b.-f.) system is defined, as described in [11]. The same coordinate systems were used in the previous method (2.4), and the definition of such systems and the derivation of the corresponding transition matrices were explained in detail. Thus, if $A_{i,i-1}$ designates the transition matrix from the (i-1)-th to the i-th coordinate system, then

$$A_{i,i-1} = \begin{bmatrix} \cos\theta_i & \sin\theta_i & 0 \\ -\sin\theta_i\cos\alpha_i & \cos\theta_i\cos\alpha_i & \sin\alpha_i \\ \sin\theta_i\sin\alpha_i & -\cos\theta_i\sin\alpha_i & \cos\alpha_i \end{bmatrix} = A^{-1}_{i-1,i} \qquad (2.5.2)$$

The transition matrix from the i-th system to the base system is obtained as

$$A_i = A_{o,i} = A_{o,1} \cdots A_{i-2,i-1} A_{i-1,i} \qquad (2.5.3)$$

The body-fixed coordinate systems were introduced because it is appropriate to express the required dynamical equations in such systems.

Kinematical relations. In order to avoid complex expressions and derivations, this method uses the recursive expressions for segment velocities and accelerations already mentioned. These expressions are written here in b.-f. coordinate systems.

$\vec{v}_{Ci}$, the vector of the i-th segment center of gravity (c.o.g.) velocity

$\vec{w}_{Ci}$, the vector of the i-th segment center of gravity acceleration

$\vec{v}_i$, the vector of the i-th coordinate system origin (O_i) velocity

$\vec{w}_i$, the vector of the i-th coordinate system origin (O_i) acceleration

$\vec{\omega}_i$, the vector of the i-th segment angular velocity

$\vec{\varepsilon}_i$, the vector of the i-th segment angular acceleration

$\vec{e}_i$, the unit vector of the $O_i z_i$-axis of the i-th coordinate system, i.e. unit vector of the axis of rotation or translation in joint S_i

$\vec{\ell}_i = \overrightarrow{O_{i-1}O_i}$ (Fig. 2.16a)

$\vec{r}_i = \overrightarrow{O_{i-1}C_i}$; C_i, the i-th segment center of gravity (Fig. 2.16a).

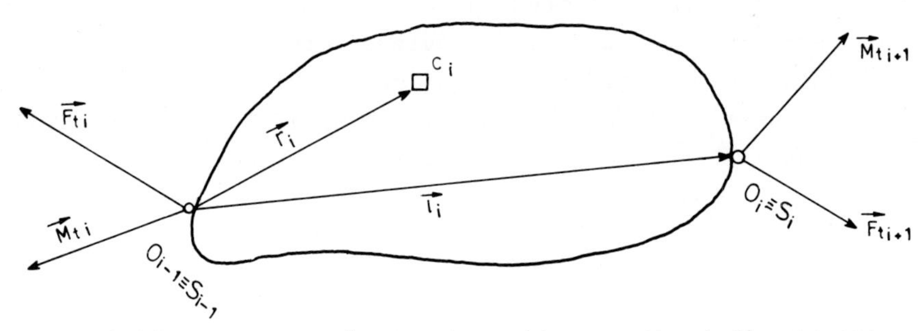

Fig. 2.16a. Forces and moments acting on the i-th segment

We introduce the following notation: $\vec{a}_i$ denotes some vector of the i-th segment, expressed in the base system or in the external system; $\vec{\tilde{a}}_i$ denotes the same vector, but expressed in the i-th b.-f. system.

Now the following recurrent expressions can be written:

$$\vec{\tilde{w}}_{Ci} = \vec{\tilde{\omega}}_i \times (\vec{\tilde{\omega}}_i \times \vec{\tilde{r}}_i) + \vec{\tilde{\varepsilon}}_i \times \vec{\tilde{r}}_i + \vec{\tilde{w}}_i \tag{2.5.4a}$$

and for the (i+1)-th segment:

$$\vec{\tilde{w}}_{C_{i+1}} = \vec{\tilde{\omega}}_{i+1} \times (\vec{\tilde{\omega}}_{i+1} \times \vec{\tilde{r}}_{i+1}) + \vec{\tilde{\varepsilon}}_{i+1} \times \vec{\tilde{r}}_{i+1} + \vec{\tilde{w}}_{i+1} \tag{2.5.4b}$$

If S_i is a rotational joint

$$\vec{\tilde{\omega}}_{i+1} = A_{i+1,i}(\vec{\tilde{\omega}}_i + \vec{\tilde{e}}_i \dot{q}_{i+1}) \tag{2.5.5}$$

$$\vec{\tilde{\varepsilon}}_{i+1} = A_{i+1,i}(\vec{\tilde{\varepsilon}}_i + \vec{\tilde{e}}_i \ddot{q}_{i+1} + \vec{\tilde{\omega}}_i \times \vec{\tilde{e}}_i \dot{q}_{i+1}) \tag{2.5.6}$$

$$\vec{\tilde{w}}_{i+1} = \vec{\tilde{\varepsilon}}_{i+1} \times \vec{\tilde{\ell}}_{i+1} + \vec{\tilde{\omega}}_{i+1} \times (\vec{\tilde{\omega}}_{i+1} \times \vec{\tilde{\ell}}_{i+1}) + A_{i+1,i}\vec{\tilde{w}}_i \tag{2.5.7}$$

If S_i is a linear joint,

$$\vec{\tilde{\omega}}_{i+1} = A_{i+1,i}\vec{\tilde{\omega}}_i \tag{2.5.8}$$

$$\vec{\tilde{\varepsilon}}_{i+1} = A_{i+1,i}\vec{\tilde{\varepsilon}}_i \tag{2.5.9}$$

$$\vec{\tilde{w}}_{i+1} = \vec{\tilde{\varepsilon}}_{i+1} \times \vec{\tilde{\ell}}_{i+1} + \vec{\tilde{\omega}}_{i+1} \times (\vec{\tilde{\omega}}_{i+1} \times \vec{\tilde{\ell}}_{i+1}) + + 2\vec{\tilde{\omega}}_{i+1} \times (A_{i+1,i}\vec{\tilde{e}}_i \dot{q}_{i+1}) + A_{i+1,i}(\vec{\tilde{w}}_i + \vec{\tilde{e}}_i \ddot{q}_{i+1}) \tag{2.5.10}$$

It should be mentioned that the vectors $\vec{\tilde{\ell}}_j$, $\vec{\tilde{r}}_j$ are constant and represent the characteristic of the segment itself (Fig. 2.16a). Vector $\vec{\tilde{e}}_i = \{0,\ 0,\ 1\}$ is also constant.

Mechanism dynamics. Let us consider one mechanism segment, the i-th one (Fig. 2.16a). Let $\vec{M}_{ti}$ be the total moment, acting in the joint S_{i-1} on the i-th segment, and $\vec{F}_{ti}$ be the total force, acting on the same segment in that joint.

Further, let $\vec{\tilde{F}}_i^{res}$ be the total resultant force acting on the i-th segment, and $\vec{\tilde{M}}_i^{res}$ the resultant moment relative to the segment c.o.g. Now the theorem about the center of gravity motion and the Euler equations, applied to the i-th segment and written in the b.-f. system, facilitate the calculation of $\vec{\tilde{F}}_i^{res}$, $\vec{\tilde{M}}_i^{res}$.

$$\vec{\tilde{F}}_i^{res} = m_i \vec{\tilde{w}}_{Ci} \tag{2.5.11}$$

$$\vec{\tilde{M}}_i^{res} = \tilde{J}_e \vec{\tilde{\varepsilon}}_i - (\tilde{J}_i \vec{\tilde{\omega}}_i) \times \vec{\tilde{\omega}}_i, \tag{2.5.12}$$

where m_i is the i-th segment mass, and $\tilde{J}_i$ the i-th segment tensor of inertia with respect to the c.o.g.-fixed system.

Finally, the relation between the resultant forces and moments and the forces and moments in the joints is given by

$$\vec{\tilde{F}}_{ti} = \vec{\tilde{F}}_i^{res} - m_i\vec{\tilde{g}}_i + A_{i,i+1}\vec{\tilde{F}}_{ti+1} \tag{2.5.13}$$

$$\vec{\tilde{M}}_{ti}=A_{i,i+1}\vec{\tilde{M}}_{ti+1}+\vec{\tilde{\ell}}_i\times A_{i,i+1}\vec{\tilde{F}}_{ti+1}+\vec{\tilde{M}}_i^{res}+\vec{\tilde{r}}_i\times(\vec{\tilde{F}}_i^{res}-m_i\vec{\tilde{g}}) \tag{2.5.14}$$

where $\vec{\tilde{g}}_i = A_i^{-1}\vec{g}$, and $\vec{g} = \{0, \; 0, \; -9,81\}$ is the gravitational acceleration vector (this numerical value holds if the z-axis of the base system is vertical).

Let us also find the relation between the forces and moments in the joints, $\vec{F}_{ti}$, $\vec{M}_{ti}$, and the drives in the same joints. Let us suppose that S_{i-1} is a linear joint. Then, a driving force $\vec{P}_i = P_i^F\vec{e}_{i-1}$ is acting in the joint and the total force in the joint, $\vec{F}_{ti}$, is

$$\vec{F}_{ti} = \vec{P}_i + \vec{F}_{Ri}, \tag{2.5.15}$$

where $\vec{F}_{Ri}$ is the reaction force in the joint S_{i-1}, whereby $\vec{F}_{Ri} \perp \vec{e}_{i-1}$. By acalar multiplication of (2.5.15) by $\vec{e}_{i-1}$,

$$P_i^F = \vec{F}_{ti}\vec{e}_{i-1}. \tag{2.5.16}$$

If S_{i-1} is a rotational joint, then

$$\vec{M}_{ti} = \vec{P}_i + \vec{M}_{Ri}, \quad \vec{M}_{Ri} \perp \vec{e}_{i-1}, \tag{2.5.17}$$

where $\vec{P}_i = P_i^M\vec{e}_{i-1}$ is the driving torque in the joint and $\vec{M}_{Ri}$ is the reaction moment. By scalar multiplication by $\vec{e}_{i-1}$,

$$P_i^M = \vec{M}_{ti}\vec{e}_{i-1} \tag{2.5.18}$$

Now the algorithm for calculating the drives P_i, i=1,...,n for known q_i, $\dot{q}_i$, $\ddot{q}_i$, i=1,...,n can be explained in principle. By knowing the motion (q_i, $\dot{q}_i$, $\ddot{q}_i$, i=1,...,n) and applying the recursive expressions (2.5.4) - (2.5.10), all kinematical values $\vec{v}_i$, $\vec{w}_i$, $\vec{\omega}_i$, $\vec{\varepsilon}_i$, $\vec{w}_{Ci}$, i=1,...,n, can be calculated. The initial conditions for this kinematical "forward" recursion are determined by the prescribed motion of the mechanism base. We apply the "backward" recursion, using expressions (2.5.11), (2.5.12), to determine $\vec{F}_i^{res}$, $\vec{M}_i^{res}$ and the "backward" recursion expressions (2.5.13), (2.5.14) to determine $\vec{M}_{ti}$, $\vec{F}_{ti}$, and finally, the expressions (2.5.16), (2.5.18) to determine the drives in the joints. The initial conditions for this dynamical "backward" recursion are $\vec{M}_{tn+1} = \vec{M}_{end}$, $\vec{F}_{tn+1} = \vec{F}_{end}$ (Fig. 2.16b). $\vec{F}_{end}$, $\vec{M}_{end}$ equal zero if the last segment (the n-th) is free. In fact, they represent the force and the couple of connection between the last segment (the manipulator gripper) and the object.

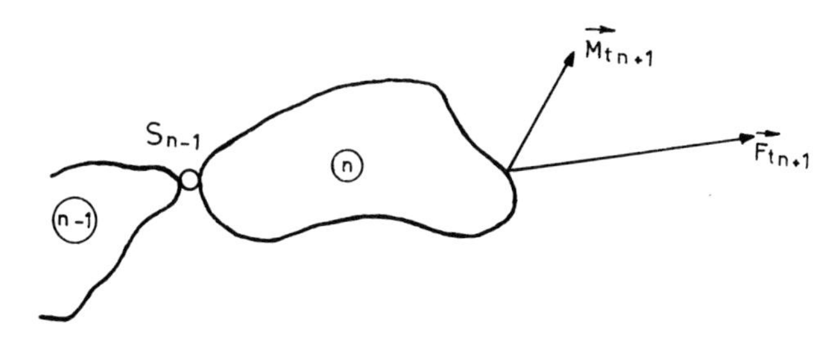

Fig. 2.16b. Boundary conditions on the last segment

2.6. The Method of Euler's Angles

This is another method utilizing kinetostatics as a mechanical approach [2]. Like the method of "block" matrices 2.4., this method is also less computer oriented. However, the analytically derived model is written in compact matrix form convenient for the usage of digital computers. This method is also presented in [6, 13].

Starting essentials. The method considers a mechanism with 3n degrees of freedom and describes it by a 3n-dimensional vector of generalized coordinates q. The kinetostatic method is applied to give a system of equations in the form

$$W\ddot{q} = P + U, \tag{2.6.1}$$

where P is the vector of driving torques in the joints; the functions f and g defined in 2.2. are thus realized as

$$\ddot{q} = f(q, \dot{q}, P, \text{configuration}) = W^{-1}(P+U) \tag{2.6.2}$$

$$P = g(q, \dot{q}, \ddot{q}, \text{configuration}) = W\ddot{q} - U \tag{2.6.3}$$

Mechanism configuration. The method considers an open-chain type mechanism with possible branching (Fig. 2.17a).

For ease in defining the configuration, let us introduce the following definitions and notation (Fig. 2.17a):

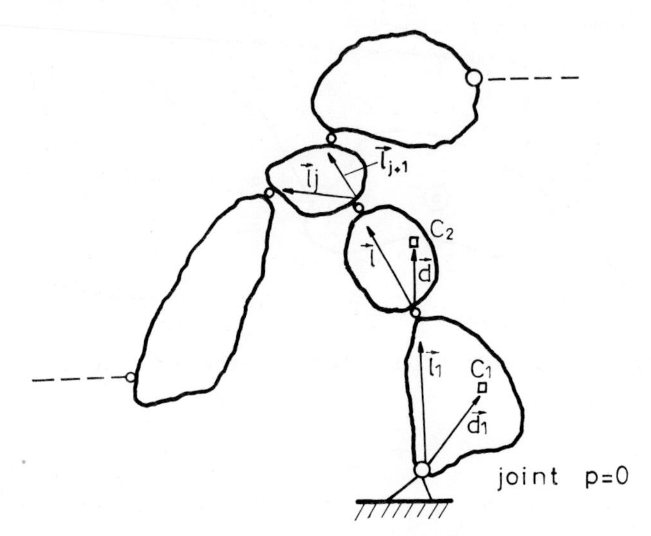

Fig. 2.17a. Open branching chain

- There are n rigid bodies subscripted by $i=1,\ldots,n$.
- The rigid bodies (segments) are interconnected by joints with three rotational degrees of freedom (spheric joints are included).
- The segments are only simply interconnected.
- Each segment "i" has its mass m_i and its inertia tensor $\tilde{J}_i$ with respect to the corresponding body-fixed system. A distance $\vec{d}_i$ from the first joint to the center of gravity C_i is also defined. The first joint of a segment is defined as the joint closest to the fixed support point.
- Since the segments are simply interconnected, there are as many joints as there are segments. Here, the fixed point $p = 0$ is included as the joint between the first segment and the fixed basis. The joints are denoted by S_p, $p = 0, 1,\ldots,m$; $(m = n - 1)$.
- Let us define a length $\vec{\ell}$ between the first and the other joints on the same segment. Each segment has none, one or more lengths $\vec{\ell}$. So

there are m lengths $\vec{\ell}$ in the whole mechanism; let us denote them by $\vec{\ell}_j$, j=1,...,m. Their indices are not related to the index of the segment on which they are but apply to the mechanism as a whole.

Let us now introduce structural matrices to be used in the algorithm for describing and prescribing the structure of the mechanism:

1. The matrix B^* is defined as having the elements β_{ip} equal to unity if the segment "i" contributes to the moment about the joint "p"; otherwise they are equal to zero. Hence the matrix describes the role of the segments with respect to the joint moments.

2. The matrix D^* (three-dimensional) is defined as having the elements δ_{ijp} equal to unity if the length $\vec{\ell}_j$ lies on the positive path from the joint "p" to the first joint of the segment "i"; otherwise they are equal to zero. The matrix describes the role of $\vec{\ell}_j$ in deriving the moment about the joint "p" due to the segment "i".

3. The matrix Γ^* is defined as having the elements γ_{ij} equal to one if the length $\vec{\ell}_j$ lies on the segment "i"; otherwise they are equal to zero. Hence the matrix connects a segment and its lengths $\vec{\ell}$.

Coordinate systems. Let us define a body-fixed (b.-f.) cartesian coordinate system for each segment. The origin of the coordinate system will be in the center of gravity (c.o.g.) of the segment, and the axes will be set arbitrarily. Let us also define a fixed external cartesian coordinate system as haivng its origin in the point of mechanism-to-ground contact (the joint p = 0) and with vertical z-axis.

Let us introduce the following notation: $\vec{a}_i$ represents a vector characteristic of the i-th segment, expressed in the fixed external system; $\vec{\tilde{a}}_i$ represents the same vector expressed in the i-th b.-f. system.

Let us also introduce matrix notation: a represents a 3×1 matrix corresponding to the vector $\vec{a}$, for each vector in the text.

In addition to these, the so-called coordinate systems of joints will also be used. Each joint "p" will have the index of the first and second segments it connects, defined by the numbers (p, 1) and (p, 2). The origin of the coordinate system of a joint is put in the joint. The first axis is attached to the first segment and the directional cosines of that axis relative to the segment will be determined by $(\ell, m, n)_{p,1}$. The second axis is attached to the second segment, and the directional cosines of the axis relative to the segment will be determined by $(\ell, m, n)_{p,2}$. If the hardware exes are in the joint then the joint system axes are set along the hardware axes. If there are no hardware axes (for instance a spheric joint), then, the first two axes of the joint system are connected to the corresponding segments arbitrarly. The third axis is defined as being perpendicular to the first two. Of course, such a system is not orthogonal.

Generalized coordinates. Let us introduce 3 generalized coordinates for each of n segments (a total of 3n coordinates). The generalized coordinates for the i-th segment will be defined as three Euler angles of the b.-f. coordinate system relative to the fixed external system (Fig. 2.17b), i.e., θ_i, ψ_i, φ_i. So, the vector of generalized coordinates will be

$$q = [\theta_1\psi_1\varphi_1 \cdots \theta_n\psi_n\varphi_n]^T \tag{2.6.4}$$

Let us also define the subvector for each segment:

$$\eta_i = [\theta_i\psi_i\varphi_i]^T, \tag{2.6.5}$$

which determines the position of a segment relative to the external space.

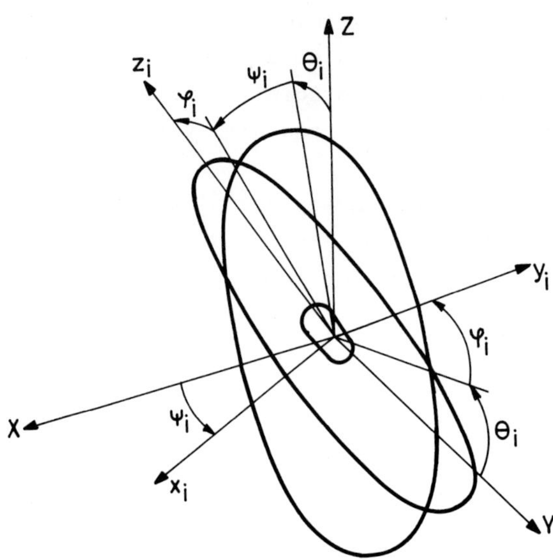

Fig. 2.17b. Set of Euler angles of the i-th segment

We now have

$$q = \begin{bmatrix} \eta_1 \\ \vdots \\ \eta_n \end{bmatrix} \tag{2.6.6}$$

The characteristic of the introduction of generalized coordinates in this method is the fact that these coordinates are "external", i.e., they determine the angular position of the body relative to the external fixed space. In the methods described so far, the generalized coordinates were "internal", i.e., they determined the relative position of two segments. The introduction of "external" coordinates is convenient for certain classes of tasks, particularly for locomotion tasks.

Transition matrices. With the generalized coordinates defined in this way, let us now form transition matrices from b.-f. systems to the fixed external system. Let us define the transition matrix A_i for the i-th segment

$$\vec{a}_i = A_i \vec{\tilde{a}}_i \tag{2.6.7}$$

The transition matrix A_i is obtained in the form

$$A_i = A_i^\theta A_i^\psi A_i^\varphi \triangleq (A^\theta A^\psi A^\varphi)_i \tag{2.6.8}$$

where

$$A_i^\theta = \begin{bmatrix} 1 & 0 & 0 \\ 0 & \cos\theta_i & -\sin\theta_i \\ 0 & \sin\theta_i & \cos\theta_i \end{bmatrix}$$

$$A_i^\psi = \begin{bmatrix} \cos\psi_i & 0 & \sin\psi_i \\ 0 & 1 & 0 \\ -\sin\psi_i & 0 & \cos\psi_i \end{bmatrix} \tag{2.6.9}$$

$$A_i^\varphi = \begin{bmatrix} 1 & 0 & 0 \\ 0 & \cos\varphi_i & -\sin\varphi_i \\ 0 & \sin\varphi_i & \cos\varphi_i \end{bmatrix}$$

The inverse matrix is obtained as

$$A_i^{-1} = A_i^{\varphi^T} A_i^{\psi^T} A_i^{\theta^T} \tag{2.6.10}$$

A point to be noted is that the transition matrix A_i depends only on the Euler angles of the i-th segment.

Angular motion of segments. As already stated, we will use matrix notation. Let us consider the i-th segment. The following is obtained for the angular velocity

$$\tilde{\omega}_i = \Pi_i \dot{\eta}_i, \tag{2.6.11}$$

where

$$\Pi_i = \begin{bmatrix} \cos\psi_i & 0 & 1 \\ \sin\psi_i \sin\varphi_i & \cos\varphi_i & 0 \\ \sin\psi_i \cos\varphi_i & -\sin\varphi_i & 0 \end{bmatrix} \tag{2.6.12}$$

Let us now write the angular acceleration $\vec{\varepsilon}_i$. The following may be obtained from (2.6.11)

$$\tilde{\varepsilon}_i = \dot{\tilde{\omega}}_i = \Pi_i \ddot{\eta}_i + {}^1\Pi_i \, {}^1\dot{\eta}_i, \tag{2.6.13}$$

where

$${}^1\Pi_i = \begin{bmatrix} 0 & 0 & -\sin\psi_i \\ -\sin\varphi_i & \sin\psi_i \cos\varphi_i & \cos\psi_i \sin\varphi_i \\ -\cos\varphi_i & -\sin\psi_i \sin\varphi_i & \cos\psi_i \cos\varphi_i \end{bmatrix}; \quad {}^1\dot{\eta}_i = \begin{bmatrix} \dot{\psi}_i \dot{\varphi}_i \\ \dot{\varphi}_i \dot{\theta}_i \\ \dot{\theta}_i \dot{\psi}_i \end{bmatrix} \tag{2.6.14}$$

The moment of momentum of a segment is expressed as follows

$$\tilde{\mathcal{G}}_i = \tilde{J}_i \tilde{\omega}_i, \tag{2.6.15}$$

where the inertia tensor is

$$\tilde{J}_i = \begin{bmatrix} I_{x_i x_i} & -I_{x_i y_i} & -I_{x_i z_i} \\ -I_{y_i x_i} & I_{y_i y_i} & -I_{y_i z_i} \\ -I_{z_i x_i} & -I_{z_i y_i} & I_{z_i z_i} \end{bmatrix} \tag{2.6.16}$$

The external moment that causes the angular motion of the segment relative to its center of gravity is obtained as the time derivative of the moment of momentum

$$\tilde{M}_i = \frac{d}{dt}\tilde{\mathcal{G}}_i + \tilde{\underline{\underline{\omega}}}_i\tilde{\mathcal{G}}_i, \tag{2.6.17}$$

where

$$\tilde{\underline{\underline{\omega}}}_i = \begin{bmatrix} 0 & -\omega_{z_i} & \omega_{y_i} \\ \omega_{z_i} & 0 & -\omega_{x_i} \\ -\omega_{y_i} & \omega_{x_i} & 0 \end{bmatrix} \tag{2.6.18}$$

is used for performing vector multiplication by the vector $\vec{\tilde{\omega}}_i = \{\omega_{x_i}, \omega_{y_i}, \omega_{z_i}\}$ but in matrix calculus.

(2.6.17) yields

$$\tilde{M}_i = \tilde{J}_i\tilde{\varepsilon}_i + \tilde{\underline{\underline{\omega}}}_i\tilde{J}_i\tilde{\omega}_i \tag{2.6.19}$$

or

$$\tilde{M}_i = \tilde{J}_i\tilde{\varepsilon}_i + {}^1\tilde{J}_i\,{}^2\tilde{\omega}_i + {}^2\tilde{J}_i\,{}^1\tilde{\omega}_i \tag{2.6.20}$$

$${}^1\tilde{J}_i = \begin{bmatrix} 0 & -I_{z_iy_i} & I_{y_iz_i} \\ I_{z_ix_i} & 0 & -I_{x_iz_i} \\ -I_{y_ix_i} & I_{x_iy_i} & 0 \end{bmatrix}; \quad {}^2\tilde{\omega}_i = \begin{bmatrix} \omega_{x_i}^2 \\ \omega_{y_i}^2 \\ \omega_{z_i}^2 \end{bmatrix}$$

$${}^2\tilde{J}_i = \begin{bmatrix} I_{z_iz_i} - I_{y_iy_i} & I_{y_ix_i} & -I_{z_ix_i} \\ -I_{x_iy_i} & I_{x_ix_i} - I_{z_iz_i} & I_{z_iy_i} \\ I_{x_iz_i} & -I_{y_iz_i} & I_{y_iy_i} - I_{x_ix_i} \end{bmatrix}; \quad {}^1\tilde{\omega}_i = \begin{bmatrix} \omega_{y_i}\omega_{z_i} \\ \omega_{z_i}\omega_{x_i} \\ \omega_{x_i}\omega_{y_i} \end{bmatrix} \tag{2.6.21}$$

Starting with expression (2.6.11), the following relations may be derived

$$\begin{aligned} {}^2\tilde{\omega}_i &= {}^2\Pi_i\,{}^2\dot{\eta}_i + {}^3\Pi_i\,{}^1\dot{\eta}_i \\ {}^1\tilde{\omega}_i &= {}^4\Pi_i\,{}^2\dot{\eta}_i + {}^5\Pi_i\,{}^1\dot{\eta}_i \end{aligned} \tag{2.6.22}$$

in which

$$
{}^{2}\Pi_i = \begin{bmatrix} \cos^2\psi_i & 0 & 1 \\ \sin^2\psi_i \sin^2\varphi_i & \cos^2\varphi_i & 0 \\ \sin^2\psi_i \cos^2\varphi_i & \sin^2\varphi_i & 0 \end{bmatrix};
$$

$$
{}^{3}\Pi_i = \begin{bmatrix} 0 & 2\cos\psi_i & 0 \\ 0 & 0 & 2\sin\psi_i \sin\varphi_i \cos\varphi_i \\ 0 & 0 & -2\sin\psi_i \sin\varphi_i \cos\varphi_i \end{bmatrix}
$$

$$
{}^{4}\Pi_i = \begin{bmatrix} \sin^2\psi_i \sin\varphi_i \cos\varphi_i & -\sin\varphi_i \cos\varphi_i & 0 \\ \sin\psi_i \cos\psi_i \cos\varphi_i & 0 & 0 \\ \sin\psi_i \cos\psi_i \sin\varphi_i & 0 & 0 \end{bmatrix} \tag{2.6.23}
$$

$$
{}^{5}\Pi_i = \begin{bmatrix} 0 & 0 & \sin\psi_i (\cos^2\varphi_i - \sin^2\varphi_i) \\ -\sin\varphi_i & \sin\psi_i \cos\varphi_i & -\cos\psi_i \sin\varphi_i \\ \cos\varphi_i & \sin\psi_i \sin\varphi_i & \cos\psi_i \cos\varphi_i \end{bmatrix}
$$

and

$$
{}^{2\cdot}\eta_i = \begin{bmatrix} \dot{\theta}_i^2 \\ \dot{\psi}_i^2 \\ \dot{\varphi}_i^2 \end{bmatrix} \tag{2.6.24}
$$

By substituting (2.6.22) in (2.6.20), we obtain

$$
\tilde{M}_i = \tilde{J}_i \Pi_i \ddot{\eta}_i + ({}^{1}\tilde{J}_i\, {}^{2}\Pi_i + {}^{2}\tilde{J}_i\, {}^{4}\Pi_i)\, {}^{2\cdot}\eta_i + (\tilde{J}_i\, {}^{1}\Pi_i + {}^{1}\tilde{J}_i\, {}^{3}\Pi_i + {}^{2}\tilde{J}_i\, {}^{5}\Pi_i)\, {}^{1\cdot}\eta_i \tag{2.6.25}
$$

The moment $\tilde{M}_i$ expressed in the b.-f. coordinate system of the i-th segment may also be expressed in the fixed external system

$$
M_i = A_i \tilde{M}_i \tag{2.6.26}
$$

By combining with (2.6.25), we obtain the form

$$
M_i = D_i \ddot{\eta}_i + E_i\, {}^{2\cdot}\eta_i + F_i\, {}^{1\cdot}\eta_i \tag{2.6.27}
$$

where

$$D_i = A_i\tilde{J}_i\Pi_i$$

$$E_i = A_i({}^1\tilde{J}_i{}^2\Pi_i + {}^2\tilde{J}_i{}^4\Pi_i) \tag{2.6.28}$$

$$F_i = A_i(\tilde{J}_i{}^1\Pi_i + {}^1\tilde{J}_i{}^3\Pi_i + {}^2\tilde{J}_i{}^5\Pi_i)$$

If the axes of the b.-f. coordinate system are in the directions of the principal inertia axes of the segment, then ${}^1\tilde{J}_i = 0$, while $\tilde{J}_i$ and ${}^2\tilde{J}_i$ take the diagonal form.

Linear motion of segments. The linear motion of a segment, the i-th, is described by considering the position of its center of gravity as given by different vectors $\vec{\ell}$ and a vector $\vec{d}_i$. $\vec{\ell}$ on a segment is defined as the vector extending from the first joint to the second one or some other joint on the same segment. Each segment has none, one or more lengths $\vec{\ell}$. Their subscripts are independent of the subscripts of their segments and are denoted by $\vec{\ell}_j$, i=1,...,m.

The vector $\vec{d}_i$, i=1,...,n, is characteristic of the i-th segment and represents the vector extending from the first joint to the center of gravity C_i. Those vectors $\vec{d}$ and $\vec{\ell}$ which are defined on one segment are time-invariant when expressed in the b.-f. system attached to that segment and time-varying when expressed in the external system.

Let us consider the i-th segment and denote by $\vec{\tilde{\ell}}$ one of the vectors $\vec{\ell}$ which is defined on the i-th segment and expressed in the b.-f. coordinate system. The index of the vector has been omitted since it is independent of the index "i".

Let us note that $\vec{\tilde{d}}_i$, $\vec{\tilde{\ell}}$ are constant vectors given as input data.

We now have

$$\vec{\ell} = A_i\vec{\tilde{\ell}} \tag{2.6.29}$$

and

$$\vec{d}_i = A_i\vec{\tilde{d}}_i \tag{2.6.30}$$

namely, in matrix form

$$\ell = A_i \tilde{\ell}; \quad d_i = A_i \tilde{d}_i$$

By differentiating (2.6.29), we obtain

$$\ddot{\ell} = \frac{d^2 A_i}{dt^2} \tilde{\ell} \tag{2.6.31}$$

By utilizing (2.6.8) and (2.6.9), the above expression yields

$$\begin{aligned}\ddot{\ell} = (&{}^2A_i^\theta A_i^\psi A_i^\varphi \dot{\theta}_i^2 + A_i^\theta\, {}^2A_i^\psi A_i^\varphi \dot{\psi}_i^2 + A_i^\theta A_i^\psi\, {}^2A_i^\varphi \dot{\varphi}_i^2 + \\ &+ 2A_i^\theta\, {}^1A_i^\psi\, {}^1A_i^\varphi \dot{\psi}_i \dot{\varphi}_i + 2\,{}^1A_i^\theta A_i^\psi\, {}^1A_i^\varphi \dot{\theta}_i \dot{\varphi}_i + 2\,{}^1A_i^\theta\, {}^1A_i^\psi A_i^\varphi \dot{\theta}_i \dot{\psi}_i + \\ &+ {}^1A_i^\theta A_i^\psi A_i^\varphi \ddot{\theta}_i + A_i^\theta\, {}^1A_i^\psi A_i^\varphi \ddot{\psi}_i + A_i^\theta A_i^\psi\, {}^1A_i^\varphi \ddot{\varphi}_i)\tilde{\ell}\end{aligned} \tag{2.6.32}$$

with

$${}^1A_i^\theta = \frac{dA_i^\theta}{d\theta_i} = \begin{bmatrix} 0 & 0 & 0 \\ 0 & -\sin\theta_i & -\cos\theta_i \\ 0 & \cos\theta_i & -\sin\theta_i \end{bmatrix}; \quad {}^2A_i^\theta = \frac{d^2 A_i^\theta}{d\theta_i^2} = \begin{bmatrix} 0 & 0 & 0 \\ 0 & -\cos\theta_i & \sin\theta_i \\ 0 & -\sin\theta_i & -\cos\theta_i \end{bmatrix}$$

$${}^1A_i^\psi = \frac{dA_i^\psi}{d\psi_i} = \begin{bmatrix} -\sin\psi_i & 0 & \cos\psi_i \\ 0 & 0 & 0 \\ -\cos\psi_i & 0 & -\sin\psi_i \end{bmatrix}; \quad {}^2A_i^\psi = \frac{d^2 A_i^\psi}{d\psi_i^2} = \begin{bmatrix} -\cos\psi_i & 0 & -\sin\psi_i \\ 0 & 0 & 0 \\ \sin\psi_i & 0 & -\cos\psi_i \end{bmatrix}$$

$${}^1A_i^\varphi = \frac{dA_i^\varphi}{d\varphi_i} = \begin{bmatrix} 0 & 0 & 0 \\ 0 & -\sin\varphi_i & -\cos\varphi_i \\ 0 & \cos\varphi_i & -\sin\varphi_i \end{bmatrix}; \quad {}^2A_i^\varphi = \frac{d^2 A_i^\varphi}{d\varphi_i^2} = \begin{bmatrix} 0 & 0 & 0 \\ 0 & -\cos\varphi_i & \sin\varphi_i \\ 0 & -\sin\varphi_i & -\cos\varphi_i \end{bmatrix}$$

(2.6.33)

Expression (2.6.32) may be written in a more compact form

$$\ddot{\ell} = H_i \Lambda \ddot{\eta}_i + B_i \Lambda\, {}^2\dot{\eta}_i + C_i \Lambda\, {}^1\dot{\eta}_i \tag{2.6.34}$$

where

$$\begin{aligned} H_i &= [{}^1A_i^\theta A_i^\psi A_i^\varphi \,\vdots\, A_i^\theta\, {}^1A_i^\psi A_i^\varphi \,\vdots\, A_i^\theta A_i^\psi\, {}^1A_i^\varphi] \\ B_i &= [{}^2A_i^\theta A_i^\psi A_i^\varphi \,\vdots\, A_i^\theta\, {}^2A_i^\psi A_i^\varphi \,\vdots\, A_i^\theta A_i^\psi\, {}^2A_i^\varphi] \\ C_i &= [A_i^\theta\, {}^1A_i^\psi\, {}^1A_i^\varphi \,\vdots\, {}^1A_i^\theta A_i^\psi\, {}^1A_i^\varphi \,\vdots\, {}^1A_i^\theta\, {}^1A_i^\psi A_i^\varphi] \end{aligned} \tag{2.6.35}$$

and

$$\Lambda = \begin{bmatrix} \tilde{\ell} & 0 & 0 \\ 0 & \tilde{\ell} & 0 \\ 0 & 0 & \tilde{\ell} \end{bmatrix} \tag{2.6.36}$$

Let us now differentiate expression (2.6.30)

$$\ddot{d}_i = \frac{d^2 A_i}{dt^2} \tilde{d}_i \tag{2.6.37}$$

Analogous to the vector ℓ, we obtain

$$\ddot{d}_i = H_i \Delta_i \ddot{\eta}_i + B_i \Delta_i \overset{2\cdot}{\eta}_i + C_i \Delta_i \overset{1\cdot}{\eta}_i \tag{2.6.38}$$

with

$$\Delta_i = \begin{bmatrix} \tilde{d}_i & 0 & 0 \\ 0 & \tilde{d}_i & 0 \\ 0 & 0 & \tilde{d}_i \end{bmatrix} \tag{2.6.39}$$

Let us consider again the i-th segment and denote by $\vec{r}_{ip}$ the vector of the distance from the joint S_p to the center of gravity C_i of the i-th segment. Using structural matrices, we may write

$$r_{ip} = \beta_{ip} d_i + \sum_{j=1}^{m} \delta_{ijp} \ell_j \tag{2.6.40}$$

The distance from the support point, i.e., from the joint $p = 0$ to the center of gravity C_i, is obtained as

$$r_{io} = d_i + \sum_{j=1}^{m} \delta_{ijo} \ell_j, \tag{2.6.41}$$

since β_{io} is always equal to unity.

The acceleration $\vec{w}_i$ of the center of gravity C_i of the i-th segment is obtained by differentiating the expression (2.6.41)

$$w_i = \ddot{r}_{io} = \ddot{d}_i + \sum_{j=1}^{m} \delta_{ijo} \ddot{\ell}_j. \tag{2.6.42}$$

By substituting (2.6.34) and (2.6.38) into (2.6.42), we obtain

$$\begin{aligned} w_i = {} & H_i \Delta_i \ddot{\eta}_i + B_i \Delta_i \overset{2\cdot}{\eta}_i + C_i \Delta_i \overset{1\cdot}{\eta}_i + \sum_{j=1}^{m} \delta_{ijo} H_{(j)} \Lambda_j \ddot{\eta}_{(j)} + \\ & + \sum_{j=1}^{m} \delta_{ijo} B_{(j)} \Lambda_j \overset{2\cdot}{\eta}_{(j)} + \sum_{j=1}^{m} \delta_{ijo} C_{(j)} \Lambda_j \overset{1\cdot}{\eta}_{(j)} . \end{aligned} \tag{2.6.43}$$

Let us explain some of the notation in expression (2.6.43). Λ_j stands for the matrix of form (2.6.36) corresponding to the vector ℓ_j. The subscript (j) of matrices H, B, C, η does not denote the matrices corresponding to the j-th segment; it denotes such matrices which correspond to that segment on which the vector ℓ_j is located.

The sums over the subscript j may be written in a more convenient form if we introduce a structural matrix Γ^*. Then

$$
\begin{aligned}
H_{(j)}\Lambda_j\ddot{\eta}_{(j)} &= \sum_{k=1}^{n}\gamma_{kj}H_k\Lambda_j\ddot{\eta}_k,\\
B_{(j)}\Lambda_j\overset{2\cdot}{\eta}_{(j)} &= \sum_{k=1}^{n}\gamma_{kj}B_k\Lambda_j\overset{2\cdot}{\eta}_k,\\
C_{(j)}\Lambda_j\overset{1\cdot}{\eta}_{(j)} &= \sum_{k=1}^{n}\gamma_{kj}C_k\Lambda_j\overset{1\cdot}{\eta}_k.
\end{aligned}
\tag{2.6.44}
$$

The expression for acceleration (2.6.43) now takes the following form

$$
\begin{aligned}
w_i = {} & H_i\Delta_i\ddot{\eta}_i + B_i\Delta_i\overset{2\cdot}{\eta}_i + C_i\Delta_i\overset{1\cdot}{\eta}_i + \\
& + \sum_{j=1}^{m}\delta_{ijo}\sum_{k=1}^{n}\gamma_{kj}H_k\Lambda_j\ddot{\eta}_k + \\
& + \sum_{j=1}^{m}\delta_{ijo}\sum_{k=1}^{n}\gamma_{kj}B_k\Lambda_j\overset{2\cdot}{\eta}_k + \\
& + \sum_{j=1}^{m}\delta_{ijo}\sum_{k=1}^{n}\gamma_{kj}C_k\Lambda_j\overset{1\cdot}{\eta}_k.
\end{aligned}
\tag{2.6.45}
$$

The moment relative to the joint S_p, necessary to produce the linear acceleration of a segment "i", will be

$$\vec{M}_{ip} = \vec{r}_{ip} \times m_i\vec{w}_i \tag{2.6.46}$$

or, in matrix form,

$$M_{ip} = \underline{\underline{r}}_{ip}m_iw_i, \tag{2.6.47}$$

where $\underline{\underline{r}}_{ip}$ is a matrix analogous to (2.6.18), but corresponding to the vector $\vec{r}_{ip}$.

Starting from (2.6.40), we may write

$$\underline{\underline{r}}_{ip} = \beta_{ip}\underline{\underline{d}}_i + \sum_{j=1}^{m}\delta_{ijp}\underline{\underline{\ell}}_j. \tag{2.6.48}$$

By using (2.6.45) in the expression for the moment (2.6.47), we obtain

$$
\begin{aligned}
M_{ip} = {} & m_i \underline{\underline{r}}_{ip} H_i \Delta_i \ddot{\eta}_i + m_i \underline{\underline{r}}_{ip} B_i \Delta_i {}^{2\cdot}\!\eta_i + m_i \underline{\underline{r}}_{ip} C_i \Delta_i {}^{1\cdot}\!\eta_i + \\
& + m_i \underline{\underline{r}}_{ip} \sum_{j=1}^{m} \delta_{ijo} \sum_{k=1}^{n} \gamma_{kj} H_k \Lambda_j \ddot{\eta}_k + \\
& + m_i \underline{\underline{r}}_{ip} \sum_{j=1}^{m} \delta_{ijo} \sum_{k=1}^{n} \gamma_{kj} B_k \Lambda_j {}^{2\cdot}\!\eta_k + \\
& + m_i \underline{\underline{r}}_{ip} \sum_{j=1}^{m} \delta_{ijo} \sum_{k=1}^{n} \gamma_{kj} C_k \Lambda_j {}^{1\cdot}\!\eta_k .
\end{aligned}
\qquad (2.6.49)
$$

In this equation the matrix $\underline{\underline{r}}_{ip}$ should be equal to zero for a segment not contributing to the moment relative to the joint p. With this additional condition the moment expression (2.6.49) is valid for any segment "i". It is the moment relative to the joint S_p, necessary for the linear motion of a segment.

<u>Joint moments</u>. The total moment relative to a joint p will be equal to the sum of all the moments necessary for angular and linear acceleration of the segments.

The moment due to gravity force of the segment "i", relative to the joint S_p, will be

$$-\underline{\underline{r}}_{ip} m_i g,$$

where $g = [0 \quad 0 \quad +9.81]^T$.

Hence, the total moment relative to a joint S_p including the gravity compensation will be

$$M_{S_p} = \sum_{i=1}^{n} M_{ip} + \sum_{i=1}^{n} \beta_{ip} M_i + \sum_{i=1}^{n} m_i \underline{\underline{r}}_{ip} g \qquad (2.6.50)$$

In this expression, M_i is to be taken according to expression (2.6.27), and M_{ip} according to (2.6.49). After substitution, the moment (2.6.50) takes the following form

$$
\begin{aligned}
M_{S_p} = {} & \sum_{i=1}^{n} m_i \underline{\underline{r}}_{ip} \sum_{j=1}^{m} \delta_{ijo} \sum_{k=1}^{n} \gamma_{kj} H_k \Lambda_j \ddot{\eta}_k + \sum_{i=1}^{n} m_i \underline{\underline{r}}_{ip} H_i \Delta_i \ddot{\eta}_i + \\
& + \sum_{i=1}^{n} m_i \underline{\underline{r}}_{ip} \sum_{j=1}^{m} \delta_{ijo} \sum_{k=1}^{n} \gamma_{kj} B_k \Lambda_j {}^{2\cdot}\!\eta_k + \sum_{i=1}^{n} m_i \underline{\underline{r}}_{ip} B_i \Delta_i {}^{2\cdot}\!\eta_i +
\end{aligned}
$$

$$+ \sum_{i=1}^{n} m_i \underline{\underline{r}}_{ip} \sum_{j=1}^{m} \delta_{ijo} \sum_{k=1}^{n} \gamma_{kj} C_k \Lambda_j \overset{1\bullet}{\eta}_k + \sum_{i=1}^{n} m_i \underline{\underline{r}}_{ip} C_i \Delta_i \overset{1\bullet}{\eta}_i +$$

$$+ \sum_{i=1}^{n} \beta_{ip} D_i \ddot{\eta}_i + \sum_{i=1}^{n} \beta_{ip} E_i \overset{2\bullet}{\eta}_i + \sum_{i=1}^{n} \beta_{ip} F_i \overset{1\bullet}{\eta}_i + \sum_{i=1}^{n} m_i \underline{\underline{r}}_{ip} g. \tag{2.6.51}$$

The terms with triple sums may be transformed by rearranging the order of summation and introducing the short notation

$$V_{jp} = \sum_{i=1}^{n} \delta_{ijo} m_i \underline{\underline{r}}_{ip}. \tag{2.6.52}$$

The remaining six terms may be transformed by changing the order of indices and summation. The resultant form for the total moment relative to a joint S_p will then be as follows.

$$\begin{aligned} M_{S_p} = & \sum_{k=1}^{n} (m_k \underline{\underline{r}}_{kp} H_k \Delta_k + \sum_{j=1}^{m} \gamma_{kj} V_{jp} H_k \Lambda_j + \beta_{kp} D_k) \ddot{\eta}_k + \\ & + \sum_{k=1}^{n} (m_k \underline{\underline{r}}_{kp} B_k \Delta_k + \sum_{j=1}^{m} \gamma_{kj} V_{jp} B_k \Lambda_j + \beta_{kp} E_k) \overset{2\bullet}{\eta}_k + \\ & + \sum_{k=1}^{n} (m_k \underline{\underline{r}}_{kp} C_k \Delta_k + \sum_{j=1}^{m} \gamma_{kj} V_{jp} C_k \Lambda_j + \beta_{kp} F_k) \overset{1\bullet}{\eta}_k + \sum_{i=1}^{n} m_i \underline{\underline{r}}_{ip} g. \end{aligned} \tag{2.6.53}$$

This holds for any joint "p".

The joint moments may be referred to some coordinate system that is different from the fixed external coordinate system. So, for example, joint coordinate systems, mentioned already, are used. The introduction of the joint coordinate system allows the moments to be expressed in terms of actual axes of rotation in the joint (Fig. 2.18).

The first axis of such a coordinate system is fixed to the first segment of the joint and has directional cosines $(\ell, m, n)_{p,1}$ relative to the b.-f. system of that segment. The second axis is fixed to the second segment of the joint and has directional cosines $(\ell, m, n)_{p,2}$ relative to the b.-f. system of that segment. Let us denote by (p, 1) the index corresponding to the first of the segments connected by the joint, and by (p, 2) the index of the second segment of the joint. In addition, let $(\alpha_{11}, \alpha_{12}, \alpha_{13})_p$, $(\alpha_{21}, \alpha_{22}, \alpha_{23})_p$ and $(\alpha_{31}, \alpha_{32}, \alpha_{33})_p$ be directional cosines of the first, second and third axes of the joint system, respectively, relative to the external coordinate system. Then

$$[\ell \quad m \quad n]_{p,1}(A^{\varphi}\ A^{\psi}\ A^{\theta})_{(p,1)} = [\alpha_{11}\ \alpha_{12}\ \alpha_{13}]_p,$$
$$[\ell \quad m \quad n]_{p,2}(A^{\varphi}\ A^{\psi}\ A^{\theta})_{(p,2)} = [\alpha_{21}\ \alpha_{22}\ \alpha_{23}]_p, \tag{2.6.54}$$

while the cosines of the third axis are obtained from the orthogonality conditions

$$[\alpha_{31}\ \alpha_{32}\ \alpha_{33}]_p = \bar{a}^1[(\alpha_{12}\,\alpha_{23} - \alpha_{22}\,\alpha_{13}) \vdots (\alpha_{13}\,\alpha_{21} - \alpha_{23}\,\alpha_{11}) \vdots (\alpha_{11}\,\alpha_{22} - \alpha_{21}\,\alpha_{12})]_p, \quad a^2 = (\alpha_{12}\,\alpha_{23} - \alpha_{22}\,\alpha_{13})^2 + (\alpha_{13}\,\alpha_{21} - \alpha_{23}\,\alpha_{11})^2 + (\alpha_{11}\,\alpha_{22} - \alpha_{21}\,\alpha_{12})^2. \tag{2.6.55}$$

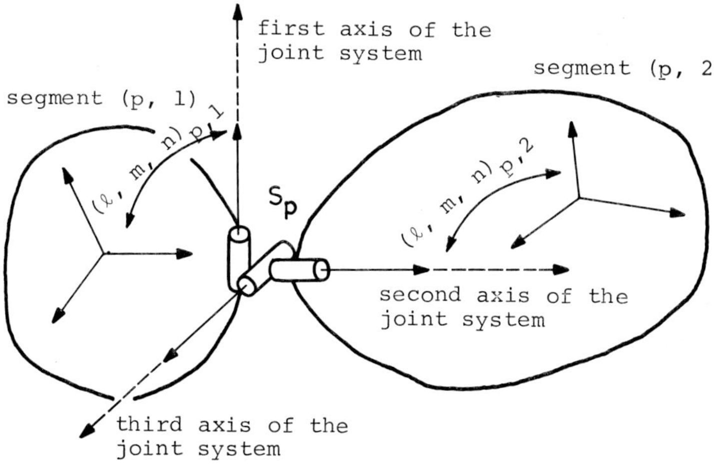

Fig. 2.18. Scheme of a joint with hardware axis, and the corresponding joint coordinate system

Directional cosines α_{ij}; i=1, 2, 3; j=1, 2, 3 assembled into a matrix $[\alpha]_p$, are used to transform the moments relative to a joint. The resultant moment, expressed in the joint coordinate system will be

$$\bar{M}_{S_p} = [\alpha]^{-1} M_{S_p}, \tag{2.6.56}$$

where the components of the vector $\bar{M}_{S_p} = [{}^1M\ {}^2M\ {}^3M]^T_{S_p}$ are the moments relative to joint axes.

To keep the expressions short, let us introduce the following notation

$$[\alpha]_p^{-1}(m_k \underline{\underline{r}}_{kp} H_k \Delta_k + \sum_{j=1}^{m} \gamma_{kj} V_{jp} H_k \Lambda_j + \beta_{kp} D_k) = W_{pk},$$

$$[\alpha]_p^{-1}(m_k \underline{\underline{r}}_{kp} B_k \Delta_k + \sum_{j=1}^{m} \gamma_{kj} V_{jp} B_k \Lambda_j + \beta_{kp} E_k) = N_{pk}, \quad (2.6.57a)$$

$$[\alpha]_p^{-1}(m_k \underline{\underline{r}}_{kp} C_k \Delta_k + \sum_{j=1}^{m} \gamma_{kj} V_{jp} C_k \Lambda_j + \beta_{kp} F_k) = R_{pk},$$

$$\sum_{i=1}^{n} m_i \underline{\underline{r}}_{ip} g = G_p.$$

The moment expressions for all the joints may now be written in matrix form. (2.6.53), (2.6.56), and (2.6.57a) yield:

$$\begin{bmatrix} \text{-----} \\ \bar{M}_{S_p} \\ \text{-----} \end{bmatrix} = \sum_{k=1}^{n} \begin{bmatrix} \text{-----} \\ W_{pk} \\ \text{-----} \end{bmatrix} \ddot{\eta}_k + \sum_{k=1}^{n} \begin{bmatrix} \text{-----} \\ N_{pk} \\ \text{-----} \end{bmatrix} {}^{2\bullet}\eta_k +$$

$$+ \sum_{k=1}^{n} \begin{bmatrix} \text{-----} \\ R_{pk} \\ \text{-----} \end{bmatrix} {}^{1\bullet}\eta_k + \begin{bmatrix} \text{----} \\ G_p \\ \text{----} \end{bmatrix}. \quad (2.6.57b)$$

To avoid the sums, i.e., to allow even shorter expressions, let us introduce the following notation

$$\begin{bmatrix} & \vdots & \\ \cdots & W_{pk} & \cdots \\ & \vdots & \end{bmatrix} = W, \quad \begin{bmatrix} & \vdots & \\ \cdots & N_{pk} & \cdots \\ & \vdots & \end{bmatrix} = N, \quad \begin{bmatrix} & \vdots & \\ \cdots & R_{pk} & \cdots \\ & \vdots & \end{bmatrix} = R,$$

$$\begin{bmatrix} \text{----} \\ \ddot{\eta}_k \\ \text{----} \end{bmatrix} = \ddot{q}; \quad \begin{bmatrix} \text{-----} \\ {}^{2\bullet}\eta_k \\ \text{-----} \end{bmatrix} = {}^{2\bullet}q, \quad \begin{bmatrix} \text{-----} \\ {}^{1\bullet}\eta_k \\ \text{-----} \end{bmatrix} = {}^{1\bullet}q, \quad \begin{bmatrix} \text{----} \\ G_p \\ \text{----} \end{bmatrix} = G. \quad (2.6.58)$$

In addition, let us introduce joint torques and denote by $\vec{P}_p$ (or, in matrix form P_p) the vector of the torque moment in a joint, which is to produce the particular motion. Then

$$\bar{P}_p = \bar{M}_{S_p} \quad (2.6.59)$$

The system of equations (2.6.57b) may now be written in the form

$$P = W\ddot{q} + N\,{}^{2\cdot}q + R\,{}^{1\cdot}q + G, \tag{2.6.60}$$

where

$$P = \begin{bmatrix} ---- \\ \bar{P}_p \\ ---- \end{bmatrix} \tag{2.6.61}$$

is the vector of joint torques, expressed by actual axes of rotation.

The system of equations (2.6.60) describing mechanism dynamics may be written in the form (2.6.1), i.e.,

$$W\ddot{q} = P + U, \tag{2.6.62}$$

with

$$U = -N\,{}^{2\cdot}q - R\,{}^{1\cdot}q - G. \tag{2.6.63}$$

Methods Based on the Lagrange's Equations

2.7. Method of Lagrange's Equations

This method uses the second-order Lagrange's equations as a mechanical approach and is very computer-oriented [8, 14, 15, 16, 17].

Starting postulates. The method considers an active mechanism with 2n degrees of freedom and describes it by a 2n-dimensional vector of the generalized coordinates q. Starting from Lagrange's equations a system of 2n second-order differential equations is formed:

$$W\ddot{q} = P + U, \tag{2.7.1}$$

where P is a 2n vector of the drives and matrices W and U (dimensions $2n\times 2n$ and $2n\times 1$ respectively), depend upon q, $\dot{q}$. Now the functions f and g, defined by (2.2.1) and (2.2.2) are obtained as

$$\ddot{q} = f(q, \dot{q}, P, \text{configuration}) = W^{-1}(P+U), \tag{2.7.2}$$

$$P = g(q, \dot{q}, \ddot{q}, \text{configuration}) = W\ddot{q} - U. \quad (2.7.3)$$

Mechanism configuration. The method considers a mechanism of the open chain type, formed by n rigid bodies of arbitrary form, without branching (Fig. 2.19). The rigid bodies, i.e., mechanism segments, are interconnected by means of joints with two degrees of freedom (d.o.f.) each, one translational and one rotational (Fig. 2.20a).

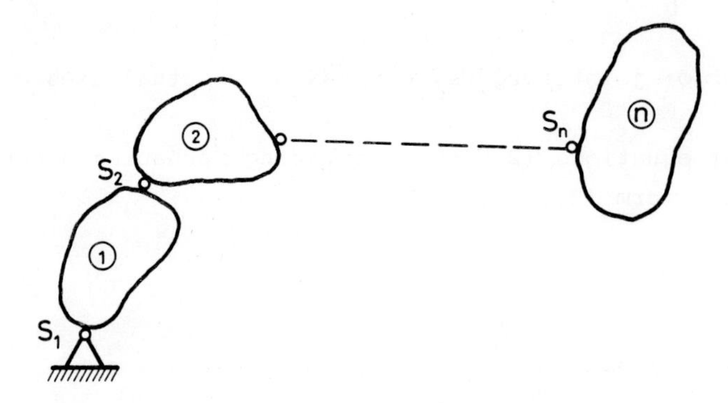

Fig. 2.19. Open kinematical chain without branching

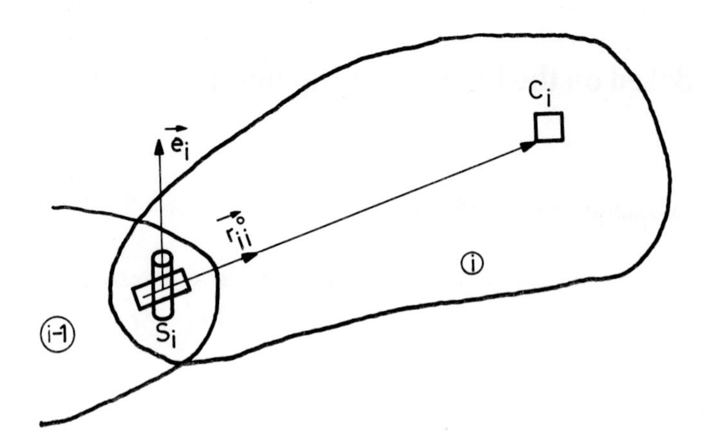

Fig. 2.20a. Joint with two d.o.f.

Rotation in the joint S_i is performed around the axis determined by the unit vector $\vec{e}_i$, and translation along the axis connecting the joint S_i and the center of gravity (c.o.g.) C_i (this axis is defined by the unit vector $\vec{r}_{ii}^{\,o}$). The point S_i and the axis $\vec{e}_i$ are immobile relative to the (i-1)-th segment and the axis $\vec{r}_{ii}^{\,o}$ is immobile relative to the i-th segment.

Consideration of the joints of such type narrows the generality to a certain extent, expecially because the translation axis is not placed arbitrary but along the straight line $\overrightarrow{S_iC_i}$. However, in the majority of practical cases this condition is justified (Fig. 2.20b); so our particular case may be considered sufficiently general for practical purposes.

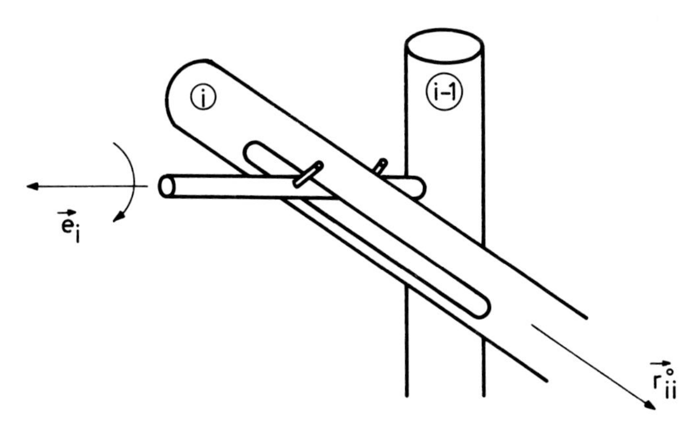

Fig. 2.20b. One practical realization of a joint with two degrees of freedom

Drives in the joint. In the mechanism joints driving forces and torques are acting; in the i-th joint the force and driving torque

$$\vec{F}_i = F_{Pi}\cdot\vec{r}^{\,o}_{ii} \tag{2.7.4a}$$

$$\vec{M}_i = M_{Pi}\cdot\vec{e}_i \tag{2.7.4b}$$

act respectively.

The vector of the drives is now of the form:

$$P = [M_{P1}F_{P1} \cdots M_{Pn}F_{Pn}]^T \tag{2.7.5}$$

Generalized coordinates. In each mechanism joint two generalized coordinates are chosen, so that they correspond to the degrees of freedom in the joint. Let consider the i-th joint S_i (Fig. 2.21) and introduce the notations according to Fig. 2.21.

Let introduce the generalized corrdinates in joint S_i as the angle of rotation θ_i around the axis $\vec{e}_i$ and the intensity u_i of the vector $\overrightarrow{S_iC_i}$. Angle θ_i can be considered as the angle between the projections of vectors $-\vec{r}_{i-1,i}$ and $\vec{r}^{\,o}_{ii}$ onto the plane perpendicular to $\vec{e}_i$, and u_i is the length $u_i = \overline{S_iC_i}$, i.e. $u_i = |\overrightarrow{S_iC_i}|$.

Thus, there are 2n generalized coordinates and the vector of generalized coordinates has the form

$$q = [\theta_1 u_1 \cdots \theta_n u_n]^T. \tag{2.7.6}$$

In the case of "specificity", i.e. $\vec{r}^{\,o}_{ii} || \vec{e}_i$ or $\vec{r}_{i-1,i} || \vec{e}_i$, one proceeds as in the method of general theorems of mechanics (2.3.).

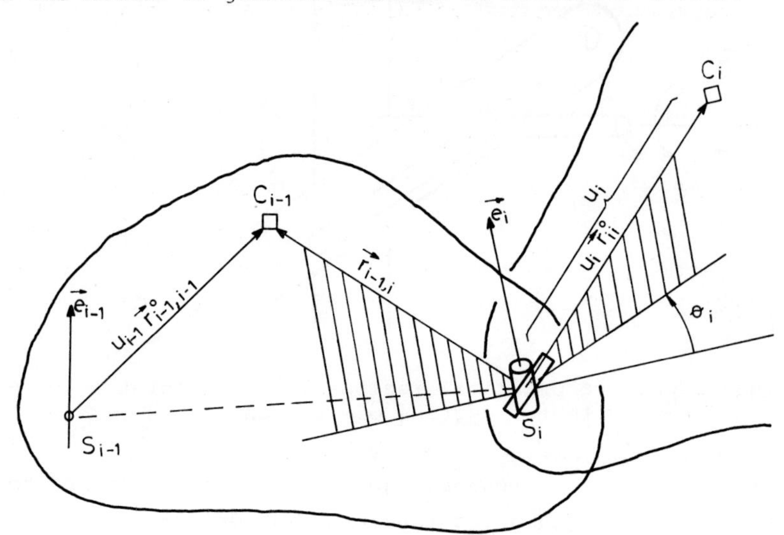

Fig. 2.21. A joint with the corresponding generalized coordinates

Coordinate systems and transition matrices. Let introduce the body-fixed (b.-f.) coordinate systems and the immobile external system, as in the method of general theorems; i.e., let us connect to each segment a system with its origin in the segment center of gravity (c.o.g.). Let us also introduce the same notation: $\vec{a}_i$ denotes a vector, characteristic of the i-th segment or joint, expressed in the external coordinate system; $\vec{\tilde{a}}_i$ is the same vector expressed in the i-th segment b.-f. system; $\vec{a}_{\sim i}$ is the same vector in the (i-1)-th b.-f. system. Further, let us likewise define the transition matrices form the i-th b.-f. system to the external one (matrix A_i) as

$$\vec{a}_i = A_i \vec{\tilde{a}}_i \tag{2.7.7}$$

and from the i-th system to the (i-1)-th (matrix $A_{i-1,i}$) as

$$\vec{\underset{\sim}{a}}_i = A_{i-1,i} \vec{\tilde{a}}_i \tag{2.7.8}$$

and conversely ($A_{i,i-1}$),

$$\vec{\tilde{a}}_i = A_{i,i-1}\underset{\sim}{\vec{a}}_i = A^{-1}_{i-1,i}\underset{\sim}{\vec{a}}_i \tag{2.7.9}$$

Let us note that the vectors $\vec{\tilde{r}}^{\,o}_{ii}$, $\vec{\tilde{r}}_{i-1,i}$, $\vec{\tilde{e}}_i$, $\underset{\sim}{\vec{e}}_i$ are constant and determine the geometry of joints and segments; so they must be prescribed for each segment and joint. Perhaps it should be explained, that $\vec{\tilde{e}}_i$ is constant, although axis $\vec{e}_i$ is moving relative to the i-th segment and its corresponding b.-f. system; $\vec{\tilde{e}}_i$ is constant, because this motion is linear.

The transition matrices are obtained recursively. In each iteration a new segment is added to the chain and its transition matrix is calculated, by using the transition matrix of the preceding segment. Thus, in the i-th iteration A_i is calculated, knowing A_{i-1}. Here, this procedure will differ to some extent from the procedure applied in the method of general theorems. Namely, the first thing to be calculated is the relative transition matrix $A_{i-1,i}$ and then:

$$A_i = A_{i-1}A_{i-1,i}. \tag{2.7.10}$$

Let consider a joint S_i and let us define the vectors

$$\underset{\sim}{\vec{a}}_i = \frac{-\underset{\sim}{\vec{e}}_i \times (\vec{r}_{i-1,i} \times \underset{\sim}{\vec{e}}_i)}{|\underset{\sim}{\vec{e}}_i \times (\vec{\tilde{r}}_{i-1,i} \times \underset{\sim}{\vec{e}}_i)|} \qquad \vec{\tilde{a}}_i = \frac{\vec{\tilde{e}}_i \times (\vec{\tilde{r}}^{\,o}_{ii} \times \vec{\tilde{e}}_i)}{|\vec{\tilde{e}}_i \times (\vec{\tilde{r}}^{\,o}_{ii} \times \vec{\tilde{e}}_i)|}, \tag{2.7.11}$$

(a) (b)

which are perpendicular to $\underset{\sim}{\vec{e}}_i$ and $\vec{\tilde{e}}_i$ respectively. The vectors (2.7.11) are unit vectors of the "a"-axis and (2.7.11b) holds for the case $\theta_i=0$ (Fig. 2.22).

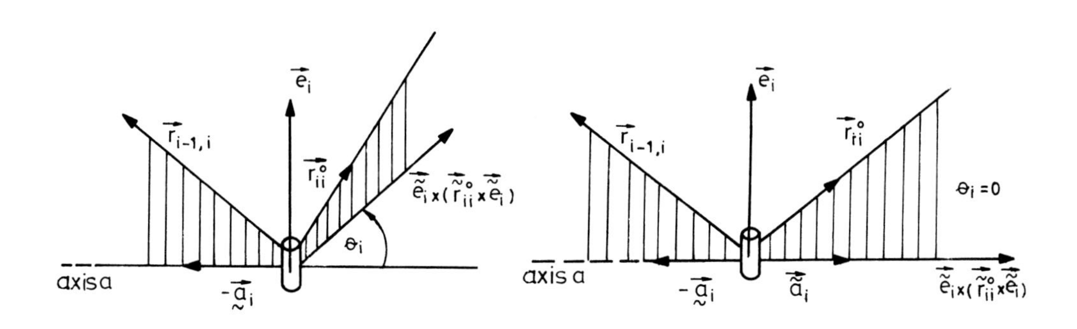

Fig. 2.22. Determining the transition matrix

By introducing the vector $\vec{b}_i = \vec{e}_i \times \vec{a}_i$, a triple of linearly independent vectors (on the (i-1)-th segment) is obtained: $\{\vec{e}_i, \vec{a}_i, \vec{b}_i\}$, and by introducing $\tilde{\vec{b}}_i = \tilde{\vec{e}}_i \times \tilde{\vec{a}}_i$, likewise a linearly independent triple (on the i-th segment): $\{\tilde{\vec{e}}_i, \tilde{\vec{a}}_i, \tilde{\vec{b}}_i\}$.

Let us denote by $A^o_{i-1,i}$ the transition matrix, corresponding to $\theta_i = 0$. Then (2.7.11b) holds and so:

$$\vec{e}_i = A^o_{i-1,i}\tilde{\vec{e}}_i; \qquad \vec{a}_i = A^o_{i-1,i}\tilde{\vec{a}}_i; \qquad \vec{b}_i = A^o_{i-1,i}\tilde{\vec{b}}_i. \tag{2.7.12}$$

Let us introduce matrix notation and let e_i denote a 3×1 matrix corresponding to the vector $\vec{e}_i$, and likewise for all other vectors in the text.

Now the relations (2.7.12) can be written together:

$$[e_i \; a_i \; b_i] = A^o_{i-1,i}[\tilde{e}_i \; \tilde{a}_i \; \tilde{b}_i], \tag{2.7.13}$$

i.e.,

$$A^o_{i-1,i} = [e_i \; a_i \; b_i] \; [\tilde{e}_i \; \tilde{a}_i \; \tilde{b}_i]^{-1}, \tag{2.7.14}$$

by the means of which the transformation matrix $A^o_{i-1,i}$ for $\theta_i = 0$ has been calculated. Let us denote the columns of the matrix obtained by v_{i1}, v_{i2}, v_{i3}; i.e.

$$A^o_{i-1,i} = [v_{i1} \; v_{i2} \; v_{i3}]. \tag{2.7.15}$$

Now, by "turning" around $\vec{e}_i$, according to Rodrigue´s formula, we obtain:

$$\vec{V}_{ij} = \vec{v}_{ij}\cos\theta_i + (1-\cos\theta_i)(\vec{e}_i \cdot \vec{v}_{ij})\vec{e}_i + \vec{e}_i \times \vec{v}_{ij}\sin\theta_i \quad j=1,2,3, \tag{2.7.16}$$

where $\vec{e}_i$ is known.

Now the transition matrix, corresponding to the angle θ_i will be:

$$A_{i-1,i} = [V_{i1} \; V_{i2} \; V_{i3}] \tag{2.7.17}$$

In the case of "specificity", one proceeds as in the method of general theorems (2.3).

Algorithm input data. The following input values are prescribed by the method:

- mechanism configuration, i.e.,

n = number of segments (and joints),

$\vec{\tilde{e}}_i$, $\vec{e}_{\sim i}$; $i=1,\dots,n$,

$\vec{\tilde{r}}^{o}_{ii}$, $\vec{\tilde{r}}_{i,i+1}$; $i=1,\dots,n$; $\vec{r}_{o1}$,

$\vec{\tilde{r}}^{*}_{ii}$, i.e. $\vec{\tilde{r}}^{*}_{i,i+1}$ (in cases of "specificity"),

m_i, $\tilde{J}_i$, $i=1,\dots,n$ (i.e. mass and tensor of inertia with respect to b.-f. system),

- initial state,

$$q(t_o) = [\theta_1(t_o)u_1(t_o) \cdots \theta_n(t_o)u_n(t_o)]^T,$$

$$\dot{q}(t_o) = [\dot{\theta}_1(t_o)\dot{u}_1(t_o) \cdots \dot{\theta}_n(t_o)\dot{u}_n(t_o)]^T.$$

- in the case of solving the direct problem of dynamics

$$\ddot{q}(t_k), \quad k=0,1,\dots,k_{end},$$

- in the case of solving the inverse problem,

$$P(t_k), \quad k=0,1,\dots,k_{end}.$$

Kinematical relations. If $\vec{v}_i$ is the center of gravity velocity of the i-th segment, and $\vec{\omega}_i$ the angular velocity of the same segment, then the recursions follow:

$$\vec{\omega}_i = \vec{\omega}_{i-1} + \dot{\theta}_i\vec{e}_i, \tag{2.7.18}$$

$$\vec{v}_i = \vec{v}_{i-1} - \vec{\omega}_{i-1} \times \vec{r}_{i-1,i} + \vec{\omega}_i \times \vec{r}^{o}_{ii}\cdot u_i + \dot{u}_i\vec{r}^{o}_{ii} \tag{2.7.19}$$

From (2.7.18) one obtains:

$$\vec{\omega}_i = \sum_{j=1}^{i} \vec{e}_j\dot{\theta}_j, \tag{2.7.20}$$

and by introducing notations, it follows that

$$\vec{\omega}_i = \sum_{j=1}^{i} \vec{\beta}_j\dot{\theta}_j, \tag{2.7.21}$$

where

$$\vec{\beta}_j = \vec{e}_j = A_{j-1}\vec{e}_{\sim j}. \tag{2.7.22}$$

From (2.7.19) and (2.7.20) and by introducing notation, it follows that

$$\vec{v}_i = \sum_{j=1}^{i} \vec{\alpha}_j^i \dot{\theta}_j + \sum_{j=1}^{i} \vec{\gamma}_j \dot{u}_j, \tag{2.7.23}$$

where

$$\vec{\alpha}_j^i = \begin{cases} -\sum_{k=j+1}^{i} \vec{R}_{kj} + \sum_{k=j}^{i} \vec{H}_{kj}, & j<i \\ \\ \vec{H}_{ii}, & j=i, \end{cases} \tag{2.7.24}$$

$$\vec{R}_{kj} = \vec{\beta}_j \times \vec{\delta}_k; \qquad \vec{H}_{kj} = \vec{\beta}_j \times \vec{\gamma}_k \cdot u_k, \tag{2.7.25}$$

$$\vec{\beta}_j = \vec{e}_j = A_{j-1} \underset{\sim}{e}_j, \quad \vec{\delta}_k = \vec{r}_{k-1,k} = A_{k-1} \vec{\tilde{r}}_{k-1,k}, \quad \vec{\gamma}_k = \vec{r}^{\,o}_{kk} = A_k \vec{\tilde{r}}^{\,o}_{kk}. \tag{2.7.26}$$

Further, from (2.7.21),

$$\vec{\tilde{\omega}}_i = A_i^{-1} \vec{\omega}_i = \sum_{j=1}^{i} \vec{B}_j^i \dot{\theta}_j, \tag{2.7.27}$$

where

$$\vec{B}_j^i = A_i^{-1} \vec{\beta}_j \quad (\vec{B}_i^i = \vec{\tilde{e}}_i). \tag{2.7.28}$$

Forming Lagrange´s equations. For the sake of using Lagrange´s equations, it is necessary to first form the expression for the system´s kinetic energy. Total kinetic energy T is equal to the sum of the segment kinetic energies:

$$T = \sum_{i=1}^{n} T_i \tag{2.7.29}$$

and for the i-th segment, according to Kenig´s theorem, it is

$$T_i = \frac{1}{2} m_i \vec{v}_i^2 + \frac{1}{2} \vec{\tilde{\omega}}_i \tilde{J}_i \vec{\tilde{\omega}}_i \tag{2.7.30}$$

The mechanism dynamics will be described by a system of 2n Lagrange equations of the form

$$\left.\begin{aligned} \frac{d}{dt}\left(\frac{\partial T}{\partial \dot{\theta}_i}\right) - \frac{\partial T}{\partial \theta_i} &= Q_i^{\theta} \\ \frac{d}{dt}\left(\frac{\partial T}{\partial \dot{u}_i}\right) - \frac{\partial T}{\partial u_i} &= Q_i^{u} \end{aligned}\right\} \quad ; \quad i=1,\dots,n \tag{2.7.31}$$

where Q_i^θ and Q_i^u are generalized forces corresponding to the coordinates θ_i and u_i, respectively.

From (2.7.23) it follows that

$$m_i\vec{v}_i^2 = m_i \sum_{p=1}^{i} \sum_{q=1}^{i} (\vec{\alpha}_p^i \vec{\alpha}_q^i \dot{\theta}_p \dot{\theta}_q + 2\vec{\alpha}_p^i \vec{\gamma}_q \dot{\theta}_p \dot{u}_q + \vec{\gamma}_p \vec{\gamma}_q \dot{u}_p \dot{u}_q) \tag{2.7.32}$$

and from (2.7.27),

$$\tilde{\vec{\omega}}_i \tilde{J}_i \tilde{\vec{\omega}}_i = \sum_{p=1}^{i} \sum_{q=1}^{i} \vec{B}_p^i \tilde{J}_i \vec{B}_q^i \dot{\theta}_p \dot{\theta}_q . \tag{2.7.33}$$

Thus we have the expression for the kinetic energy (2.7.29), (2.7.30).

Let us now find the partial derivatives $\frac{\partial T}{\partial \dot{\theta}_s}$, $\frac{\partial T}{\partial \dot{u}_s}$, s=1,...,n. From (2.7.29), (2.7.30) it follows that

$$\frac{\partial T}{\partial \dot{\theta}_s} = \frac{1}{2} \sum_{i=1}^{n} (\frac{\partial}{\partial \dot{\theta}_s} m_i\vec{v}_i^2 + \frac{\partial}{\partial \dot{\theta}_s} \tilde{\vec{\omega}}_i \tilde{J}_i \tilde{\vec{\omega}}_i) \tag{2.7.34}$$

and from (2.7.32) and (2.7.33),

$$\frac{\partial T}{\partial \dot{\theta}_s} = \frac{1}{2} \sum_{i=s}^{n} \sum_{r=1}^{i} [(\vec{B}_r^i \tilde{J}_i \vec{B}_s^i + \vec{B}_s^i \tilde{J}_i \vec{B}_r^i + 2m_i \vec{\alpha}_r^i \vec{\alpha}_s^i)\dot{\theta}_r + 2m_i \vec{\alpha}_s^i \vec{\gamma}_r \dot{u}_r] . \tag{2.7.35}$$

Differentiating (2.7.35) with respect to time qives

$$\begin{aligned} \frac{d}{dt}(\frac{\partial T}{\partial \dot{\theta}_s}) = \frac{1}{2} \sum_{i=s}^{n} \sum_{r=1}^{i} [(\dot{\vec{B}}_r^i \tilde{J}_i \vec{B}_s^i + \vec{B}_r^i \tilde{J}_i \dot{\vec{B}}_s^i + \dot{\vec{B}}_s^i \tilde{J}_i \vec{B}_r^i + \vec{B}_s^i \tilde{J}_i \dot{\vec{B}}_r^i + \\ + 2m_i \dot{\vec{\alpha}}_r^i \vec{\alpha}_s^i + 2m_i \vec{\alpha}_r^i \dot{\vec{\alpha}}_s^i)\dot{\theta}_r + (\vec{B}_r^i \tilde{J}_i \vec{B}_s^i + \vec{B}_s^i \tilde{J}_i \vec{B}_r^i + \\ + 2m_i \vec{\alpha}_r^i \vec{\alpha}_s^i)\ddot{\theta}_r + (2m_i \dot{\vec{\alpha}}_s^i \vec{\gamma}_r + 2m_i \vec{\alpha}_s^i \dot{\vec{\gamma}}_r)\dot{u}_r + 2m_i \vec{\alpha}_s^i \vec{\gamma}_r \cdot \ddot{u}_r] . \end{aligned} \tag{2.7.36}$$

Let us now find the time derivatives of the vector coefficients. From (2.7.28),

$$\dot{\vec{B}}_j^i = -A_i^{-1}\dot{A}_i A_i^{-1}\vec{\beta}_j + A_i^{-1}\dot{\vec{\beta}}_j \tag{2.7.37}$$

and from (2.7.22)

$$\dot{\vec{\beta}}_j = \dot{\vec{e}}_j = \dot{A}_{j-1}\underset{\sim}{e}_j . \tag{2.7.38}$$

From (2.7.24),

$$\dot{\vec{\alpha}}^i_j = \begin{cases} -\sum_{k=j+1}^{i} \dot{\vec{R}}_{kj} + \sum_{k=j}^{i} \dot{\vec{H}}_{kj}, & j<i \\ \dot{\vec{H}}_{ii}, & j=i \end{cases} \tag{2.7.39}$$

and from (2.7.25), (2.7.26),

$$\begin{aligned} \dot{\vec{R}}_{kj} &= \dot{\vec{\beta}}_j \times \vec{\delta}_k + \vec{\beta}_j \times \dot{\vec{\delta}}_k, \\ \dot{\vec{H}}_{kj} &= \dot{\vec{\beta}}_j \times \vec{\gamma}_k u_k + \vec{\beta}_j \times \dot{\vec{\gamma}}_k u_k + \vec{\beta}_j \times \vec{\gamma}_k \dot{u}_k, \end{aligned} \tag{2.7.40}$$

$$\begin{aligned} \dot{\vec{\beta}}_j &= \dot{\vec{e}}_j = \dot{A}_{j-1}\underset{\sim}{e}_j, \\ \dot{\vec{\delta}}_k &= \dot{\vec{r}}_{k-1,k} = \dot{A}_{k-1}\tilde{r}_{k-1,k}, \\ \dot{\vec{\gamma}}_k &= \dot{\vec{r}}^{\,o}_{kk} = \dot{A}_k\tilde{r}^{\,o}_{kk}. \end{aligned} \tag{2.7.41}$$

Now,

$$\frac{\partial T}{\partial \dot{u}_s} = \frac{1}{2}\sum_{i=1}^{n}\left(\frac{\partial}{\partial \dot{u}_s} m_i v_i^2 + \frac{\partial}{\partial \dot{u}_s} \tilde{\vec{\omega}}_i \tilde{J}_i \tilde{\vec{\omega}}_i\right). \tag{2.7.42}$$

From (2.7.32) and (2.7.33) it follows that

$$\frac{\partial T}{\partial \dot{u}_s} = \frac{1}{2}\sum_{i=s}^{n}\sum_{r=1}^{i}(m_i\vec{\alpha}^i_r\vec{\gamma}_s\dot{\theta}_r + m_i\vec{\gamma}_r\vec{\gamma}_s\dot{u}_r). \tag{2.7.43}$$

By differentiating with respect to time,

$$\begin{aligned} \frac{d}{dt}\left(\frac{\partial T}{\partial \dot{u}_s}\right) = \frac{1}{2}\Big[\sum_{i=s}^{n}\sum_{r=1}^{i}&(m_i\dot{\vec{\alpha}}^i_r\vec{\gamma}_s + m_i\vec{\alpha}^i_r\dot{\vec{\gamma}}_s)\dot{\theta}_r + \\ &+ m_i\vec{\alpha}^i_r\vec{\gamma}_s\ddot{\theta}_r + (m_i\dot{\vec{\gamma}}_r\vec{\gamma}_s + m_i\vec{\gamma}_r\dot{\vec{\gamma}}_s)\dot{u}_r + m_i\vec{\gamma}_r\vec{\gamma}_s\ddot{u}_r\Big], \end{aligned} \tag{2.7.44}$$

where $\dot{\vec{\alpha}}^i_j$ and $\dot{\vec{\gamma}}_j$ are determined by expressions (2.7.39), (2.7.40), (2.7.41).

Thus, the problem has been reduced to that of obtaining the transition matrices and their time derivatives, i.e. A_p, $\dot{A}_p$, p=1,...,n. This will be dealt with later.

For the purpose of forming Lagrange´s equations it is also necessary to determine the partial derivatives $\frac{\partial T}{\partial \theta_s}$, $\frac{\partial T}{\partial u_s}$, s=1,...,n. From (2.7.29), (2.7.30) it follows that

$$\frac{\partial T}{\partial \theta_s} = \frac{1}{2} \sum_{i=1}^{n} \frac{\partial}{\partial \theta_s} m_i \vec{v}_i^2 + \frac{\partial}{\partial \theta_s} \tilde{\vec{\omega}}_i \tilde{J}_i \tilde{\vec{\omega}}_i . \tag{2.7.45}$$

From (2.7.32) and (2.7.33) one finds

$$\frac{\partial}{\partial \theta_s} m_i \vec{v}_i^2 = \begin{cases} 0, & s>i \\ m_i \sum_{p=1}^{i} \sum_{q=1}^{i} \Big[\Big(\frac{\partial \vec{\alpha}_p^i}{\partial \theta_s} \vec{\alpha}_q^i + \vec{\alpha}_p^i \frac{\partial \vec{\alpha}_q^i}{\partial \theta_s} \Big) \dot{\theta}_p \dot{\theta}_q + \\ \quad + \Big(2 \frac{\partial \vec{\alpha}_p^i}{\partial \theta_s} \vec{\gamma}_q + 2 \vec{\alpha}_p^i \frac{\partial \vec{\gamma}_q}{\partial \theta_s} \Big) \dot{\theta}_p \dot{u}_q + \\ \quad + \Big(\frac{\partial \vec{\gamma}_p}{\partial \theta_s} \vec{\gamma}_q + \vec{\gamma}_p \frac{\partial \vec{\gamma}_q}{\partial \theta_s} \Big) \dot{u}_p \dot{u}_q \Big], & s \le i, \end{cases} \tag{2.7.46}$$

$$\frac{\partial}{\partial \theta_s} \vec{\omega}_i \tilde{J}_i \vec{\tilde{\omega}}_i = \begin{cases} 0, & s>i \\ \sum_{p=1}^{i} \sum_{q=1}^{i} \Big(\frac{\partial \vec{B}_p^i}{\partial \theta_s} \tilde{J}_i \vec{B}_q^i + \vec{B}_p^i \tilde{J}_i \frac{\partial \vec{B}_q^i}{\partial \theta_s} \Big) \dot{\theta}_p \dot{\theta}_q , & s \le i. \end{cases} \tag{2.7.47}$$

Let us determine the partial derivatives of the vector coefficients. From (2.7.28) we find

$$\frac{\partial \vec{B}_j^i}{\partial \theta_s} = -A_i^{-1} \frac{\partial A_i}{\partial \theta_s} A_i^{-1} \vec{\beta}_j + A_i^{-1} \frac{\partial \vec{\beta}_j}{\partial \theta_s} \tag{2.7.48}$$

and from (2.7.22),

$$\frac{\partial \vec{\beta}_j}{\partial \theta_s} = \frac{\partial \vec{e}_j}{\partial \theta_s} = \frac{\partial A_{j-1}}{\partial \theta_s} \vec{\underset{\sim}{e}}_j . \tag{2.7.49}$$

From (2.7.24) it follows that

$$\frac{\partial \vec{\alpha}_j^i}{\partial \theta_s} = \begin{cases} -\sum_{k=j+1}^{i} \frac{\partial \vec{R}_{kj}}{\partial \theta_s} + \sum_{k=j}^{i} \frac{\partial \vec{H}_{kj}}{\partial \theta_s} , & j<i \\ \frac{\partial \vec{H}_{ii}}{\partial \theta_s}, & j=i \end{cases} \tag{2.7.50}$$

and from (2.7.25), (2.7.26),

$$\frac{\partial \vec{R}_{kj}}{\partial \theta_s} = \frac{\partial \vec{\beta}_j}{\partial \theta_s} \times \vec{\delta}_k + \vec{\beta}_j \times \frac{\partial \vec{\delta}_k}{\partial \theta_s}; \quad \frac{\partial \vec{H}_{kj}}{\partial \theta_s} = \frac{\partial \vec{\beta}_j}{\partial \theta_s} \times \vec{\gamma}_k u_k + \vec{\beta}_j \times \frac{\partial \vec{\gamma}_k}{\partial \theta_s} u_k, \tag{2.7.51}$$

$$\left.\begin{aligned} \frac{\partial \vec{\beta}_j}{\partial \theta_s} &= \frac{\partial \vec{e}_j}{\partial \theta_s} = \frac{\partial A_{j-1}}{\partial \theta_s} \underset{\sim}{\vec{e}}_j, \\ \frac{\partial \vec{\delta}_k}{\partial \theta_s} &= \frac{\partial \vec{r}_{k-1,k}}{\partial \theta_s} = \frac{\partial A_{k-1}}{\partial \theta_s} \tilde{\vec{r}}_{k-1,k}, \\ \frac{\partial \vec{\gamma}_k}{\partial \theta_s} &= \frac{\partial \vec{r}^{\,o}_{kk}}{\partial \theta_s} = \frac{\partial A_k}{\partial \theta_s} \tilde{\vec{r}}^{\,o}_{kk}. \end{aligned}\right\} \tag{2.7.52}$$

For the partial derivative of the kinetic energy with respect to the coordinate u_s from (2.7.29), (2.7.30), (2.7.32) and (2.7.33) we find

$$\frac{\partial T}{\partial u_s} = \frac{1}{2} \sum_{i=1}^{n} \frac{\partial}{\partial u_s} m_i \vec{v}_i^{\,2} + \frac{\partial}{\partial u_s} \tilde{\vec{\omega}}_i \tilde{J}_i \tilde{\vec{\omega}}_i, \tag{2.7.53}$$

$$\frac{\partial}{\partial u_s} \tilde{\vec{\omega}}_i \tilde{J}_i \tilde{\vec{\omega}}_i = 0, \tag{2.7.54a}$$

and

$$\frac{\partial}{\partial u_s} m_i \vec{v}_i^{\,2} = \begin{cases} 0, & s>i \\ \sum_{p=1}^{i} \sum_{q=1}^{i} \left(\frac{\partial \vec{\alpha}^i_p}{\partial u_s} \vec{\alpha}^i_q \dot{\theta}_p \dot{\theta}_q + \vec{\alpha}^i_p \frac{\partial \vec{\alpha}^i_q}{\partial u_s} \dot{\theta}_p \dot{\theta}_q + 2 \frac{\partial \vec{\alpha}^i_p}{\partial u_s} \vec{\gamma}_q \dot{\theta}_p \cdot \dot{u}_q\right) \end{cases} \tag{2.7.54b}$$

since $\frac{\partial \vec{B}^i_j}{\partial u_s} = 0$ for any s, i, j, and $\frac{\partial \vec{\gamma}_j}{\partial \theta_s} = 0$ for any s, j. From (2.7.24) it follows that

$$\frac{\partial \vec{\alpha}^i_j}{\partial u_s} = \begin{cases} -\sum_{k=j+1}^{i} \frac{\partial \vec{R}_{kj}}{\partial u_s} + \sum_{k=j}^{i} \frac{\partial \vec{H}_{kj}}{\partial u_s}, & j<i \\ \frac{\partial \vec{H}_{ii}}{\partial u_s}, & j=i \end{cases} \tag{2.7.55}$$

and from (2.7.25), (2.7.26),

$$\frac{\partial \vec{R}_{kj}}{\partial u_s} = 0; \quad \frac{\partial \vec{H}_{kj}}{\partial u_s} = \begin{cases} 0, & k \neq s \\ \vec{\beta}_j \times \vec{\gamma}_k = A_{j-1} \underset{\sim}{\vec{e}}_j \times A_k \tilde{\vec{r}}^{\,o}_{kk}, & k=s. \end{cases} \tag{2.7.56}$$

Thus, the problem of determining the partial derivatives of the kinetic energy with respect to generalized coordinates has been reduced to obtaining the partial derivatives of the transition matrices i.e. $\frac{\partial A_p}{\partial \theta_s}$, for every p, s.

On the whole, it can be concluded, that in order to formulate the left-hand sides of the Lagrange´s equations (2.7.31) it is sufficient to calculate A_p, $\dot{A}_p$, $\frac{\partial A_p}{\partial \theta_s}$, for every p, s, i.e. to calculate the transition matrices and their partial and time derivatives. This will be discussed next.

The recursive procedure for calculating the transition matrices has already been described in detail, i.e. A_i is calculated starting from the known A_{i-1}. Likewise, the calculation of the derivatives of the transition matrices is also recursive, i.e. $\dot{A}_i$ will be calculated starting from the known $\dot{A}_{i-1}$, and $\frac{\partial A_i}{\partial \theta_s}$ from the known $\frac{\partial A_{i-1}}{\partial \theta_s}$.

Let us first obtain the derivative $\frac{\partial A_i}{\partial \theta_s}$. The relation (2.7.10) is

$$A_i = A_{i-1}A_{i-1,i}. \tag{2.7.57}$$

In determining the partial derivative, from (2.7.57) we obtain

$$\frac{\partial A_i}{\partial \theta_s} = \begin{cases} \frac{\partial A_{i-1}}{\partial \theta_s} A_{i-1,i}, & s<i \\ A_{i-1}\frac{\partial A_{i-1,i}}{\partial \theta_i}, & s=i \\ 0, & s>i \end{cases} \tag{2.7.58}$$

where the property that the relative transition matrix $A_{i-1,i}$ depends on θ_i only (for each i), has been used. As A_{i-1}, $\frac{\partial A_{i-1}}{\partial \theta_s}$ are known, and the procedure for calculating $A_{i-1,i}$ has been described, only calculation of $\frac{\partial A_{i-1,i}}{\partial \theta_i}$ remains. From (2.7.17) it follows that

$$\frac{\partial A_{i-1,i}}{\partial \theta_i} = \begin{bmatrix} \frac{\partial V_{i1}}{\partial \theta_i} & \frac{\partial V_{i2}}{\partial \theta_i} & \frac{\partial V_{i3}}{\partial \theta_i} \end{bmatrix} \tag{2.7.59}$$

and from (2.7.16),

$$\frac{\partial \vec{v}_{ij}}{\partial \theta_i} = -\vec{v}_{ij}\sin\theta_i + \sin\theta_i(\vec{e}_i \cdot \vec{v}_{ij})\vec{e}_i + \vec{e}_i \times \vec{v}_{ij}\cos\theta_i, \quad j=1,2,3, \tag{2.7.60}$$

where $\vec{v}_{ij}$ has been determined in (2.7.15) and $\vec{e}_i$ is known and constant.

Thus, the problem of calculatiing the partial derivatives of the transitionmatrices $\frac{\partial A_i}{\partial \theta_s}$ has been solved. We now find the derivative $\dot{A}_i$. By differentiating the relation (2.7.57) with respect to time we get

$$\dot{A}_i = \dot{A}_{i-1}A_{i-1,i} + A_{i-i}\dot{A}_{i-1,i}. \tag{2.7.61}$$

As A_{i-1}, $\dot{A}_{i-1}$ are known, and the procedure for calculating A_{i-1} is described, the calculation of $\dot{A}_{i-1,i}$ still remains. Using the fact that $A_{i-1,i}$ depends on θ_i only, we find that

$$\dot{A}_{i-1,i} = \frac{\partial A_{i-1,i}}{\partial \theta_i}\dot{\theta}_i \tag{2.7.62}$$

The calculation of $\frac{\partial A_{i-1,i}}{\partial \theta_i}$ has been described, starting from relation (2.7.59).

Thus, the calculation of the time derivatives $\dot{A}_i$ of the transition matrices has been accomplished.

The problem of formulating the left-hand sides of Lagrange´s equations (2.7.31) has been solved in principle. On the right-hand side of the equations the generalized forces appear.

Calculation of the generalized forces. Let us designate by Q_i^θ, Q_i^u the generalized forces corresponding to the generalized coordinates θ_i, u_i, respectively. The expressions for the generalized forces will be derived by means of the virtual displacements method.

Let us first find Q_i^θ. Let us allow the coordinate θ_i to have some virtual displacement $\delta\theta_i$ (Fig. 2.23a), keeping all the other coordinates constant. Further, let us find the expression for the virtual work of the active forces over that displacement. In this mechanical system the active forces are the drives and the gravity forces.

Over the virtual displacement $\delta\theta_i$ the work is performed by the driving torque M_{P_i} and the gravity forces of the segments i,i+1,...,n (Fig. 2.23a):

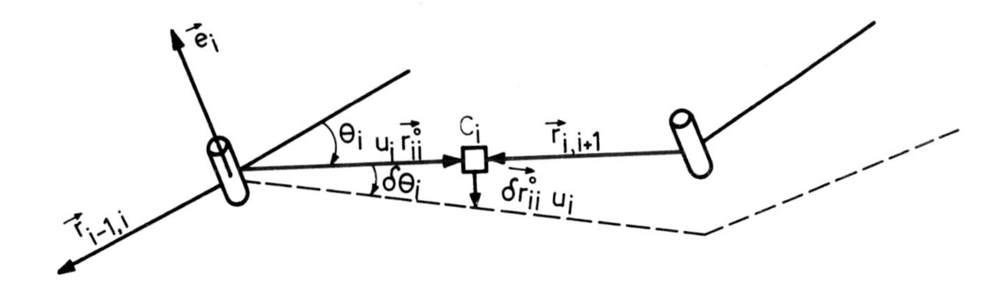

Fig. 2.23a. Revolute virtual displacement in joint S_i

Work of the driving torque is

$$\delta A^P = M_{P_i} \delta\theta_i . \tag{2.7.63}$$

Work of the gravity forces is

$$\delta A^g = \delta A^o + \delta A^1 + \dots + \delta A^{n-1}, \tag{2.7.64}$$

where δA^k is the work of the gravity forces of segment i+k.

For segment (i+k), using notation from Fig. 2.23b, the work done by the gravity forces has the form

$$\delta A^k = m_{i+k} \vec{g} \cdot \delta \vec{r}^{\,i}_k , \tag{2.7.65}$$

where $\vec{g}$ is the gravitational acceleration, $\vec{g} = \{0, \; 0, \; -9,81\}$.

The vector $\vec{r}^{\,i}_k$ is determined as

$$\vec{r}^{\,i}_k = \sum_{\ell=0}^{k} \vec{r}^{\,o}_{i+\ell,\, i+\ell} u_{i+\ell} - \sum_{\ell=0}^{k-1} \vec{r}_{i+\ell,\, i+\ell+1} \tag{2.7.66}$$

and then,

$$\delta \vec{r}^{\,i}_k = \vec{e}_i \times \vec{r}^{\,i}_k \cdot \delta\theta_i \tag{2.7.67}$$

Substituting (2.7.67) into (2.7.65) we find

$$\delta A^k = m_{i+k} \, [\vec{g}, \, \vec{e}_i, \, \vec{r}^{\,i}_k] \delta\theta_i , \tag{2.7.68}$$

where the expression in the square parenthesis represents the vector box product.

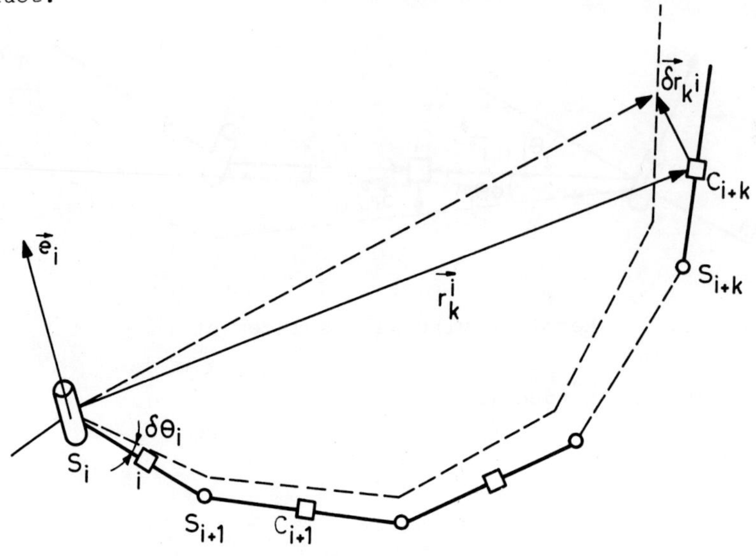

Fig. 2.23b. Determining the work of the gravity forces

Now the total work of the gravity forces, by (2.7.64) and (2.7.68), is equal to

$$\delta A^g = \sum_{k=0}^{n-i} m_{i+k} \, [\vec{g}, \vec{e}_i, \vec{r}_k^i]\delta\theta_i \tag{2.7.69}$$

and the total work of the active forces, according to (2.7.63), (2.7.69) is

$$\delta A^\theta = \delta A^P + \delta A^g = \{M_{P_i} + \sum_{k=0}^{n-i} m_{i+k} \, [\vec{g}, \vec{e}_i, \vec{r}_k^i]\}\delta\theta_i. \tag{2.7.70}$$

Now, by definition, the generalized force Q_i^θ is

$$Q_i^\theta = M_{P_i} + \sum_{k=0}^{n-i} m_{i+k} \, [\vec{g}, \vec{e}_i, \vec{r}_k^i], \tag{2.7.71}$$

where $\vec{r}_k^i$ was given by (2.7.66).

Let us now determine the generalized force Q_i^u. Let us allow the corresponding coordinate u_i to have a virtual displacement δu_i, keeping all other coordinates constant (Fig. 2.24). Over that displacement, work will be done by the driving force F_{P_i} and the gravity forces of segments i, i+1,...,n.

$$\delta A^P = \vec{F}_i \cdot \delta\vec{r}_i = F_{P_i} \cdot \delta u_i, \qquad (2.7.72)$$

Since

$$\delta\vec{r}_i = \vec{r}^{\,o}_{ii} \delta u_i . \qquad (2.7.73)$$

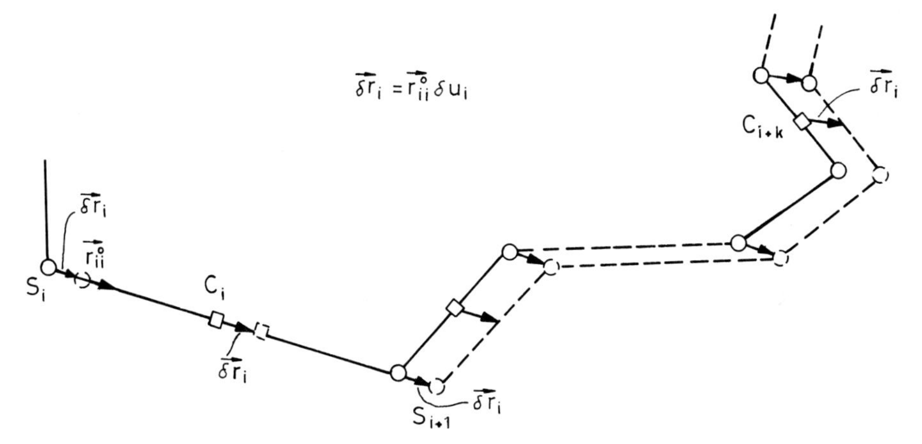

Fig. 2.24. Linear virtual displacement in joint S_i

Note that with the virtual displacement δu_i the whole part of the chain (from S_i to the free end) is moved linearly by $\delta\vec{r}_i = \vec{r}^{\,o}_{ii}\delta u_i$. Thus, the work done by gravity is

$$\delta A^g = \sum_{k=0}^{n-1} m_{i+k} \vec{g} \cdot \delta\vec{r}_i \qquad (2.7.74)$$

or, by (2.7.73),

$$\delta A^g = \vec{r}^{\,o}_{ii} \vec{g} \left(\sum_{k=0}^{n-1} m_{i+k} \right) \delta u_i . \qquad (2.7.75)$$

By (2.7.72) and (2.7.75), the total work of the active forces over the virtual displacement δu_i is now

$$\delta A^u = \delta A^P + \delta A^g = \{F_{P_i} + \vec{r}^{\,o}_{ii} \vec{g} \sum_{k=0}^{n-i} m_{i+k}\} \delta u_i \qquad (2.7.76)$$

By definition, the generalized force Q^u_i is

$$Q^u_i = F_{P_i} + \vec{r}^{\,o}_{ii} \vec{g} \sum_{k=0}^{n-i} m_{i+k} \qquad (2.7.77)$$

So, the formulation of the right-hand sides of Lagrange´s equations (2.7.31) has been achieved.

Let us introduce the vector of generalized forces of dimension 2n:

$$Q = [Q_1^\theta \; Q_1^u \; \cdots \; Q_n^\theta \; Q_n^u]^T. \tag{2.7.78}$$

By considering expressions (2.7.71) and (2.7.77) we note that the vector Q can be written in the form

$$Q = P + Y, \tag{2.7.79}$$

where P is the drives vector given by (2.7.5) and the vector Y is

$$\begin{aligned} Y &= [Y_1^\theta \; Y_1^u \; \cdots \; Y_n^\theta \; Y_n^u]^T, \\ Y_i^\theta &= \sum_{k=0}^{n-i} m_{i+k} [\vec{g}, \vec{e}_i, \vec{r}_k^{\,i}], \\ Y_i^u &= \vec{r}_{ii}^{\,o} \vec{g} \sum_{k=0}^{n-i} m_{i+k}. \end{aligned} \tag{2.7.80}$$

Computer matrix algorithm. In order to obtain a computer-oriented method for automatic formulation of the active mechanism mathematical model, it is necessary, besides using the recursive kinematical relations (2.7.18) - (2.7.23) and recursive calculation of the transition matrices and their derivatives, to also introduce matrix notation and recursive formulation of the equation system (2.7.1).

Let us write the relation for the center of gravity velocity (2.7.23) in matrix form:

$$v_i = \left[\alpha_1^i \gamma_1 \; \cdots \; \alpha_i^i \gamma_i \; 00 \; \cdots \; 00\right] \begin{bmatrix} \dot\theta_1 \\ \dot u_1 \\ \vdots \\ \dot\theta_i \\ \dot u_i \\ \vdots \\ \dot\theta_n \\ \dot u_n \end{bmatrix} \tag{2.7.81}$$

By introducing the vector of generlaized coordinates q (2.7.6) and the designations for the 3×2n matrix

$$M^{(i)} = [\alpha_1^i \gamma_1 \; \cdots \; \alpha_i^i \gamma_i \;\; 00 \; \cdots \; 00] \tag{2.7.82}$$

(2.7.81) acquires the form

$$v_i = M^{(i)}\dot{q}. \tag{2.7.83}$$

For the angular velocity, we find from (2.7.27) that

$$\tilde{\omega}_i = \tilde{N}^{(i)}\dot{q}, \tag{2.7.84}$$

where the 3×2n matrix is

$$\tilde{N}^{(i)} = [B_1^i 0 \cdots B_i^i 0 \; 00 \cdots 00]. \tag{2.7.85}$$

Let us write the kinetic energy of the i-th segment in the matrix form

$$T_i = \frac{1}{2} m_i v_i^T v_i + \frac{1}{2} \tilde{\omega}_i^T \tilde{J}_i \tilde{\omega}_i \tag{2.7.86}$$

and, by substituting (2.7.83) and (2.7.84), in the form

$$T_i = \frac{1}{2} \dot{q}^T (m_i M^{(i)^T} M^{(i)} + \tilde{N}^{(i)^T} \tilde{J}_i \tilde{N}^{(i)})\dot{q} \tag{2.7.87}$$

or

$$T_i = \frac{1}{2} \dot{q}^T W_i \dot{q}, \tag{2.7.88}$$

where the 2n×2n matrix is

$$W_i = m_i M^{(i)^T} M^{(i)} + \tilde{N}^{(i)^T} \tilde{J}_i \tilde{N}^{(i)}. \tag{2.7.89}$$

Now the total kinetic energy of the chain is

$$T = \sum_{i=1}^{n} T_i = \frac{1}{2} \dot{q}^T (\sum_{i=1}^{n} W_i)\dot{q}, \text{ or} \tag{2.7.90}$$

$$T = \frac{1}{2} \dot{q}^T W \dot{q}, \tag{2.7.91}$$

where the 2n×2n matrix is

$$W = \sum_{i=1}^{n} W_i. \tag{2.7.92}$$

The system of Lagrange equations (2.7.31) can be written in the matrix notation

$$\frac{d}{dt}(\frac{\partial T}{\partial \dot{q}}) - \frac{\partial T}{\partial q} = Q, \tag{2.7.93}$$

where the vector Q of generalized forces is given by (2.7.78). Substituting (2.7.91) into (2.7.93), we get

$$W\ddot{q} + \dot{W}\dot{q} - \frac{\partial T}{\partial q} = Q. \tag{2.7.94}$$

If we write the generalized forces Q in the form $Q = P + Y$ according to (2.7.79) and (2.7.80), and introduce the notation

$$U = Y + \frac{\partial T}{\partial q} - \dot{W}\dot{q}, \tag{2.7.95}$$

from (2.7.94) we find

$$W\ddot{q} = P + U, \tag{2.7.96}$$

i.e. the system of Lagrange´s equations in the form (2.7.1).

Before deriving of the recursion algorithm, let us discuss the meaning of the indices in the parenthesis (for instance $M^{(i)}$, $\tilde{N}^{(i)}$). Such matrices do not have an index in the real algorithm. The upper index in parenthesis (i) indicates that only the values which these matrices acquire in the i-th iteration are considered.

The algorithm for formulating the system (2.7.94), i.e. the calculation of the corresponding matrices, is recursive. In each iteration a new segment is added to the system-chain. Thus, in the i-th iteration, the i-th segment is added.

The matrix W is calculated recursively according to (2.7.92):

$$W^{(i)} = W^{(i-1)} + W_i; \qquad W = W^{(n)}, \tag{2.7.97}$$

where W_i is calculated according to (2.7.89).

Matrix $\dot{W}$ is also calculated recursively:

$$\dot{W}^{(i)} = \dot{W}^{(i-1)} + \dot{W}_i, \tag{2.7.98}$$

where, from (2.7.89), it follows that

$$\dot{W}_i = m_i(\dot{M}^{(i)^T}M^{(i)} + M^{(i)^T}\dot{M}^{(i)}) + \dot{\tilde{N}}^{(i)^T}\tilde{J}_i\tilde{N}^{(i)} + \tilde{N}^{(i)^T}\tilde{J}_i\dot{\tilde{N}}^{(i)}. \tag{2.7.99}$$

By (2.7.82) and (2.7.85), the matrices $\dot{M}^{(i)}$ and $\dot{\tilde{N}}^{(i)}$ have the form

$$\dot{M}^{(i)} = [\dot{\alpha}_1^i\dot{\gamma}_1 \cdots \dot{\alpha}_i^i\dot{\gamma}_i \; 00 \cdots 00],$$

$$\dot{\tilde{N}}^{(i)} = [\dot{B}_1^i \; 0 \cdots \dot{B}_i^i 0 \;\; 00 \cdots 00]. \tag{2.7.100}$$

$\dot{B}_j^i$ has been determined by (2.7.37), (2.7.38), $\dot{\alpha}_j^i$ by (2.7.39)-(2.7.41), and $\dot{\gamma}_j$ by (2.7.41). By examining the enumerated expressions, it can be seen that it is appropriate to introduce into the algorithm the matrices N, $\dot{N}$ for preserving the coefficients β_j and $\dot{\beta}_j$, as well the matrices L, $\dot{L}$ for preserving the coefficients δ_j and $\dot{\delta}_j$. As these matrices do not enter directly into the calculations but serve only to preserve the vector coefficients, their dimensions can be 3×n. Thus

$$N^{(i)} = [\beta_1 \cdots \beta_i 0 \cdots 0],$$

$$\dot{N}^{(i)} = [\dot{\beta}_1 \cdots \dot{\beta}_i 0 \cdots 0], \tag{2.7.101}$$

$$L^{(i)} = [\delta_1 \cdots \delta_i 0 \cdots 0],$$

$$\dot{L}^{(i)} = [\dot{\delta}_1 \cdots \dot{\delta}_i 0 \cdots 0]. \tag{2.7.102}$$

In each iteration the recursive modifications and supplementations of the matrices N, $\dot{N}$, $\tilde{N}$, $\dot{\tilde{N}}$, M, $\dot{M}$, L, $\dot{L}$ are carried out. The formulae for doing this are derived from (2.7.22), (2.7.24), (2.7.26), (2.7.28).

Let us consider the (i+1)-th iteration, in which the (i+1)-th segment is added to the chain.

The supplementations are performed in the matrices N, $\dot{N}$, according to the expressions (2.7.22) and (2.7.38), i.e.

$$\beta_{i+1} = A_i \underset{\sim}{e}_{i+1}; \qquad \dot{\beta}_{i+1} = \dot{A}_i \underset{\sim}{e}_{i+1} \tag{2.7.103}$$

and the matrices $N^{(i+1)}$, $\dot{N}^{(i+1)}$ are obtained instead of $N^{(i)}$, $\dot{N}^{(i)}$.

The supplementations are performed in the matrices L, $\dot{L}$, according to (2.7.26) and (2.7.41), i.e.

$$\delta_{i+1} = A_i \tilde{r}_{i,i+1}; \qquad \dot{\delta}_{i+1} = \dot{A}_i \tilde{r}_{i,i+1} \tag{2.7.104}$$

and $L^{(i+1)}$, $\dot{L}^{(i+1)}$ are obtained.

In the matrices $\tilde{N}$, $\dot{\tilde{N}}$ the modifications are performed according to (2.7.28) and (2.7.37), i.e.

$$\left.\begin{aligned} B_j^{i+1} &= A_{i+1}^{-1}\beta_j; \\ \dot{B}_j^{i+1} &= -A_{i+1}^{-1}\dot{A}_{i+1}A_{i+1}^{-1}\beta_j + A_{i+1}\dot{\beta}_j \end{aligned}\right\} ;\quad j=1,\dots,i+1 \tag{2.7.105}$$

and $\tilde{N}^{(i+1)}$, $\dot{\tilde{N}}^{(i+1)}$ are obtained.

In matrices M, $\dot{M}$ the modifications and supplementations are performed according to (2.7.24) - (2.7.26) and (2.7.39) - (2.7.41). i.e.,

$$\begin{cases} \alpha_j^{i+1} = \alpha_j^i - R_{i+1,j} + H_{i+1,j}, & j=1,\dots,i \\ \alpha_{i+1}^{i+1} = H_{i+1,i+1}, \end{cases} \tag{2.7.106}$$

$$\gamma_{i+1} = A_{i+1}\tilde{r}^o_{i+1,i+1}$$

and

$$\begin{cases} \dot{\alpha}_j^{i+1} = \dot{\alpha}_j^i - \dot{R}_{i+1,j} + \dot{H}_{i+1,j} & j=1,\dots,i \\ \dot{\alpha}_{i+1}^{i+1} = \dot{H}_{i+1,i+1}, \end{cases} \tag{2.7.107}$$

$$\dot{\gamma}_{i+1} = \dot{A}_{i+1}\tilde{r}^o_{i+1,i+1},$$

where R_{kj}, H_{kj}, $\dot{R}_{kj}$, $\dot{H}_{kj}$ are given by expressions (2.7.25) - (2.7.26) and (2.7.40) - (2.7.41).

Thus, the calculation of the matrices W, $\dot{W}$, appearing in (2.7.94) has been done.

Now, in order to formulate the system (2.7.94) it is necessary to calculate the vector

$$\frac{\partial T}{\partial q} \triangleq \left[\frac{\partial T}{\partial \theta_1}\ \frac{\partial T}{\partial u_1}\ \cdots\ \frac{\partial T}{\partial \theta_n}\ \frac{\partial T}{\partial u_n}\right]^T \tag{2.7.108}$$

using the expressions (2.7.45) - (2.7.47) and (2.7.53) - (2.7.54b). In the algorithm, the summations are, of course, carried out in such may as to avoid double summation. As it is here, it is necessary to know the partial derivatives of the coefficients β_j, B_j^i, γ_j, α_j^i; so, the matrices for preserving them are also introduced into the algorithm. For placing the coefficients $\frac{\partial \beta_j}{\partial \theta_s}$ a 3-dimensional matrix has been introduced, designated by $\frac{\partial N}{\partial \theta}$ (in the i-th iteration, $\frac{\partial N^{(i)}}{\partial \theta}$) and shown in Fig. 2.25a. When adding the (i+1)-th segment to the chain, i.e. in the (i+1)-th iteration, the matrix is supplemented. Starting from (2.7.49), the supplementing formula is obtained:

$$\left.\begin{aligned} &\frac{\partial \beta_{i+1}}{\partial \theta_s} = \frac{\partial A_i}{\partial \theta_s} \underset{\sim}{e}_{i+1}, \quad s=1,\ldots,i \\ &\frac{\partial \beta_{i+1}}{\partial \theta_{i+1}} = 0 \end{aligned}\right\} \tag{2.7.109}$$

In order to store the coefficients $\frac{\partial B_j^i}{\partial \theta_s}$, the three-dimensional matrix $\frac{\partial \tilde{N}}{\partial \theta}$ is introduced (Fig. 2.25b). The modifications of and supplements to this matrix are carried out in the (i+1)-th iteration. These are based on (2.7.48) applied to the (i+1)-th iteration.

Further, in order to store the coefficients $\frac{\partial \alpha_j^i}{\partial \theta_s}$, $\frac{\partial \gamma_j}{\partial \theta_s}$, the three - dimensional matrix $\frac{\partial M}{\partial \theta}$ is introduced. The structure of this matrix is similar to that in Fig. 2.25b. with the vector $\frac{\partial B_j^i}{\partial \theta_s}$ replaced by $\frac{\partial \alpha_j^i}{\partial \theta_s}$ and the zero vector $\vec{0}$ replaced by $\frac{\partial \gamma_j}{\partial \theta_s}$. Modifications of and supplements to this matrix are performed in the (i+1)-th iteration, based on relations (2.7.50) - (2.7.52), i.e.:

$$\left.\begin{aligned} &\left.\begin{aligned} &\frac{\partial \alpha_j^{i+1}}{\partial \theta_s} = \frac{\partial \alpha_j^i}{\partial \theta_s} - \frac{\partial R_{i+1,j}}{\partial \theta_s} + \frac{\partial H_{i+1,j}}{\partial \theta_s}; \quad j=1,\ldots,i \\ &\frac{\partial \alpha_{i+1}^{i+1}}{\partial \theta_s} = \frac{\partial H_{i+1,i+1}}{\partial \theta_s} \end{aligned}\right\} s=1,\ldots,i \\ &\frac{\partial \alpha_j^{i+1}}{\partial \theta_{i+1}} = - \frac{\partial R_{i+1,j}}{\partial \theta_{i+1}} + \frac{\partial H_{i+1,j}}{\partial \theta_{i+1}}; \quad j=1,\ldots,i \\ &\frac{\partial \alpha_{i+1}^{i+1}}{\partial \theta_{i+1}} = \frac{\partial H_{i+1,i+1}}{\partial \theta_{i+1}} \end{aligned}\right\} \tag{2.7.110}$$

where $\frac{\partial R_{kj}}{\partial \theta_s}$, $\frac{\partial H_{kj}}{\partial \theta_s}$ are given by expressions (2.7.51) - (2.7.52), and for the vector γ_{i+1},

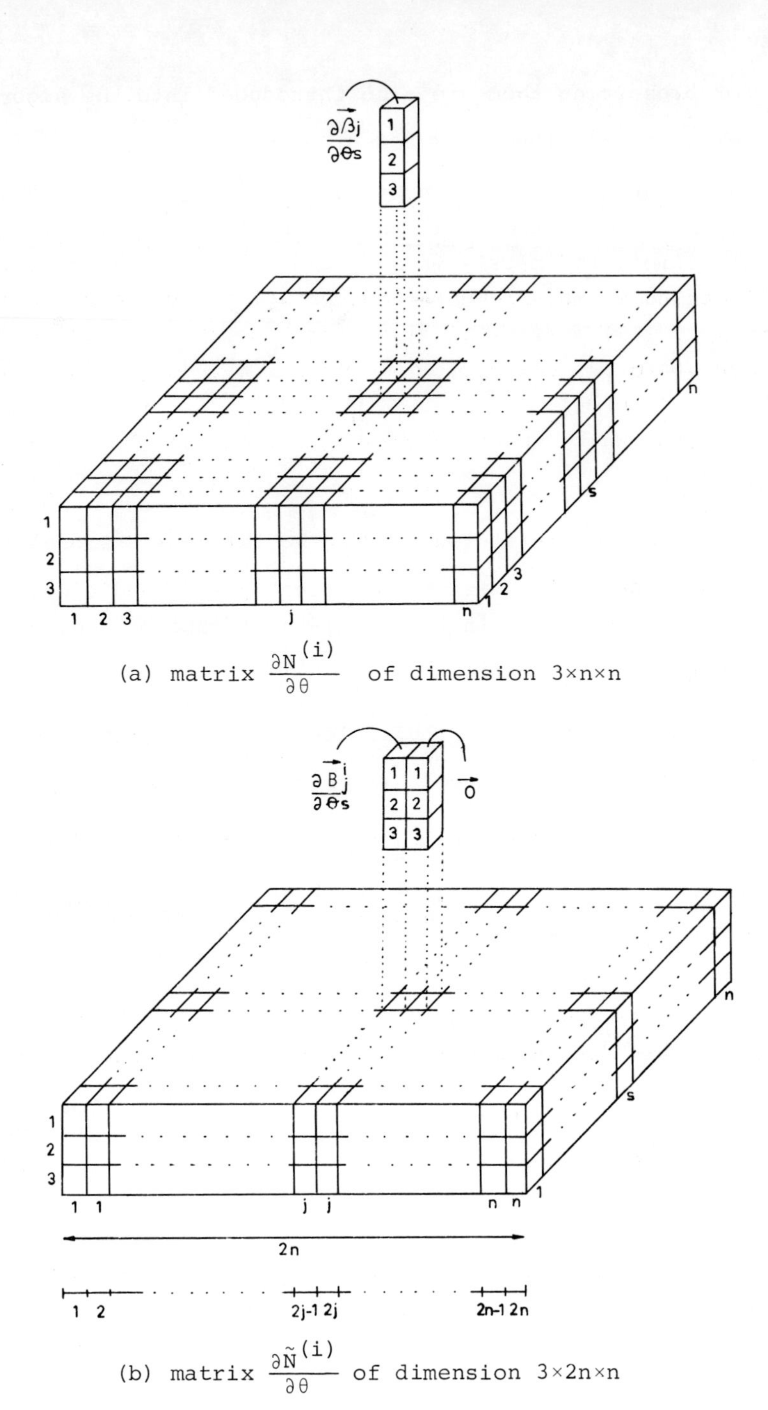

(a) matrix $\frac{\partial N^{(i)}}{\partial \theta}$ of dimension 3×n×n

(b) matrix $\frac{\partial \tilde{N}^{(i)}}{\partial \theta}$ of dimension 3×2n×n

Fig. 2.25. Structure of the matrices $\frac{\partial N}{\partial \theta}$, $\frac{\partial \tilde{N}}{\partial \theta}$

$$\frac{\partial \gamma_{i+1}}{\partial \theta_s} = \frac{\partial A_{i+1}}{\partial \theta_s} \tilde{r}^o_{i+1,i+1} \tag{2.7.111}$$

The coefficients $\frac{\partial \alpha^i_j}{\partial u_s}$ are stored in the three-dimensional matrix $\frac{\partial M}{\partial u}$, the structure of which is analogous to the structure of the matrix $\frac{\partial M}{\partial \theta}$. Modifications and supplements are carried out using expressions similar to (2.7.110) and (2.7.111) except that the derivatives are taken with respect to the coordinate u_s, taking care about (2.7.56)

Finally, for storing the coefficients $\frac{\partial \delta_j}{\partial \theta_s}$, there is the three-dimensional matrix $\frac{\partial L}{\partial \theta}$ with a structure like that in Fig. 2.25a except that the vector $\frac{\partial \beta_j}{\partial \theta_s}$ is replaced by $\frac{\partial \delta_j}{\partial \theta_s}$. Starting from (2.7.52), the expression for supplementing the matrix in the (i+1)-th iteration is derived

$$\left.\begin{aligned} &\frac{\partial \delta_{i+1}}{\partial \theta_s} = \frac{\partial A_i}{\partial \theta_s} \tilde{r}_{i,i+1}; \quad s=1,\dots,i \\ &\frac{\partial \delta_{i+1}}{\partial \theta_{i+1}} = 0 \end{aligned}\right\} \tag{2.7.112}$$

We described the calculation of the vector $\frac{\partial T}{\partial q}$ in eq. (2.7.94).

The generalized forces Q are calculated in the form (2.7.79), using (2.7.80).

Thus, the system (2.7.94) has been formed and, by introducing the substitution (2.7.95), it reduces to the form (2.7.96), i.e. (2.7.1).

The description of the method of Lagrange´s equations is therefore complete. This method can be represented as a whole by means of a block-sheme, as in Fig. 2.26.

Case of joints with one degree of freedom (d.o.f.). The method of Lagrange´s equations treats a mechanism with two-d.o.f. joints. If there is a one-d.o.f. joint in the mechanism, for instance a rotational one, then the linear coordinate in that joint is kept constant. The same holds for the linear joint with one d.o.f. However, in this case a 2n-order system is still formed and solved, i.e., the system order is higher than the real number of d.o.f. With minor modifications, this method can be adapted to one-degree-of-freedom joints, and for separa-

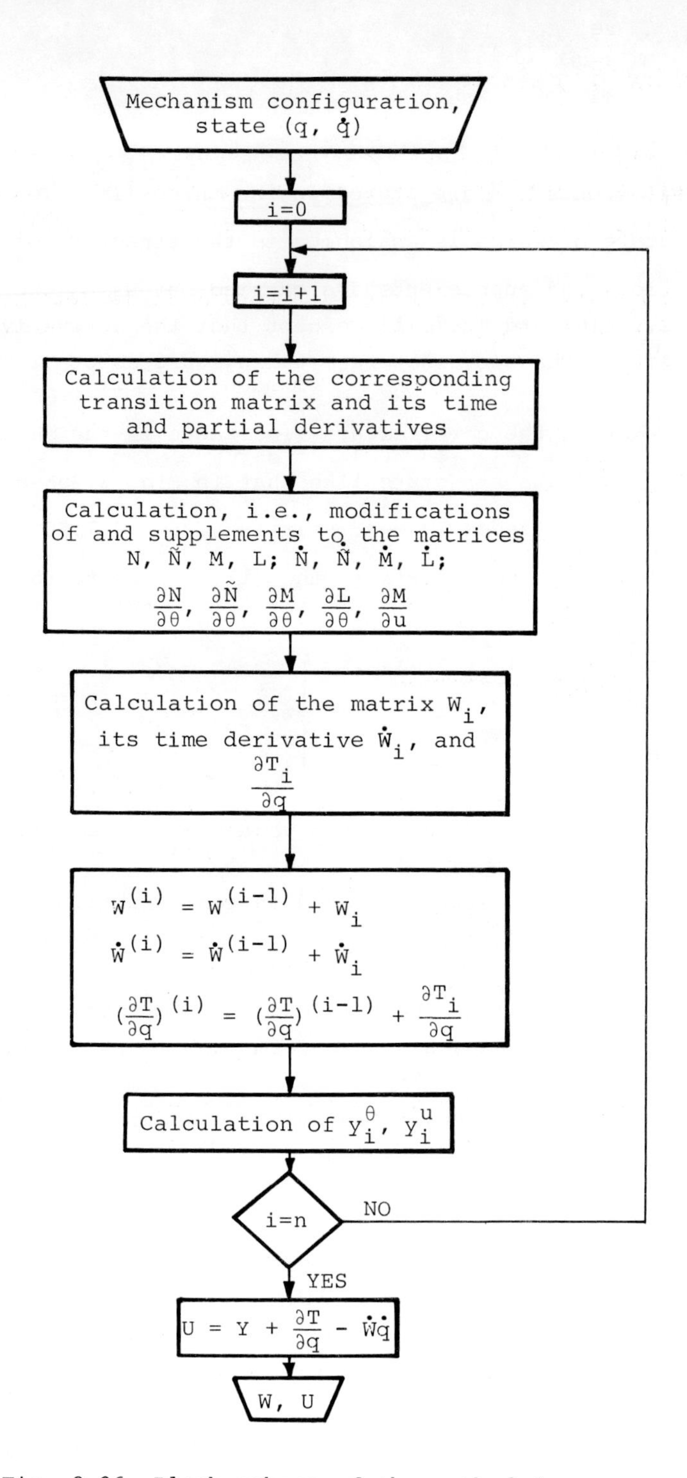

Fig. 2.26. Block-scheme of the method for equations system forming (2.7.1)

tion of complex joints into series-connected simple joints.

Let us consider an open chain without branching (Fig. 2.19), interconnected by joints having one rotational (Fig. 2.27a) or one linear degree of freedom (Fig. 2.27b). The mechanism has in all n degrees of freedom.

The axis of rotation or translation in a joint S_i is determined by means of a unit vector $\vec{e}_i$.

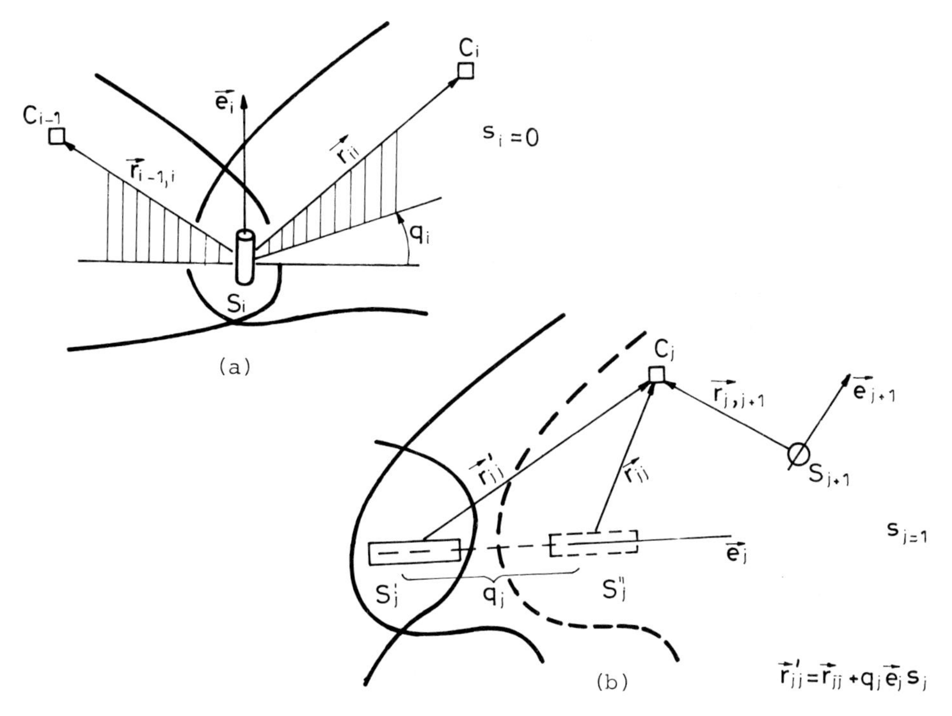

Fig. 2.27. (a) Mechanism rotation joint
(b) Mechanism linear joint

The drives in the joints are given in the form

$$\vec{P}_i = P_i\vec{e}_i. \tag{2.7.113}$$

Generalized coordinates $q = [q_1 \cdots q_n]^T$ are introduced in the form of rotation angles and displacements in the joints, as in Fig. 2.27a, b.

The transition matrices are calculated as in the method of general theorems, described in 2.3.

For such a mechanism, the recursive expressions for the center of gravity velocities and the angular velocities of segments acquire the form

$$\vec{\omega}_i = \vec{\omega}_{i-1} + \dot{q}_i\vec{e}_i(1-s_i), \tag{2.7.114}$$

$$\vec{v}_i = \vec{v}_{i-1} - \vec{\omega}_{i-1} \times \vec{r}_{i-1,i} + \vec{\omega}_i \times \vec{r}\,'_{ii} + \dot{q}_i\vec{e}_i s_i, \tag{2.7.115}$$

where

$$s_i = \begin{cases} 0, & \text{if the joint } S_i \text{ is rotational,} \\ 1, & \text{if the joint } S_i \text{ is linear.} \end{cases}$$

From these expressions,

$$\vec{\omega}_i = \sum_{j=1}^{i} \vec{\beta}_j \dot{q}_j, \tag{2.7.116}$$

where

$$\vec{\beta}_j = (1-s_j)\vec{e}_j = (1-s_j)A_{j-1}\underset{\sim}{e}_j. \tag{2.7.117}$$

One also finds that

$$\vec{v}_i = \sum_{j=1}^{i} \vec{\alpha}_j^i \dot{q}_j, \tag{2.7.118}$$

where

$$\vec{\alpha}_j^i = \begin{cases} \vec{\gamma}_j - \sum\limits_{k=j+1}^{i} \vec{R}_{kj} + \sum\limits_{k=j}^{i} \vec{H}_{kj}; & j<i, \\ \vec{\gamma}_j + \vec{H}_{ii}; & j=i, \end{cases} \tag{2.7.119}$$

$$\vec{\gamma}_j = \vec{e}_j s_j, \quad \vec{R}_{kj} = \vec{\beta}_j \times \vec{r}_{k-1,k}, \quad \vec{H}_{kj} = \vec{\beta}_j \times (\vec{\gamma}_k q_k + \vec{r}_{kk}). \tag{2.7.120}$$

From (2.7.116) it also follows that

$$\vec{\tilde{\omega}} = A_i^{-1}\vec{\omega}_i = \sum_{j=1}^{i} \vec{B}_j^i \dot{q}_j, \tag{2.7.121}$$

where

$$\vec{B}_j^i = A_i^{-1}\vec{\beta}_j. \tag{2.7.122}$$

The analogy is now clear as are the differences from the method of treating joints with two degrees of freedom.

Relations (2.7.121) and (2.7.118) can be written in the matrix notation

$$\tilde{\omega}_i = \tilde{N}^{(i)}\dot{q} \tag{2.7.123}$$

and

$$v_i = M^{(i)}\dot{q}, \tag{2.7.124}$$

where the 3×n matrices $\tilde{N}^{(i)}$ and $M^{(i)}$ are of the form

$$\tilde{N}^{(i)} = [B_1^i \cdots B_i^i \; 0 \cdots 0], \tag{2.7.125}$$

$$M^{(i)} = [\alpha_1^i \cdots \alpha_i^i \; 0 \cdots 0]. \tag{2.7.126}$$

Just as before, the kinetic energy of the whole mechanism is written in the form

$$T = \sum_{i=1}^{n} \frac{1}{2} m_i v_i^T v_i + \frac{1}{2}\tilde{\omega}_i^T \tilde{J}_i \tilde{\omega}_i = \frac{1}{2} \dot{q}^T W \dot{q}, \tag{2.7.127}$$

where the n×n matrix is

$$W = \sum_{i=1}^{n} W_i; \quad W_i = m_i M^{(i)^T} M^{(i)} + \tilde{N}^{(i)^T} \tilde{J}_i \tilde{N}^{(i)}. \tag{2.7.128}$$

The generalized forces are derived analogously as before by means of the method of virtual displacements:

$$\begin{aligned} Q_i^{(rot)} &= P_i + \sum_{k=0}^{n-i} [m_{i+k}\vec{g}, \vec{e}_i, \vec{r}_k^{\,i}], \\ \vec{r}_k^{\,i} &= \sum_{\ell=0}^{k} \vec{r}\,'_{i+\ell,i+\ell} - \sum_{\ell=0}^{k-1} \vec{r}_{i+\ell,i+\ell+1} \end{aligned} \tag{2.7.129}$$

if S_i is a rotational joint, and

$$Q_i^{(trans)} = P_i + \vec{e}_i \vec{g} \sum_{k=0}^{n-i} m_{i+k} \tag{2.7.130}$$

if S_i is a linear joint.

By introducing matrix notation, the equation for generalized forces acquires the form

$$Q = P + Y, \tag{2.7.131}$$

where $Q = [Q_1 \cdots Q_n]$ is an n-dimensional vector of generalized forces and $P = [P_1 \cdots P_n]$ is an n-dimensional vector of the drives in the joints.

By substituting (2.7.127) and (2.7.131) into the Lagrange´s equations

$$\frac{d}{dt}\left(\frac{\partial T}{\partial \dot{q}}\right) - \frac{\partial T}{\partial q} = Q, \tag{2.7.132}$$

a system of equations

$$W\ddot{q} = P + U \tag{2.7.133}$$

is obtained, where $U = Y + \frac{\partial T}{\partial q} - \dot{W}\dot{q}$, the necessary partial and time derivatives, $\frac{\partial T}{\partial q}$ and $\dot{W}$ being calculated as in the method considering two-d.o.f. joints.

Thus the system (2.7.133) of order n is obtained, of order i.e. equal to the number of mechanism degrees of freedom.

Methods Based on the Gauss' Principle and Appel's Equations

2.8. Method of Gauss' Principle

This rather computer-oriented method differs in many aspects from the methods described earlier. For the description of the mechanism dynamics, other methods use mainly the system of second-order differential equations derived in some way. At a particular time instant, such a system represents a system of algebraic equations with respect to generalized accelerations; so the functions f and g, defined in Paragraph 2.2., can be easily derived. The method described in this Paragraph [18, 19] is based on Gauss´ principle of minimal compulsion and, according to that principle, on the minimization of a scalar function. So the inverse problem of dynamics (i.e. the function f) is solved by means of numerical minimization. In order to solve the direct problem (i.e. the function g) we use the necessary conditions of minimum to derive the equation system from which the drives are obtained. Consequently, the functions f and g are realized in rather different ways.

This is due to the fact that the procedure of numerical minimization is one directional; so to solve the direct problem, we have to give up the basic idea of minimization and to return to the equation system which has a different form than in other methods.

Because of its length, the method of Gauss´ principle will be presented here in a very shortened form, so if the need arises, the reader is advised to consult [19].

At the start, the treatment of a general configuration will be presented and later elaborated for the case of the open kinematic chain without branching.

The general case

Since the method utilizes (the apparatus of) homogenous coordinates, we breifly define them and the manner of using them in the problem considered.

Let us consider the Cartesian orthogonal coordinate system Oxyz and a point with coordinates (x, y, z). The homogenous coordinates of that point are defined as a quadruplet of the numbers (x_1, x_2, x_3, x_4), satisfying

$$x = \frac{x_1}{x_4}, \quad y = \frac{x_2}{x_4}, \quad z = \frac{x_3}{x_4} \tag{2.8.1}$$

which evidently are not uniquely determined.

Let us say that the position of the poitn in the four dimensional space of homogenous coordinates is determined by means of a four-dimensional position vector

$$X = [x \quad y \quad z \quad 1]^T. \tag{2.8.2}$$

Let us now take two coordinate systems, O´x´y´z´ and O"x"y"z", and the transformation of one coordinate system into the other. Such a transformation can be presented by

$$X' = TX'', \tag{2.8.3}$$

where the transformation matrix (4×4) is of the form

$$T = \begin{bmatrix} A & \ell \\ 0 & 1 \end{bmatrix}, \qquad (2.8.4)$$

A being the 3×3 transition matrix between the two systems, defined as in the preceding methods, and ℓ representing a 3×1 matrix of the point O" coordinates in the system O´x´y´z´ (Fig. 2.28).

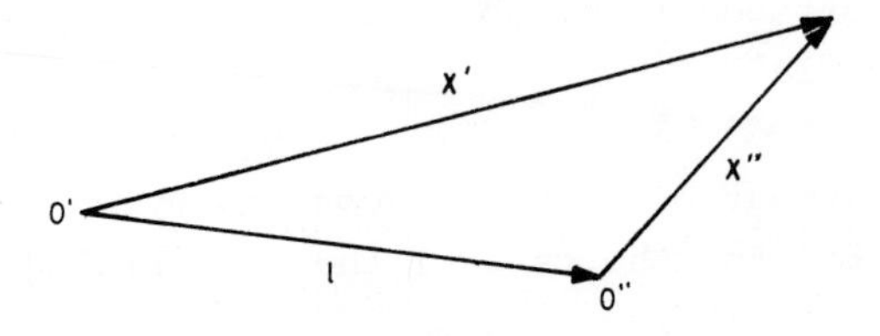

Fig. 2.28. Transformation of coordinates

It is important to notice the difference between the transformation matrix T, which transofrms the coordinates of some point from one system to another, and the transition matrix connecting the projections of a vector onto the axes of the two coordinate systems.

Let us take a vector $\vec{F} = \{F_x, F_y, F_z\}$, which is not a position vector. For such vectors, we introduce the matrix notation by the 4×1 matrix vector $F = [F_x \;\; F_y \;\; F_z \;\; 0]^T$. If

$$F' = [F_{x'} \;\; F_{y'} \;\; F_{z'} \;\; 0]^T \quad \text{and} \quad F'' = [F_{x''} \;\; F_{y''} \;\; F_{z''} \;\; 0]^T$$

then

$$F' = TF''. \qquad (2.8.5)$$

Since the fourth elements of the vectors F´ and F" are zero, for such vectors the transformation matrix T performs the same function as the transition matrix A, i.e., (2.8.5) is equivalent to

$$\vec{F}' = A\vec{F}'' \qquad (2.8.6)$$

For the position vectors, i.e., the transformation of coordinates, (2.8.2) and (2.8.3) hold.

<u>Mechanism configuration</u>. Let us consider a mechanism consisting of m rigid bodies forming, among themselves and other fixed bodies, n kinematic pairs with one d.o.f, rotational or linear (with connections by means of corresponding joints), as in Fig. 2.29.

Let us associate with each body an orthogonal Cartesian coordinate system (body-fixed or b.-f. system), and let us also define an immobile external coordinate system connected to the base. By introducing homogenous coordinates, we can describe the position of the body in space by means of a 4×4 transformation matrix T from the b.-f. system to the external one. From (2.8.4), we find that for the i-th body,

$$T_i = \begin{bmatrix} A_i & \ell_i \\ 0 & 1 \end{bmatrix}, \qquad (2.8.7)$$

where A_i is the corresponding transition matrix and ℓ_i the position vector of the origin of i-th system relative to the external system. Such a matrix T_i will often be called the position matrix of the i-th body, the matrix $\dot{T}_i$ will be called the velocity matrix and $\ddot{T}_i$ the acceleration matrix.

The relative position of two bodies, the "i-th" relative to the "j-th",

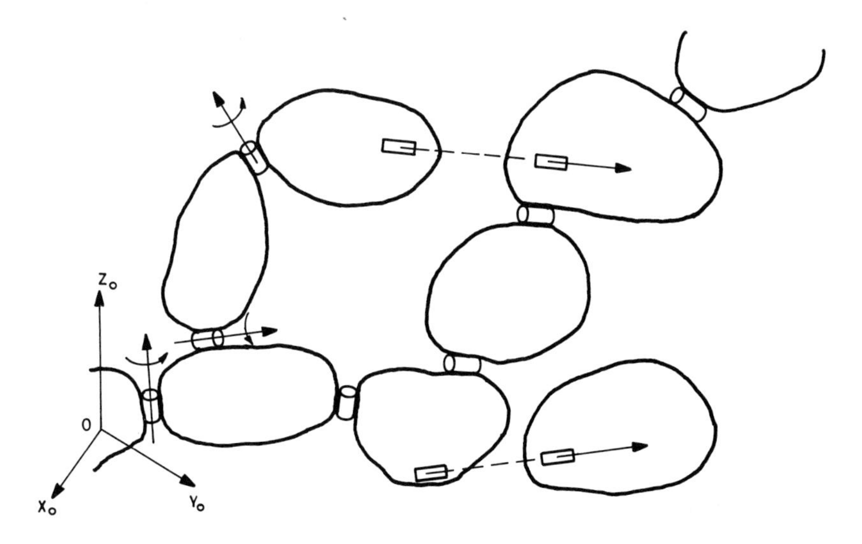

Fig. 2.29. General mechanism configuration

can be described by the corresponding transformation matrix from the i-th b.-f. system into the j-th one:

$$T_{ji} = \begin{bmatrix} A_{ji} & \underset{\sim}{\ell}_{ji} \\ 0 & 1 \end{bmatrix}, \qquad (2.8.8)$$

where A_{ji} is the corresponding transition matrix and $\underset{\sim}{\ell}_{ji}$ is the position vector of the i-th coordinate system origin expressed in the j-th system.

The transformation matrices (2.8.7) and (2.8.8) have 13 non-zero elements each but it should be noticed that these 13 elements depend on 6 variables (e.g. 3 coordinates of the b.-f. system origin and 3 Euler angles).

Let us consider two bodies "i" and "j", connected by means of a joint S_k. Let us denote the corresponding transformation matrix by ${}^kT_{ji}$, where the upper index k means connection by means of the joint S_k. Since the bodies are connected by means of a joint with one degree of freedom, the elements of matrix ${}^kT_{ji}$ depend on one variable only: a rotation angle if the joint is rotational, and a linear displacement if the joint is linear.

Internal coordinates. For the joint S_k let us define the coordinate q_k as a linear displacement in the joint if the joint is linear, and a rotation angle if it is rotational. The vector of internal coordinates is $q = [q_1 \cdots q_n]^T$.

Thus, the matrix ${}^kT_{ji}$ depends on q_k, i.e.

$${}^kT_{ji} = {}^kT_{ji}(q_k).$$

Drives in the joints. A drive is acting in each mechanism joint. In the rotational joint S_i, the driving torque of intensity P_i is acting, and if S_i is a linear joint, then a driving force of intensity P_i is acting. As a vector, the drive has the same direction and positive sense as the axis of rotation or translation in the joint. Let us introduce the n×1 matrix of the driving forces and torques $P=[P_1 \cdots P_n]$.

Basic postulates of the method. The method is based on Gauss´ principle of least compulsion. For a system of material points Gauss formed a scalar function proportional to the square of the real motion deviation from the free motion and called it the measure of compulsion Z:

$$Z = \frac{1}{2} \sum_{\nu} m_{\nu} (\ddot{\vec{r}}_{\nu} - \frac{\vec{F}_{\nu}}{m_{\nu}})^2, \tag{2.8.9}$$

where $\sum_{\nu}$ designates the sum over all material points of the system, $\vec{r}_{\nu}$ is the position vector of the point, m_{ν} is the mass, and $\vec{F}_{\nu}$ is the force acting at the point ν. Gauss´ principle now states: "At each time instant, the real motion of the system, being under the action of forces and having ideal connections, differs from all possible motions (possible in the sense of being compatible with the connections) which are performed from the same initial configuration and same initial velocities, by the fact that for the real motion the measure of the deviation from free motion, i.e. the compulsion Z, is minimal" [19].

The idea in the realization of the function f (defined by (2.2.1)) is to form the function $Z = Z(\ddot{u})$ with the known configuration, positions and velocities, as a function of the acceleration $\ddot{u}$ only. The accelerations are then calculated by numerical minimization of the function $Z(\ddot{u})$.

<u>The compulsion function</u>. For the mechanism considered, it is possible to write the compulsion function Z in the form [19]:

$$Z = \sum_{i=1}^{m} \mathrm{tr}\left(\frac{1}{2}\, \ddot{T}_i \tilde{H}_i \ddot{T}_i^T - \Phi_i \ddot{T}_i^T\right) - \sum_{j=1}^{n} P_j \ddot{q}_j, \qquad (2.8.10)$$

where tr designates the trace of the matrix. The 4×4 matrix H_i characterizes the inertial properties of the i-th body and has the structure:

$$\tilde{H}_i = \begin{bmatrix} \tilde{J}_i & \tilde{V}_i \\ \tilde{V}_i^T & m_i \end{bmatrix}, \qquad (2.8.11)$$

where $\tilde{J}_i$ is the 3×3 tensor of the inertia of the i-th body with respect to the b.-f. coordinate system, m_i is the body mass, V_i is the 3×1 vector of the body mass moment (static moment) relative to the b.-f. system (For the material point of mass m, the mass moment $\vec{V}$ is defined as $\vec{V} = m\vec{r}$, where $\vec{r}$ is the position vector relative to some system. For the rigid body, summations is taken over all body points).

The 4×4 matrix Φ_i characterizes the external forces acting on the i-th body. Let us designate by F_{Ei} the 3×1 vector of the resultant external force acting on the i-th segment, and by M_{Ei} the resultant couple of the external forces (resultant moment). Let us introduce the so-called projection matrices:

$$\Theta_1 = \begin{bmatrix} 0 & 0 & 0 & 0 \\ 0 & 0 & -1 & 0 \\ 0 & 1 & 0 & 0 \\ 0 & 0 & 0 & 0 \end{bmatrix} \Theta_2 = \begin{bmatrix} 0 & 0 & 1 & 0 \\ 0 & 0 & 0 & 0 \\ -1 & 0 & 0 & 0 \\ 0 & 0 & 0 & 0 \end{bmatrix} \Theta_3 = \begin{bmatrix} 0 & -1 & 0 & 0 \\ 1 & 0 & 0 & 0 \\ 0 & 0 & 0 & 0 \\ 0 & 0 & 0 & 0 \end{bmatrix}$$

(2.8.12)

$$\Theta_4 = \begin{bmatrix} 0 & 0 & 0 & 1 \\ 0 & 0 & 0 & 0 \\ 0 & 0 & 0 & 0 \\ 0 & 0 & 0 & 0 \end{bmatrix} \Theta_5 = \begin{bmatrix} 0 & 0 & 0 & 0 \\ 0 & 0 & 0 & 1 \\ 0 & 0 & 0 & 0 \\ 0 & 0 & 0 & 0 \end{bmatrix} \Theta_6 = \begin{bmatrix} 0 & 0 & 0 & 0 \\ 0 & 0 & 0 & 0 \\ 0 & 0 & 0 & 1 \\ 0 & 0 & 0 & 0 \end{bmatrix}$$

Now, for the sake of more condensed writing, let us introduce the following notation for the 6×1 vector

$$F_i = \begin{bmatrix} F_{Ei} \\ M_{Ei} \end{bmatrix} = [f_{i1} \; f_{i2} \; \cdots \; f_{i6}]^T. \tag{2.8.13}$$

Now let us write the matrix Φ_i in the form:

$$\Phi_i = T_i \sum_{k=1}^{6} \Theta_k f_{ik} \delta_k, \tag{2.8.14}$$

where

$$\delta_k = \begin{cases} 1/2 & \text{for } k = 1, 2, 3, \\ 1 & \text{for } k = 3, 4, 5. \end{cases} \tag{2.8.15}$$

Let us, for instance, express the gravity forces in the form of matrix Φ_i. It can be shown that

$$\Phi_i = G\tilde{H}_i = m_i g \tilde{\rho}_i, \tag{2.8.16}$$

where

$$G = \begin{bmatrix} 0 & 0 & 0 & g_1 \\ 0 & 0 & 0 & g_2 \\ 0 & 0 & 0 & g_3 \\ 0 & 0 & 0 & 0 \end{bmatrix} \qquad g = \begin{bmatrix} g_1 \\ g_2 \\ g_3 \\ 0 \end{bmatrix} \tag{2.8.17}$$

where g_1, g_2, g_3 are the projections of the gravitational acceleration on the axes of the external system (for instance $g = [0 \;\; 0 \;\; -9,81 \;\; 0]^T$

if the z-axis of the external system is directed vertically upwards), and $\tilde{\rho}_i$ is a four-dimensional position vector of the i-th segment with respect to the i-th b.-f. system.

Kinematical connections. Let us consider two bodies "i_k" and "j_k" connected by means of joint S_k. Then

$$T_{i_k} = T_{j_k} {}^k T_{ji}. \tag{2.8.18}$$

By differentiating (2.8.18) twice with respect to time, one obtains

$$\ddot{T}_{i_k} = \ddot{T}_{j_k} {}^k T_{ji} + B_k \ddot{q}_k + C_k, \tag{2.8.19}$$

where

$$B_k = T_{j_k} \frac{d\, {}^k T_{ji}}{dq_k} \qquad C_k = 2\dot{T}_{j_k} \frac{d\, {}^k T_{ji}}{dq_k} \dot{q}_k + T_{j_k} \frac{d^2\, {}^k T_{ji}}{dq_k^2} \dot{q}_k^2. \tag{2.8.20}$$

Solving the inverse problem. Let us note that the vector "u", defined in Paragraph 2.2., now posses six variables for each body. They determine the position of the body in space and the body position matrix T is dependent on them. The internal coordinates also appear as the elements of the vector "u". Thus, the total dimension of vector "u" is 6m+n. Now the compulsion function Z, defined by (2.8.10), can be considered as the function of the vector $\ddot{u}$, i.e.

$$Z = Z(\ddot{u}). \tag{2.8.21}$$

Let introduce the notation:

$$\ddot{\Delta}_k = \ddot{T}_{i_k} - \ddot{T}_{j_k} \cdot {}^k T_{ji} - B_k \ddot{q}_k - C_k. \tag{2.8.22}$$

Then the kinematical connection (2.8.19) can be written in the form:

$$\ddot{\Delta}_k(\ddot{u}) = 0, \quad k=1,\dots,n. \tag{2.8.23}$$

Thus, the calculation of $\ddot{u}$, i.e. the realization of the function f, is reduced to the task of quadratic programming, i.e. to the minimization of the compulsion function Z defined by (2.8.21) and (2.8.10), along with constraints (2.8.23), i.e. (2.8.19). This means that we look for the $\ddot{u}$ which ensures the minimum value of the function (2.8.10) and satisfy (2.8.19). As the constraints (2.8.23) are of the equality type, the minimization with the constraints can be reduced to the minimization without constraints. This is achieved by the method of the "penalty function", i.e. by introducing the function [20, 21]

$$Z^{*} = Z + \sum_{k=1}^{n} \frac{1}{2\varepsilon} \operatorname{tr}(\ddot{\Delta}_k \ddot{\Delta}_k^T) \tag{2.8.24}$$

which is minimized without constraints and has a minimum for the same value of the vector $\ddot{u}$.

Case of the kinematical chain without branching

Mechanism configuration. Let us now consider the special case, i.e. the open chain without branching (Fig. 2.30). Then, the number of segments equals the number n of joints and the number of degrees of freedom. Let us introduce the indicator s_i for the joint type as:

$$s_i = \begin{cases} 0 & \text{for the rotational joints } S_{i-1}, \\ 1 & \text{for the linear joints } S_{i-1}. \end{cases}$$

Generalized coordinates. Let us introduce n generalized coordinates, one in each joint. Let q_i be the generalized coordinate, corresponding to joint S_{i-1}. If the joint S_{i-1} is rotational and θ_i is the rotation angle, then $q_i = \theta_i$. If the joint is linear and u_i is the displacement, then $q_i = u_i$. Thus,

$$q_i = (1-s_i)\theta_i + s_i u_i .$$

i.e. the earlier introduced internal coordinates now become the generalized coordinates of the system.

Coordinate systems and transformation matrices. Let us introduce the body-fixed (b.-f.) systems. For the i-th segment (body), the corresponding b.-f. system $O_i x_i y_i z_i$ has its origin in the point S_i (Fig. 2.30). The way of setting the coordinate systems is the same as in the

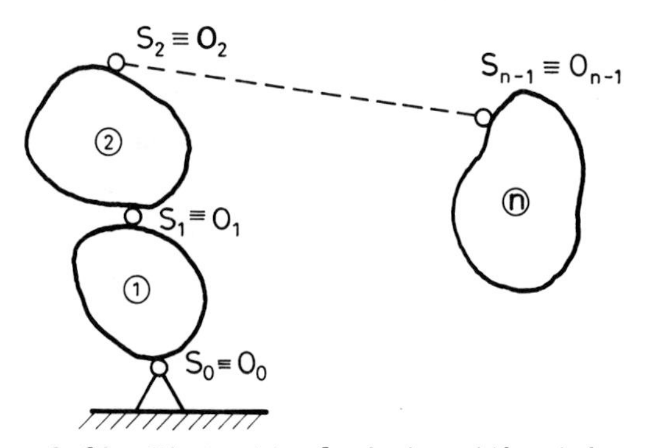

Fig. 2.30. Kinematical chain without branching

method of block-matrices, described in Paragraph 2.4. Thus, the transition matrix is

$$A_{i-1,i} = \begin{bmatrix} \cos\theta_i & -\sin\theta_i\cos\alpha_i & \sin\theta_i\sin\alpha_i \\ \sin\theta_i & \cos\theta_i\cos\alpha_i & -\cos\theta_i\sin\alpha_i \\ 0 & \sin\alpha_i & \cos\alpha_i \end{bmatrix} \qquad (2.8.25)$$

and the position vector

$$\underset{\sim}{\ell}_{i-1,i} = [a_i\cos\theta_i \quad a_i\sin\theta_i \quad u_i]^T, \qquad (2.8.26)$$

where θ_i, α_i, a_i, u_i, are the same as in the method in Paragraph 2.4.

Taking into account (2.8.8), the transformation matrix is

$$^{i-1}T_{i-1,i} = \begin{bmatrix} A_{i-1,i} & \underset{\sim}{\ell}_{i-1,i} \\ 0 & 1 \end{bmatrix} =$$

$$= \begin{bmatrix} \cos\theta_i & -\sin\theta_i\cos\alpha_i & \sin\theta_i\sin\alpha_i & a_i\cos\theta_i \\ \sin\theta_i & \cos\theta_i\cos\alpha_i & -\cos\theta_i\sin\alpha_i & a_i\sin\theta_i \\ 0 & \sin\alpha_i & \cos\alpha_i & u_i \\ 0 & 0 & 0 & 1 \end{bmatrix} \qquad (2.8.27)$$

Evidently, matrix $^{i-1}T_{i-1,i}$ depends on one variable only, the generalized coordinate q_i (having in mind the definition of the generalized coordinate). Thus,

$$^{i-1}T_{i-1,i} = {}^{i-1}T_{i-1,i}(q_i). \qquad (2.8.28)$$

Kinematical connections. It can be shown that in the case of the kinematical chain without branching, the kinematical connections (2.8.19), (2.8.22) can be written in the form

$$\ddot{T}_i = \ddot{T}_{i-1}\,{}^{i-1}T_{i-1,i} + B_{i-1}\ddot{q}_i + C_{i-1} \qquad (2.8.29)$$

i.e.

$$\ddot{\Delta}_{i-1} = \ddot{T}_i - \ddot{T}_{i-1}\,{}^{i-1}T_{i-1,i} - B_{i-1}\ddot{q}_i - C_{i-1} = 0, \qquad (2.8.30)$$

where

$$B_{i-1} = T_{i-1}\Theta\,{}^{i-1}T_{i-1,i},$$
$$C_{i-1} = 2\dot{T}_{i-1}\Theta\,{}^{i-1}T_{i-1,i}\dot{q}_i + T_{i-1}\Theta^2\,{}^{i-1}T_{i-1,i}\dot{q}_i^2, \qquad (2.8.31)$$

whereby

$$\Theta = (1-s_i)\Theta_3 + s_i\Theta_6. \qquad (2.8.32)$$

The solution of the inverse problem is found in the manner already described, the compulsion function Z being

$$Z = \sum_{i=1}^{n} \operatorname{tr}\left(\frac{1}{2}\ddot{T}_i\tilde{H}_i\ddot{T}_i^T - \Phi_i\ddot{T}_i^T\right) - \sum_{i=1}^{n} P_i\ddot{q}_i . \qquad (2.8.33)$$

Solving the direct problem. As we said in the introduction to the method, the direct problem is solved by using the conditions necessary for the minimum to derive the equations from which the drives are calculated.

Using the Lagrangian multipliers, let us form the function

$$Z^{**} = Z + \sum_{k=1}^{n} \operatorname{tr}(\lambda_k\ddot{\Delta}_{k-1})^T . \qquad (2.8.34)$$

Using also the necessary condition for the minimum, $\delta Z^{**} = 0$, the following equations can be derived:

$$\lambda_i = \lambda_{i+1}{}^{i}T^T_{i,i+1} + \Phi_i - \ddot{T}_i\tilde{H}_i, \quad i=0,1,\ldots,n, \qquad (2.8.35)$$

$$P_j = \operatorname{tr}(\lambda_j B_{j-1}), \quad j=1,\ldots,n, \qquad (2.8.36)$$

with the boundary conditions

$$\operatorname{tr}(\lambda_o\delta\ddot{T}_o^T) = 0 \quad \operatorname{tr}(\lambda_{n+1}\delta\ddot{T}_n^T) = 0, \qquad (2.8.37)$$

which, for the case under consideration of the chain on a fixed base and with a free end, become

$$\ddot{T}_o = 0, \quad \lambda_{n+1} = 0 . \qquad (2.8.38)$$

Thus, the system of recursive equations (2.8.29), (2.8.35), (2.8.36) with boundary conditions (2.8.38) can be solved.

Reduction to generalized coordinates. As a simple and suitable method, we demonstrate the reduction of the problem to generalized coordinates. Then, the vector "u", defining the position of the system (introduced in Para. 2.2), is chosen as $u = [q_1 \cdots q_n]^T$.

In the case of the chain in question, the position matrix T_i can be expressed, using (2.8.18), as

$$T_i = T_{i-1}{}^{i-1}T_{i-1,i} \qquad (2.8.39)$$

from which it follows that

$$T_i = {}^{o}T_{o,1}\,{}^{1}T_{1,2} \cdots {}^{i-1}T_{i-1,i}. \tag{2.8.40}$$

As ${}^{i-1}T_{i-1,i}$ depends on q_i only, matrix T_i can be expressed in terms of the generalized coordinates q_ℓ; $\ell = 1,\dots,i$. In order to make the compulsion function Z depend on the generalized coordinates, let us first find the derivative $\dot{T}$ from (2.8.40).

$$\dot{T}_i = \sum_{j=1}^{i} V_{ij}\dot{q}_j, \tag{2.8.41}$$

where

$$V_{ij} = {}^{o}T_{o,1}\,{}^{1}T_{1,2} \cdots {}^{j-2}T_{j-2,j-1}\,\Theta\,{}^{j-1}T_{j-1,j} \cdots {}^{i-1}T_{i-1,i}$$

$$1 \le j \le i, \quad i=1,\dots,n \tag{2.8.42}$$

and the second derivative,

$$\ddot{T}_i = \sum_{j=1}^{i} V_{ij}\ddot{q}_j + \sum_{j=1}^{i}\sum_{k=1}^{i} V_{ijk}\dot{q}_j\dot{q}_k, \tag{2.8.43}$$

where

$$V_{ijk} = {}^{o}T_{o,1}\,{}^{1}T_{1,2} \cdots \Theta\,{}^{j-1}T_{j-1,j} \cdots \Theta\,{}^{k-1}T_{k-1,k} \cdots {}^{i-1}T_{i-1,i} \tag{2.8.44}$$

Substituting (2.8.43) into expression (2.8.33) the compulsion function Z is obtained in terms of the generalized accelerations:

$$Z = Z(\ddot{q}_1,\dots,\ddot{q}_n) = Z(\ddot{u}); \quad u = q. \tag{2.8.45}$$

The form of the function Z is

$$Z = \frac{1}{2}\,\ddot{q}^T W \ddot{q} - U^T\ddot{q} - P^T\ddot{q} + \text{const} \tag{2.8.46}$$

where the elements of the matrices W and U are:

$$\begin{aligned} W_{sj} &= \sum_{i=k}^{n} \mathrm{tr}(V_{ij}H_iV_{is}^T), \quad k = \max\,(s, j) \\ U_s &= \sum_{i=s}^{n}\sum_{j=1}^{i}\sum_{k=1}^{i} \mathrm{tr}(V_{ijk}H_iV_{is}^T)\dot{q}_j\dot{q}_k - \mathrm{tr}(\Phi_iV_{is}^T) \quad s,j=1,\dots,n \end{aligned} \tag{2.8.47}$$

Solving the inverse problem, i.e. calculating $\ddot{q}$, now reduces to minimizing the function (2.8.45), (2.8.46) without constraints.

Starting from the compulsion function in the form (2.8.46), it is possible to derive the system of dynamical equations with respect to generalized coordinates. If we use the conditions $\frac{\partial Z}{\partial \ddot{q}_i} = 0$, $i=1,\ldots,n$ necessary for the minimum, the equation system follows:

$$W\ddot{q} = P + U \qquad (2.8.48)$$

2.9. The Method of Appel's Equations

This method is also computer-oriented to a great extent. It was first proposed in [8] and [16]. Appel's equations were used as a dynamical approach. This method, in its starting postulates about the configuration and kinematics it is treating, coincides in many aspects with the method of general theorems (para. 2.3). However, these aspects which coincide will be presented here again for the sake of completness of presentation and easier follow-up, but in a somewhat more compact form.

Starting postulates. The method considers a mechanism with n degrees of freedom and describes it by means of the n-dimensional vector of generalized coordinates:

$$q = [q_1 \ \cdots \ q_n]^T. \qquad (2.9.1)$$

Starting from the Appel's equations, a system of n second-order differential equations is formed:

$$W\ddot{q} = P + U, \qquad (2.9.2)$$

where P is an n-dimensional vector of the driving forces and torques and the $n \times n$ matrix W and $n \times 1$ matrix U depend on the mechanism state q, $\dot{q}$.

Now the functions f and g, defined by (2.2.1) and (2.2.2), are obtained as

$$\ddot{q} = f(P, q, \dot{q}, \text{configuration}) = W^{-1}(P+U), \qquad (2.9.3)$$

$$P = g(q, \dot{q}, \ddot{q}, \text{configuration}) = W\ddot{q} - U. \qquad (2.9.4)$$

Mechanism configuration. The method considers a mechanism of the open chain type, consisting of n rigid bodies and without branching (Fig. 2.31).

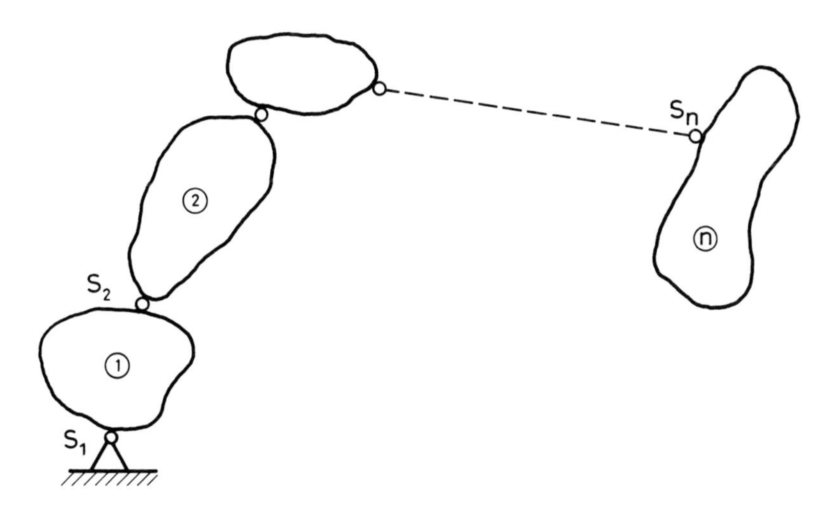

Fig. 2.31. Open kinematical chain without branching

The bodies forming the mechanism (mechanism segments) are interconnected by means of joints having one degree of freedom (d.o.f.) each. The joints are of rotational or linear type. The rotational joint S_i (Fig. 2.32a) enables rotation about the axis defined by the unit vector $\vec{e}_i$ and the linear joint S_j enables translation along the axis defined by the unit vector $\vec{e}_j$ (Fig. 2.32b).

C_i, C_j denote the centers of gravity (c.o.g.) of the segments. s_i, s_j are indicators of the joint type:

$$s_k = \begin{cases} 0, & \text{if joint } S_k \text{ is rotational} \\ 1, & \text{if joint } S_k \text{ is linear.} \end{cases}$$

<u>Drives in the joints</u>. There are driving actuators acting in the mechanism joints. Thus, in the rotational joints S_i, a driving torque is acting:

$$\vec{P}_i = P_i^M \vec{e}_i \tag{2.9.5a}$$

and in the linear joint S_j, the driving force

$$\vec{P}_j = P_j^F \vec{e}_j . \tag{2.9.5b}$$

The drives vector becomes

$$P = [P_1 \cdots P_n]^T, \tag{2.9.6}$$

where the upper indices M and F were deleted since the joint type, and consequently the type of the drive, is determined by the indicator s_k.

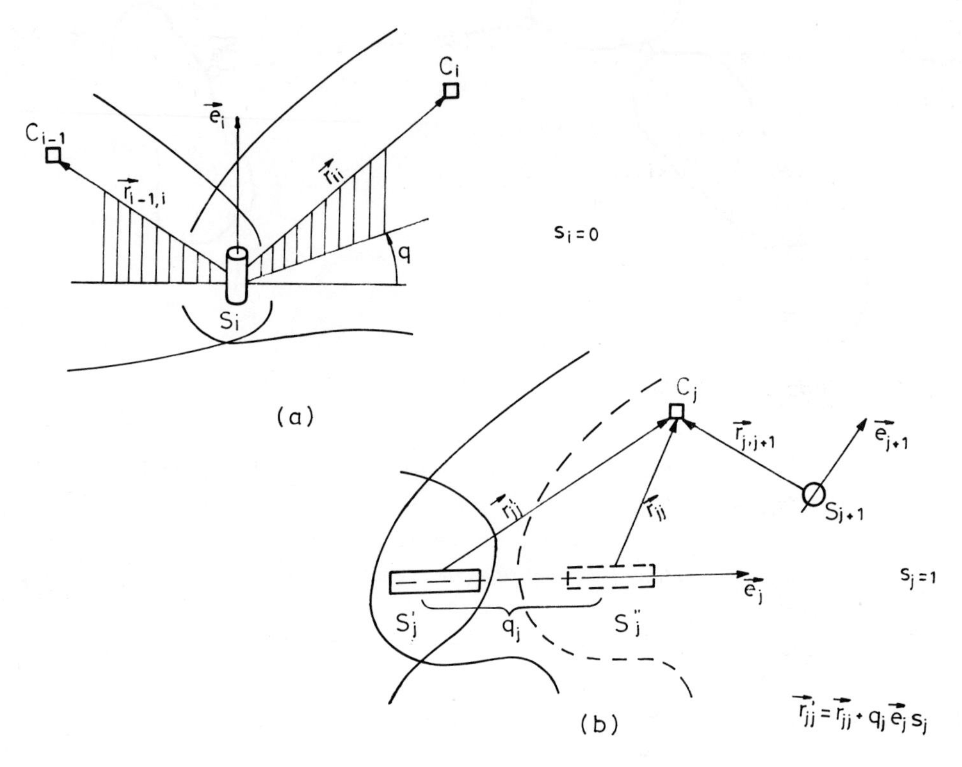

Fig. 2.32. (a) Mechanism rotational joint
(b) Mechanism linear joint

Generalized coordinates. In each joint a generalized coordinate is chosen.

For the rotational joint S_i the generalized coordinate q_i is defined as the rotation angle in the joint around axis $\vec{e}_i$ and can be considered as the angle between the projections of the vectors $-\vec{r}_{i-1,i}$ and $\vec{r}_{ii}$ onto a plane orthogonal to $\vec{e}_i$ (Fig. 2.32a). In the case of "specificity", i.e. $\vec{r}_{ii} || \vec{e}_i$ or $\vec{r}_{i-1,i} || \vec{e}_i$, let us proceed in a manner analogous to the method of general theorems (Para. 2.3).

For the linear joint S_j, the corresponding generalized coordinate q_j is defined as translation displacement along the translation axis $\vec{e}_j$, i.e. $q_j = \overline{S_j' S_j''}$ (Fig. 2.32b).

Coordinate systems and transition matrices. Let us introduce the body-fixed (b.-f.) coordinate systems and the immobile external system as in the method of general theorems (Para. 2.3). Also let us introduce the same notation: $\vec{a}_i$ denotes some vector characteristic for the i-th segment, expressed in the external coordinate system; $\vec{\tilde{a}}_i$ is the same vector expressed in the i-th b.-f. system, and $\vec{a}_{\sim i}$ in the (i-1)-th b.-f. system. Let us define the transition matrices as in 2.3.

$$\vec{a}_i = A_i \vec{\tilde{a}}_i, \tag{2.9.7}$$

$$\vec{a}_{\sim i} = A_{i-1,i} \vec{\tilde{a}}_i, \tag{2.9.8}$$

$$\vec{\tilde{a}}_i = A_{i,i-1} \vec{a}_{\sim i} = A_{i-1,i}^{-1} \vec{a}_{\sim i}. \tag{2.9.9}$$

Note that the vectors $\vec{\tilde{r}}_{ii}$, $\vec{\tilde{r}}_{i-1,i}$, $\vec{\tilde{e}}_i$, $\vec{e}_{\sim i}$ are constant.

Let us introduce matrix notation: for each vector $\vec{a}_i$, a corresponding 3×1 matrix a_i is defined.

The transition matrices can be calculated as in the method of general theorems but it is more appropriate to calculate them as in the method of Lagrange´s equations, i.e. first calculate the transition matrix $A_{i-1,i}$ and then A_i recursively by (2.7.10).

The calculation of the transition matrix for two segments connected by a linear joint is like that in the method of general theorems except that the relative transition matrix $A_{i-1,i}$ is sought.

Algorithm input data. The following input values are fed to the method:

- mechanism configuration:

  n, the number of d.o.f. (= number of segments = number of joints);

  s_i, $i=1,\ldots,n$,

  $\vec{\tilde{e}}_i$, $\vec{e}_{\sim i}$: $i=1,\ldots,n$,

  $\vec{\tilde{u}}_j$, $\vec{u}_{\sim j}$ for linear joints S_j,

  $\vec{\tilde{r}}_{ii}$, $\vec{\tilde{r}}_{i,i+1}$, $i=1,\ldots,n$; $\vec{r}_{01}$,

  $\vec{\tilde{r}}^*_{ii}$, $\vec{\tilde{r}}^*_{i,i+1}$ (in the cases of "specificity"),

  m_i, $\tilde{J}_i$, $i=1,\ldots,n$ (i.e. mass and tensor of inertia with respect to the b.-f. system),

- the initial state

$$q(t_o),\ \dot{q}(t_o)$$

- $\ddot{q}(t_k)$; $k=0,1,\ldots,k_{end}$ (in the case of solving the direct problem of dynamics),

- $P(t_k)$; $k=0,1,\ldots,k_{end}$ (in the case of solving the inverse problem).

Kinematic relations. Let us denote the velocity and acceleration of the i-th segment center of gravity by $\vec{v}_i$, $\vec{w}_i$, respectively, and the angular velocity and acceleration by $\vec{\omega}_i$, $\vec{\varepsilon}_i$ respectively. Now, the recursive expressions (2.3.14) - (2.3.18) hold but it is more suitable to write them in the b.-f. coordinate system. Thus, the velocities are

$$\vec{\tilde{\omega}}_i = A_{i,i-1}\vec{\tilde{\omega}}_{i-1} + \dot{q}_i(1-s_i)\vec{\tilde{e}}_i, \tag{2.9.10}$$

$$\vec{\tilde{v}}_i = A_{i,i-1}(\vec{\tilde{v}}_{i-1}-\vec{\tilde{\omega}}_{i-1}\times\vec{\tilde{r}}_{i-1,i}) + \vec{\tilde{\omega}}_i \times \vec{\tilde{r}}\,'_{ii} + \dot{q}_i s_i \vec{\tilde{e}}_i,$$
$$\vec{\tilde{r}}\,'_{ii} = \vec{\tilde{r}}_{ii} + q_i\vec{\tilde{e}}_i s_i, \tag{2.9.11}$$

and the accelerations are

$$\vec{\tilde{\varepsilon}}_i = A_{i,i-1}\vec{\tilde{\varepsilon}}_{i-1} + [\ddot{q}_i\vec{\tilde{e}}_i+\dot{q}_i(\vec{\tilde{\omega}}_i\times\vec{\tilde{e}}_i)](1-s_i), \tag{2.9.12}$$

$$\vec{\tilde{w}}_i = A_{i,i-1}[\vec{\tilde{w}}_{i-1}-\vec{\tilde{\varepsilon}}_{i-1}\times\vec{\tilde{r}}_{i-1,i}-\vec{\tilde{\omega}}_{i-1}\times(\vec{\tilde{\omega}}_{i-1}\times\vec{\tilde{r}}_{i-1,i})] +$$
$$+ \vec{\tilde{\varepsilon}}_i \times \vec{\tilde{r}}\,'_{ii} + \vec{\tilde{\omega}}_i \times (\vec{\tilde{\omega}}_i\times\vec{\tilde{r}}\,'_{ii})+[\ddot{q}\vec{\tilde{e}}_i+2\dot{q}_i(\vec{\tilde{\omega}}_i\times\vec{\tilde{e}}_i)]s_i, \tag{2.9.13}$$

with initial conditions

$$\vec{\tilde{v}}_o = 0, \quad \vec{\tilde{\omega}}_o = 0, \quad \vec{\tilde{w}}_o = 0, \quad \vec{\tilde{\varepsilon}}_o = 0. \tag{2.9.14}$$

Forming the system of equations. The mechanism dynamics will be described by means of a system of n differential equations in Appel´s form:

$$\frac{\partial G}{\partial \ddot{q}_i} = Q_i, \quad i=1,\ldots,n, \tag{2.9.15}$$

where the function G is the "acceleration energy" and Q_i is the generalized force corresponding to the coordinate q_i.

The system of Appel´s equations (2.9.15) can be written in matrix form:

$$\frac{\partial G}{\partial \ddot{q}} = Q, \tag{2.9.16}$$

where

$$Q = [Q_1 \cdots Q_n]^T \tag{2.9.17}$$

is the vector of generalized forces.

The function of the "acceleration energy" G for the whole chain under consideration represents the sum of the corresponding functions for each segment, i.e.,

$$G = \sum_{i=1}^{n} G_i, \tag{2.9.18}$$

and for the individual segment, the function G_i is given by means of a known expression [22]:

$$G_i = \frac{1}{2} m_i \vec{\tilde{w}}_i^2 + \frac{1}{2} \vec{\tilde{\varepsilon}}_i \tilde{J}_i \vec{\tilde{\varepsilon}}_i - [2(\tilde{J}_i \vec{\tilde{\omega}}_i) \times \vec{\tilde{\omega}}_i] \vec{\tilde{\varepsilon}}_i . \tag{2.9.19}$$

For the sake of forming the iterative matrix computer algorithm (in each iteration the next segment is considered), let us define the following matrices:

Ω is a 3×n matrix, the columns of which are the coefficients of the generalized accelerations in the expression for $\vec{\tilde{w}}_i$ (in the i-th iteration), Θ is a 3×1 matrix, containing the free term of the same expression.

Now the acceleration $\vec{\tilde{w}}_i$ can be written in the form

$$\tilde{w}_i = \Omega \ddot{q} + \Theta . \tag{2.9.20}$$

Let us introduce notation for the columns:

$$\left.\begin{aligned} \Omega &= [\beta_1^i \cdots \beta_i^i \; 0 \cdots 0] \\ \Theta &= [\delta^i] . \end{aligned}\right\} \tag{2.9.21}$$

Further, let us define the following matrices:

Γ is a 3×n matrix, the columns of which are, in the i-th iteration, the coefficients of the generalized accelerations in the expression for $\vec{\varepsilon}_i$, Φ is a 3×1 matrix, containing the free term of same expression.

Now, we write the angular acceleration in form

$$\tilde{\varepsilon}_i = \Gamma\ddot{q} + \Phi \tag{2.9.22}$$

and let us write for the columns

$$\left.\begin{aligned} \Gamma &= [\alpha_1^i \cdots \alpha_i^i \; 0 \cdots 0], \\ \Phi &= [\gamma^i]. \end{aligned}\right\} \tag{2.9.23}$$

In each iteration, a new segment is added to the chain, and modifications of and supplements to the matrices Ω, Θ, Γ, Φ are obtained so that they correspond to the new segment. The expressions for modifications and supplements are derived from the recursive expressions (2.9.10) - (2.9.13). Thus, for the i-th iteration,

$$\left.\begin{aligned} \vec{\alpha}_j^i &= A_{i,i-1}\vec{\alpha}_j^{i-1}; \quad j=1,\dots,i-1, \\ \vec{\alpha}_i^i &= \vec{e}_i(1-s_i), \end{aligned}\right\} \tag{2.9.24}$$

$$\vec{\gamma}_i = A_{i,i-1}\vec{\gamma}^{i-1} + \dot{q}_i(1-s_i)(\vec{\tilde{\omega}}_i \times \vec{\tilde{e}}_i), \tag{2.9.25}$$

$$\left.\begin{aligned} \vec{\beta}_j^i &= A_{i,i-1}\vec{\beta}_j^{i-1} - A_{i,i-1}(\vec{\alpha}_j^{i-1} \times \vec{\tilde{r}}_{i-1,i}) + \vec{\alpha}_j^i \times \vec{\tilde{r}}'_{ii}; \; j=1,\dots,i-1 \\ \vec{\beta}_i^i &= \vec{\alpha}_i^i \times \vec{\tilde{r}}'_{ii} + \vec{e}_i s_i, \end{aligned}\right\} \tag{2.9.26}$$

$$\begin{aligned} \vec{\delta}^i &= A_{i,i-1}\vec{\delta}^{i-1} - A_{i,i-1}(\vec{\gamma}^{i-1} \times \vec{\tilde{r}}_{i-1,i}) + \vec{\gamma}^i \times \vec{\tilde{r}}'_{ii} + \vec{h}, \\ \vec{h} &= -A_{i,i-1}[\vec{\tilde{\omega}}_{i-1} \times (\vec{\tilde{\omega}}_{i-1} \times \vec{\tilde{r}}_{i-1,i})] + \\ &\quad + \vec{\tilde{\omega}}_i \times (\vec{\tilde{\omega}}_i \times \vec{\tilde{r}}'_{ii}) + 2\dot{q}_i(\vec{\tilde{\omega}}_i \times \vec{\tilde{e}}_i)s_i . \end{aligned} \tag{2.9.27}$$

Substituting (2.9.20) and (2.9.22) into (2.9.19), the function G_i acquires the form

$$G_i = \frac{1}{2}\ddot{q}^T W_i \ddot{q} + V_i \ddot{q} + D_i, \tag{2.9.28}$$

where

$$W_i = m_i \Omega^T \Omega + \Gamma^T \tilde{J}_i \Gamma,$$

$$V_i = m_i \Theta^T \Omega + \Phi^T \tilde{J}_i \Gamma - 2\tilde{u}^T \Gamma, \tag{2.9.29}$$

$$D_i = \frac{1}{2} m_i \Theta^T \Theta + \frac{1}{2} \Phi^T \tilde{J}_i \Phi - 2\tilde{u}^T \Phi,$$

$$\vec{\tilde{u}} = (\tilde{J}_i \vec{\tilde{\omega}}_i) \times \vec{\tilde{\omega}}_i .$$

Taking care about (2.9.18) and (2.9.28), it is clear that the "acceleration energy" function G for the whole chain will have the form

$$G = \frac{1}{2} \ddot{q}^T W \ddot{q} + V \ddot{q} + D. \tag{2.9.30}$$

Let us substitute (2.9.30) into Appel´s equations (2.9.16):

$$\frac{\partial G}{\partial \ddot{q}} = W \ddot{q} + V^T = Q. \tag{2.9.31}$$

About the calculation of the generalized forces, more will be said somewhat later and it will be shown that they can be calculated in the form

$$Q = P + Y, \tag{2.9.32}$$

where P is the vector of the drives and Y can be calculated independently of P. Thus, (2.9.31) acquires the form

$$W \ddot{q} = P + Y - V^T \tag{2.9.33}$$

and by introducing

$$U = Y - V^T, \tag{2.9.34}$$

the form (2.9.2) is obtained, or

$$W \ddot{q} = P + U. \tag{2.9.35}$$

Let us now consider the calculation of the matrices W and V. Due to (2.9.18) and (2.9.28) it is evident that

$$W = \sum_{i=1}^{n} W_i, \quad V = \sum_{i=1}^{n} V_i, \tag{2.9.36}$$

which yields the possibility of recursively calculating the matrices W and V. In the i-th iteration this is

$$W^{(i)} = W^{(i-1)} + W_i, \qquad V^{(i)} = V^{(i-1)} + V_i, \tag{2.9.37}$$

the upper index in parenthesis indicating that a matrix with no index in the algorithm is involved, i.e., its value is calculated interatively and the upper index (i) designates the value of that matrix in the i-th iteration. Let us now consider the calculation of the generalized forces. By means of the method of virtual displacements, just as in the method of Lagrange´s equations, the expression for the generalized forces is derived. Thus, if S_i is a rotational joint, the generalized force corresponding to the generalized coordinate q_i is

$$Q_i^{(rot)} = P_i^M + \sum_{k=0}^{n-i} [m_{i+k}\vec{g}, \vec{e}_i, \vec{r}_k^{\,i}], \tag{2.9.38}$$

where the square parantheses denote the vector box product, $\vec{g}$ being the gravity acceleration vector, $\vec{g} = \{0, \;\; 0, \;\; -9.81\}$, and

$$\vec{r}_k^{\,i} = \sum_{\ell=0}^{k} \vec{r}\,'_{i+\ell,i+\ell} - \sum_{\ell=0}^{k-1} \vec{r}_{i+\ell,i+\ell+1}. \tag{2.9.39}$$

If S_j is a linear joint, then the generalized force corresponding to the coordinate q_j is

$$Q_j^{(trans)} = P_j^F + \vec{e}_j\vec{g} \sum_{k=0}^{n-j} m_{j+k}. \tag{2.9.40}$$

It should be pointed out, that the calculation of the generalized forces can be done in the external system using (2.9.38) - (2.9.40), recursively of course, but they can also be calculated in the b.-f. systems.

By considering the expressions (2.9.38) and (2.9.40), it may be noted that the expressions for the generalized forces can be written in the form

$$Q = P + Y, \tag{2.9.41}$$

where P is the drive vector given by (2.9.6) and Y is a vector given as

$$Y = [y_1 \;\cdots\; y_n]^T, \tag{2.9.42a}$$

$$y_i = \begin{cases} \sum_{k=0}^{n-i} [m_{i+k}\vec{g}, \vec{e}_i, \vec{r}_k^{\,i}], & s_i=0, \\ \vec{e}_i\vec{g} \sum_{k=0}^{n-i} m_{i+k}, & s_i=1, \end{cases} \tag{2.9.42b}$$

and that Y is calculated independently of P. Naturally, the expressions (2.9.42) are calculated recursively without duplicating the summations.

Globally, the algorithm of the method of Appel´s equations can be presented by means of the block-scheme in Fig. 2.33.

The case of joints with two degrees of freedom. The two-d.o.f. joints often appear in practical designs. In such cases there are usually one rotational and one linear d.o.f. in the one joint. The method of Appel´s equations can be slightly modified to work with such joints.

Let consider a chain of n bodies with two-d.o.f. joints permitting one translation (along the unit vector $\vec{a}_i$) and one rotation (around the unit vector $\vec{e}_i$), (Fig. 2.34). However, two cases should be distinguished. These are shown in Fig. 2.34. (a) and (b), the possible practical realizations being shown in Fig. 2.35 (a) and (b). In case (a), the axis of rotation $\vec{e}_i$ is fixed to the segment (i-1) and in case (b) it is fixed to the i-th segment, i.e., it moves translatorily relative to segment (i-1). However, due to the linear displacement, the orientation of axis $\vec{e}_i$ with respect to both segments does not change, i.e., $\vec{\tilde{e}}_i$ and $\underset{\sim}{\vec{e}}_i$ remain constant vectors in each case.

Let us introduce the indicator p_i, so that $p_i = 1$ in the case of the (a) type joint, and $p_i = 0$ in the (b) case.

Such a mechanism has 2n degrees of freedom and is described by means of a 2n-dimensional vector with generalized coordinates

$$q = [\varphi_1 x_1 \cdots \varphi_n x_n]^T. \tag{2.9.43}$$

φ_i represents the angle of rotation around axis $\vec{e}_i$ and x_i represents the translation displacement along the axis $\vec{a}_i$ (Fig. 2.34).

There are driving actuators acting in the mechanism joints. In each joint, there is a driving force for translation,

$$\vec{P}_i^F = P_i^F \cdot \vec{a}_i, \tag{2.9.44a}$$

and a driving torque for rotation,

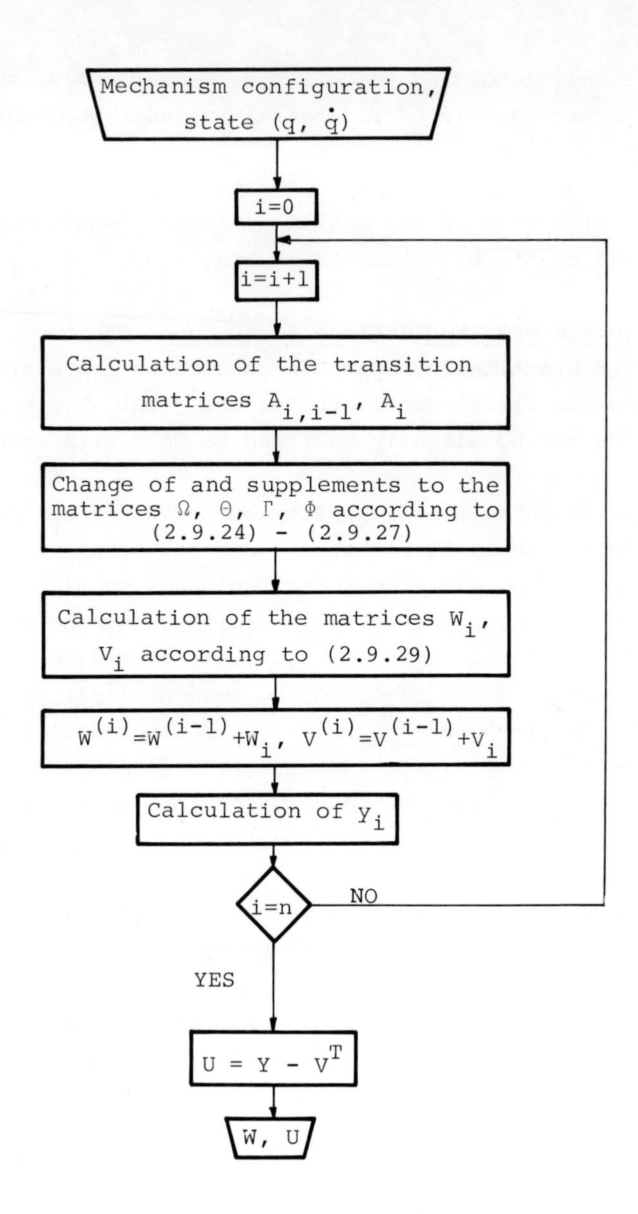

Fig. 2.33. Block scheme of the method of Appel´s equations

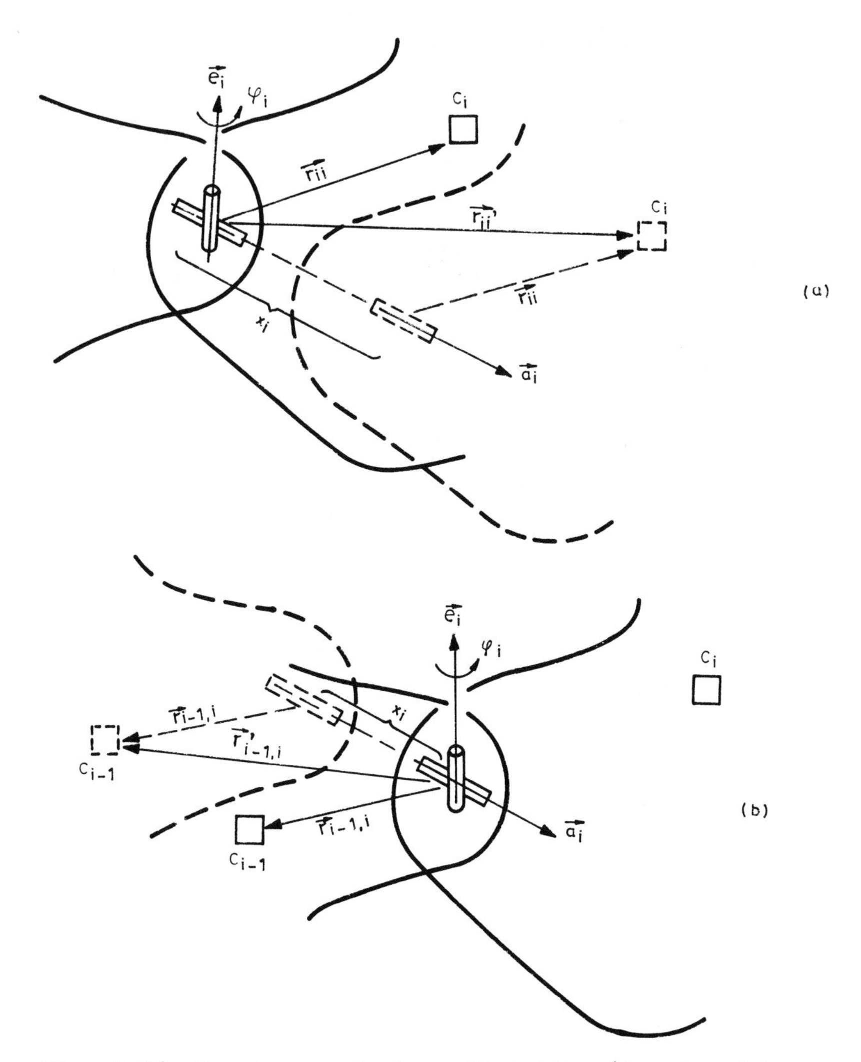

Fig. 2.34. Two cases of class IV joints (two d.o.f.)

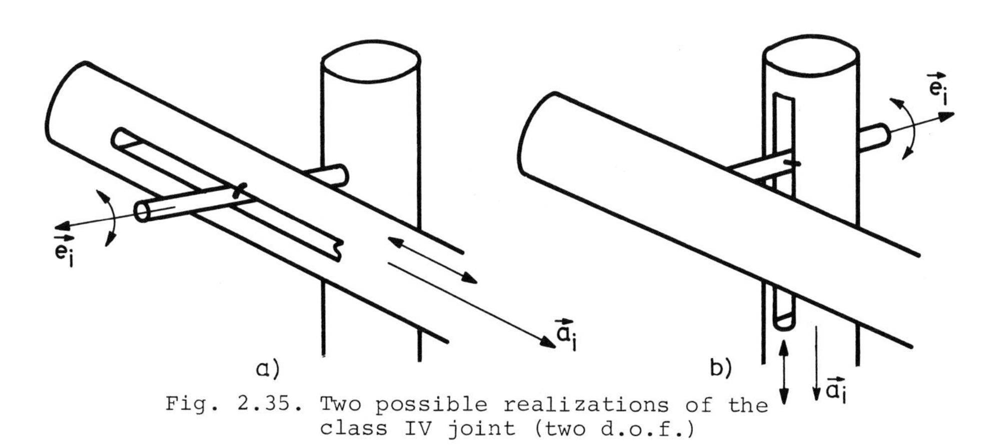

Fig. 2.35. Two possible realizations of the class IV joint (two d.o.f.)

$$\vec{P}_i^M = P_i^M \vec{e}_i . \tag{2.9.44b}$$

Further, the recursive relation for velocities and accelerations of segments are derived as in (2.9.10) - (2.9.13), but for joints with two degrees of freedom:

$$\begin{aligned}
\vec{\tilde{\omega}}_i &= A_{i,i-1}\vec{\tilde{\omega}}_{i-1} + \dot{\varphi}_i\vec{\tilde{e}}_i , \\
\vec{\tilde{v}}_i &= A_{i,i-1}[\vec{\tilde{v}}_{i-1} - \vec{\tilde{\omega}}_{i-1}\times\vec{\tilde{r}}'_{i-1,i}] + \vec{\tilde{\omega}}_i\times\vec{\tilde{r}}'_{ii} + \dot{x}_i\vec{\tilde{a}}_i , \\
\vec{\tilde{\varepsilon}}_i &= A_{i,i-1}\vec{\tilde{\varepsilon}}_{i-1} + \ddot{\varphi}_i\vec{\tilde{e}}_i + \dot{\varphi}_i\vec{\tilde{\omega}}_i\times\vec{\tilde{e}}_i , \\
\vec{\tilde{w}}_i &= A_{i,i-1}[\vec{\tilde{w}}_{i-1} - \vec{\tilde{\varepsilon}}_{i-1}\times\vec{\tilde{r}}'_{i-1,i} - \vec{\tilde{\omega}}_{i-1}\times(\vec{\tilde{\omega}}_{i-1}\times\vec{\tilde{r}}'_{i-1,i})] + \\
&\quad + \vec{\tilde{\varepsilon}}_i\times\vec{\tilde{r}}'_{ii} + \vec{\tilde{\omega}}_i\times(\vec{\tilde{\omega}}_i\times\vec{\tilde{r}}'_{ii}) + \ddot{x}_i\vec{\tilde{a}}_i + 2\dot{x}_i\vec{\tilde{\omega}}_i\times\vec{\tilde{a}}_i , \\
\vec{\tilde{r}}'_{i-1,i} &= \vec{\tilde{r}}_{i-1,i} - (1-p_i)x_i\vec{\tilde{a}}_i ; \quad \vec{\tilde{r}}'_{ii} = \vec{\tilde{r}}_{ii} + p_i x_i \vec{\tilde{a}}_i .
\end{aligned} \tag{2.9.45}$$

The procedure for forming the mathematical model, i.e., the calculation of the matrices W, U in the system (2.9.2) is analogous to the case of the one-d.o.f. joints. Consequently, we will derive only what is different from the previous method.

In the case of the mechanism possessing two-d.o.f. joints, the vector of generalized coordinates is different, i.e., has the form (2.9.43). In the case of introducing matrix calculus by the expressions (2.9.20) and (2.9.22), the matrices Ω, Γ, Θ, Φ will likewise have a different form. Ω is a $3\times 2n$ matrix of the form

$$\Omega = [\beta_1^i\eta_1^i \cdots \beta_i^i\eta_i^i \ 0 \cdots 0] \tag{2.9.46}$$

and the $3\times 2n$ matrix Γ is:

$$\Gamma = [\alpha_1^i \ 0 \cdots \alpha_i^i \ 0 \ 0 \cdots 0] . \tag{2.9.47}$$

The recursive modifications of and supplements to the matrices Ω, Γ, Θ, Φ, caried out in each iteration during the circling of the chain, are derived from (2.9.45).

Introduction of the described joints with two degrees of freedom and the corresponding new generalized coordinates will modify the expressions for the generalized forces. The vector of the generalized forces Q will have dimension 2n and be of the form

$$Q = [Q_1^{\varphi}\ Q_1^{x}\ \cdots\ Q_n^{\varphi}\ Q_n^{x}]^T, \qquad (2.9.48)$$

where Q_i^{φ} is the generalized force corresponding to the φ_i coordinate and Q_i^{x} corresponds to x_i.

By means of the virtual displacement method, as in the proceding method, the expressions for the generalized forces are found to be

$$Q_i^{\varphi} = P_i^M + \sum_{k=0}^{n-i} [m_{i+k}\vec{g},\ \vec{e}_i,\ \vec{r}_k^{\,i}], \qquad (2.9.49)$$

where

$$\vec{r}_k^{\,i} = \sum_{\ell=0}^{k} \vec{r}\,'_{i+\ell,i+\ell} - \sum_{\ell=0}^{k-1} \vec{r}\,'_{i+\ell,i+\ell+1} \qquad (2.9.50)$$

and

$$Q_i^{x} = P_i^F + \vec{a}_i\vec{g} \sum_{k=0}^{n-i} m_{i+k}. \qquad (2.9.51)$$

With these modifications, the same procedure as in the preceding case of joints with one degree of freedom, starting with Appel´s equations

$$\frac{\partial G}{\partial \ddot{q}} = Q, \qquad (2.9.52)$$

yields a system of 2n linear equations with respect to generalized accelerations:

$$W\ddot{q} = P + U \qquad (2.9.53)$$

2.10. Some Problems of Closed Chain Dynamics

In most of the methods so far described, open kinematic chains have been considered. We now show how these methods, developed for open chains, can be used for solving the dynamics in some cases of chains with an unfree end.

Let us consider the open chain without branching. Let us close the chain by imposing a constraint upon the last segment position. Two types of constraints will be considered:

(i) where the last segment is connected to a given surface by means of one of its points,

(ii) where the last segment is connected to a given line by means of one of its points.

Closed chains with such constraints have been chosen because they often appear in practical problems of industrial robotics, i.e. in some manipulation tasks.

(i) Let us consider an open chain of n rigid bodies without branching. Let the joints have one degree of freedom each (rotational or linear) and let $\vec{e}_i$ be the unit vector of the axis of rotation (if S_i is a rotational joint) or the axis of translation (if S_i is a linear joint). So, there are n degrees of freedom.

Consider a point V on the last (n-th) segment and let us denote by (x, y, z) the coordinates of this point in the immobile external system.

Let us impose the constraint that the point V is always on the surface (Fig. 2.36) determined by the equation

$$h(x, y, z) = 0 \tag{2.10.1a}$$

if the surface is immobile, or,

$$h(x, y, z, t) = 0 \tag{2.10.1b}$$

if the surface is moving according to some given law. Such a mechanism how possesses $m = n - 1$ degrees of freedom.

Let us consider the mechanism as free and substitute the action of the connection (constraint) by means of a reaction force $\vec{R}$ (Fig. 2.36). As the direction of the reaction force is determined by the gradient of the surface, only one unknown value appears, namely, the algebraic value of the reaction force with respect to the positive sense of the gradient (Fig. 2.36a).

Such a quasi-free mechanism is described by a system of n differential equations:

$$W\ddot{q} = P + U + DR. \quad (2.10.2)$$

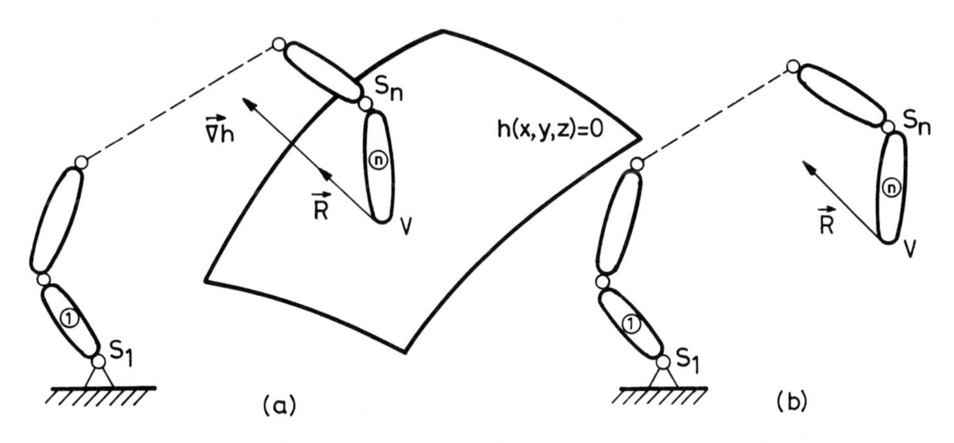

Fig. 2.36. Mechanism with unfree tip and the reaction force of the connection

The matrices W, U were defined earlier in the text, and W, U are calculated independently of the reaction R, i.e. for the free mechanism. P is an $n\times1$ vector of the driving forces and torques in the joints. R is a 3×1 matrix corresponding to the vector $\vec{R}$. The product $D\cdot R$ represents the component of the generalized forces due to the action of the force R. Starting with the procedure for determining the generalized forces (2.9), an $n\times3$ matrix D is obtained in the form

$$D = \begin{bmatrix} u_1^T \\ \vdots \\ u_n^T \end{bmatrix}, \quad (2.10.3)$$

where $\vec{u}_i = \vec{e}_i \times \vec{r}_V^{\,i}$, if S_i is a rotational joint, and $\vec{u}_i = \vec{e}_i$ if S_i is a linear joint. Here, $\vec{r}_V^{\,i} = \overrightarrow{S_iV}$. Thus, the problem of the numerical calculation of matrix D for a known mechanism state is solved.

Let us consider the point V. A function $\eta(q)$ can be derived which calculates the coordinates (x, y, z) of the point V for known generalized coordinates q. Thus $(x, y, z) = \eta(q)$. About the derivation of this function, more will be said later in this book.

Solving the direct problem means, in principle, calculating the forces for some known motion of the mechanical system. In the case of the mechanism considered, the direct problem will involve calculation of the drives P for the known mechanism motion (satisfying the imposed constraint) and the known pressure which we want to realize on the surface (i.e. the algebraic value of the reaction $\vec{R}$). In most practical problems, the pressure on the surface is given. It should be mentioned that in practical problems the motion is not given in generalized coordinates but by means of some other input values from which the motion in generalized coordinates can be calculated. However, this will be treated elsewhere in the book.

Since time is regarded in a series of time instants, let us assume that in each time instant the q, $\dot{q}$, $\ddot{q}$ and the algebraic value N of the reaction $\vec{R}$ are given.

The direction of the reaction $\vec{R}$ is determined by means of the gradient of the surface h (2.10.1). The gradient is calculated at the contact point of the surface and the mechanism (point V) for each particular time instant t_k. As the algebraic value N of the reaction $\vec{R}$ at that instant is known,

$$\vec{R}(t_k) = N(t_k)\frac{\vec{\nabla}h(x(t_k),\ y(t_k),\ z(t_k))}{|\vec{\nabla}h|} \tag{2.10.4a}$$

or

$$\vec{R}(t_k) = N(t_k)\frac{\vec{\nabla}h(x(t_k),\ y(t_k),\ z(t_k),\ t_k)}{|\vec{\nabla}h|}, \tag{2.10.4b}$$

depending on whether the surface is immobile or not.

When R is calculated, we calculate W, U, D and then the required drives at that time instant:

$$P = W\ddot{q} - U - DR \tag{2.10.5}$$

Repeating the procedure for all time instants, the drives P are obtained over the whole interval.

Solving the inverse problem. Let us now consider the drives P as known in a series of time instants; the motion and pressure on the surface has to be calculated.

Let us start from the initial time instant t_o with the initial state q^o, $\dot{q}^o$. By calcualting W, U, D, the system (2.10.2) is obtained in which P is known but $\ddot{q}$ and R are unknown. The direction of the reaction $\vec{R}$ is determined by means of the gradient of the surface. Then, the gradient and normal $\vec{n}$ to the surface are calculated as follows:

$$\vec{n} = \frac{\vec{\nabla} h(x^o, y^o, z^o)}{|\vec{\nabla} h|}, \tag{2.10.6a}$$

$$\vec{n} = \frac{\vec{\nabla} h(x^o, y^o, z^o, t)}{|\vec{\nabla} h|}. \tag{2.10.6b}$$

Note that (x^o, y^o, z^o) are easily calculated for known q^o by means of the function $(x, y, z) = \eta(q)$. Now the reaction of the connection is given by

$$\vec{R} = N\vec{n} \tag{2.10.7}$$

By substituting (2.10.7) into (2.10.2) one obtains

$$W\ddot{q} = P + U + D \cdot n \cdot N. \tag{2.10.8}$$

In (2.10.8), $\ddot{q}$ and N are unknown. The algebraic value of N should be determined so that the point V is on the surface. Discretely, this means that after integration over the interval Δt, i.e., in the next instant $t_1 = t_o + \Delta t$, generalized coordinates $q^1 = q(t_1)$ should be obtained, such that the point V : $(x^1, y^1, z^1) = \eta(q^1)$ satisfies the equation of the surface (2.10.1) at that time instant. With this in mind, let us write (2.10.8) in the form

$$\ddot{q} = W^{-1}(P+U) + W^{-1}DnN. \tag{2.10.9}$$

Further, let us suppose that the integration over the interval Δt is performed considering $\ddot{q}$ as a constant, i.e.,

$$q^1 = q(\overbrace{t_o+\Delta t}^{t_1}) = \frac{1}{2}\ddot{q}(t_o)\Delta t^2 + \dot{q}(t_o)\Delta t + q(t_o), \tag{2.10.10a}$$

$$\dot{q}^1 = \dot{q}(t_o+\Delta t) = \ddot{q}(t_o)\Delta t + \dot{q}(t_o). \tag{2.10.10b}$$

Substituting (2.10.9) into (2.10.10a) one finds that

$$q^1 = \frac{1}{2} W^{-1}(P+U)\Delta t^2 + \dot{q}^o\Delta t + q^o + \frac{1}{2} W^{-1}Dn\Delta t^2 N. \tag{2.10.11}$$

This expression can be written in the form

$$q^1 = E + HN, \tag{2.10.12}$$

where

$$E = \frac{1}{2} W^{-1}(P+U)\Delta t^2 + \dot{q}^o \Delta t + q^o,$$
$$H = \frac{1}{2} W^{-1} Dn\Delta t^2. \tag{2.10.13}$$

Matrices E and H are calculated numerically.

Now, the scalar N is determined in such way that the coordinates $(x^1, y^1, z^1) = \eta(q^1)$ satisfy (2.10.1). In fact, the following system of equations must be solved:

$$q^1 = E + HN, \tag{2.10.14a}$$

$$(x^1, y^1, z^1) = \eta(q^1), \tag{2.10.14b}$$

$$h(x^1, y^1, z^1, t_1) = 0, \tag{2.10.14c}$$

the unkonwns q^1, N, (x^1, y^1, z^1); the determination of N being most essential.

Such a problem can be solved by means of uni-dimensional search with respect to N. By some of the uni-dimensional search methods [21] such a value of N can be found. If this is substituted into (2.10.14a), then (2.10.14b) yields the coordinates x^1, y^1, z^1 satisfying (2.10.14c). The methods of uni-dimensional search are very efficient so the whole procedure does not take long.

When N is calculated in the way described, then from (2.10.9) and (2.10.10) we can calculate the state q^1, $\dot{q}^1$ at the instant t_1, and the procedure is repeated for a new instant.

(ii) Let us consider the chain as in case (i) but let us impose a different constaint on the last segment, i.e., the point V is to be on a given line. Let the line be given in space by the equations

$$h_1(x, y, z, t) = 0, \tag{2.10.15a}$$

$$h_2(x, y, z, t) = 0. \tag{2.10.15b}$$

The problem is treated in a manner similar to (i) because the constraint in question has the form of two intersected surfaces (which gives a curve). Thus, the reaction force is

$$\vec{R} = N_1 \frac{\vec{\nabla} h_1}{|\vec{\nabla} h_1|} + N_2 \frac{\vec{\nabla} h_2}{|\vec{\nabla} h_2|}, \tag{2.10.16}$$

so the problem of a two-dimensional search for N_1 and N_2 arises (with the inverse problem), which is of solving a slower procedure (Fig. 2.37).

However, in the case of solving the direct problem, when the pressure $\vec{R}$ is known, one simply proceeds in same way as in (i).

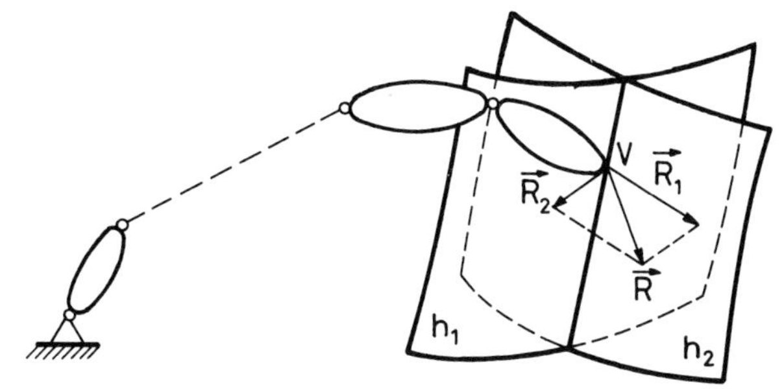

Fig. 2.37. Last segment connected to a line

CONCLUSION. In this chapter we did not compare the various methods in Terms of efficiency, speed and the like. On the one hand, this was not done because authors often do not give sufficient data, and, on the other hand, it is not customary to compare methods if they have not been tested under equal conditions. Further, it should be said that even definitive comparisons of speed would not suffice for an evaluation of efficiency. This can be explained by considering the different ways of placing the connected (b.-f.) coordinate systems. In the main, the coordinate systems are placed in two ways. The first (2.4., 2.5, 2.8.) is to place the origin in the joint and connect it to one of the segments, the position of the coordinate axes being precisely determined. This was explained in detail in 2.4. Such a method enables us to write the transition matrix between two segments in analytical form. The second way (2.3., 2.7., 2.9.) is to place the origin in the segment center of gravity, with arbitrarily positioned orthogonal exes. In this case the computer procedure for calculating the transition matrix has been defined. At first glance, it seems that calculating the transition matrix by means of the analytical expression is faster. But, each methods demands, as input data, the tensor of inertia with respect to the axes of the body-fixed system. If the second way of placing the coordinate system is used, then the system axes are usually assumed to be along the main inertia axes of the segment; so, writing the expression for the inertia tensor in easy. With the first way of placing the coordinate system, a new system is adopted with the origin in the center of gravity and with the axes parallel to the axes of the system in the joint. These axes are not in the direction of the main inertial axes of the segment (except in some special cases); so, the formation of the expression for the inertia tensor is considerably more complex. If the inertia tensor is calculated by hand at the stage of prepating the data, many advantages of the c.-a. model formation are lost, notably the possibility of making simple changes in the configuration. If the main moments of inertia (for the main axes) are still the input data and the calculation for the axes of the connected system is performed inside the algorithm, then the speed is lost because such a calculation is equivalent to calculating a new transition matrix.

References

[1] Stepanenko Yu., "Method of Analysis of Spatial Lever Mechanisms", (in Russian), Mekhanika mashin, Vol. 23, Moscow, 1970.

[2] Juričić D., Vukobratović M., "Mathematical Modelling of a Bipedal Walking System", ASME Publ. 72-WA/BHF-13.

[3] Vukobratović M., Stepanenko Yu., "Mathematical Models of General Anthropomorphic Systems", Math. Biosciences, Vol. 17, 1973.

[4] Stepanenko Yu., Vukobratović M., "Dynamics of Articulated Open--Chain Active Mechanisms", Math. Biosciences, Vol. 28, No 1/2, 1976.

[5] Vukobratović M., "Computer Method for Mathematical Modelling of Active Kinematic Chains via Generalized Coordinates", Journal of IFToMM Mechanisms and Machine Theory, Vol. 13, No 1, 1978.

[6] Vukobratović M., Legged Locomotion Robots and Anthropomorphic Mechanisms,Monograph, 1975, Institute "M.Pupin", Beograd, Yugoslavia.

[7] Orin D., Vukobratović M., R.B. Mc Ghee., G. Hartoch, "Kinematic and Kinetic Analysis of Open-Chain Linkages Utilizing Newton--Euler Methods", Math. Biosciences, Vol. 42, 1978.

[8] Vukobratović M., "Synthesis of Artificial Motion", Journal of Mechanism and Machine Theory, Vol. 13, No 1, 1978.

[9] Vukobratović M., Potkonjak V., "Contribution to the Forming of Computer Methods for Automatic Modelling of Active Spatial Mechanisms Motion", PART1: "Method of Basic Theoremes of Mechanics", Journal of Mechanisms and Machine Theory, Vol. 14, No 3, 1979.

[10] Vukobratović M., Potkonjak V., "Contribution to the Computer

Methods for Generation of Active Mechanisms via Basic Theoremes of Mechanics" (in Russian), Teknitcheskaya kibernetika ANUSSR, No 2, 1979.

[11] Paul R., Modelling, Trajectory Calculation and Servoing of a Computer Controlled Arm, (in Russian) "Nauka", Moscow, 1976.

[12] Luh, J.Y.S., Walker M.W., Paul R.P.C., "On-Line Computational Scheme for Mechanical Manipulators", Trans. of ASME Journal of Dynamic Systems, Measurement and Control, June, 1980, Vol. 102.

[13] Vukobratović M., "Computer Method for Mathematical Modelling of Active Kinematic Chains via Euler´s Angles", Journal of IFToMM Mechanisms and Machine Theory, Vol. 13, No 1, 1978.

[14] Vukobratović M., Potkonjak V., "Contribution to Automatic Forming of Active Chain Models via Lagrangian Form", Journal of Applied Mechanics, No 1, 1979.

[15] Hollerbach J.M., "A Recursive Formulation of Lagrangian Manipulator Dynamics", Proc. of JACC, Aug. 13-15, 1980, San Francisco.

[16] Potkonjak V., Vukobratović M., "Two Methods for Computer Forming of Dynamic Equations of Active Mechanisms", Journal of Mechanism and Machine Theory, Vol. 14, No 3, 1979.

[17] Vukobratović M., Nikolić I., "Further Development of Methods for Automatic Forming of Dynamic Models of Manipulators", Proc. of International Conference on Systems Engineering, Coventry, England - 1980.

[18] Popov E.P.,Vereschagin A.F., Ivkin A.M., Leskov A.G., Medvedov V.S., "Design of Robot Control Using Dynamic Models of Manipulator Devices", Proc. of VI IFAC Symp. on Automatic Control in Space, (in Russian), Erevan, USSR, 1974.

[19] Popov E.P., Vereschagin A.F. Zenkevich S.A., Manipulation Robots: Dynamics and Algorithms, (in Russian), "Nauka", Moscow, 1978.

[20] Polak Z., Numerical Methods of Optimization (in Russian), "Nauka", Moscow, 1978.

[21] Gottfred B., Weisman J., Introduction to Optimization Theory, Prentice Hall, New Jersey, 1973.

[22] Pars L., Analytical Mechanics, (in Russian), "Nauka", Moscow, 1971.

[23] Medvedov V.S., Leskov A.G., Yuschenko A.S., Control Systems of Manipulation Robots, (in Russian), "Nauka", Moscow, 1978.

Chapter 3
Simulation of Manipulator Dynamics and Adjusting to Functional Movements

The methods for computer-aided formation and solution of the mathematical models of active mechanisms (described in Chapter II and in [1 - 15]) have served as a basis for the development of the algorithm for the simulation of manipulator dynamics. We shall first indicate the main ideas and problems, and then describe the general algorithm for the simulation of dynamics [16, 17]. Later, we shall adjust this algorithm to some classes of practical tasks, i.e., to functional movements, particularly because of the efficiency of handling the algorithm. At the same time we shall analyse the need for a certain number of degrees of freedom of the manipulator as well as their use in the various categories task [18]. Something more should be said about the number of manipulator d.o.f. and about the number of d.o.f. necessary for performing a particular manipulation task. At first, we note the difference between the number of d.o.f. of the whole manipulator (considered as a dynamical system), and the number of d.o.f. of the gripper (considered as the last rigid body in the chain). These two numbers need not be equal. The number of manipulator d.o.f. depends on the number of joints and the number of d.o.f. in each joint. For instance, if a manipulator represents an open chain without branching and consists of n segments and n one-d.o.f. joints, then it has n degrees of freedom. On the other hand, the gripper, considered as a rigid body, cannot have more than six d.o.f. Even if n is less then six, those two numbers of d.o.f. need not be equal. For instance, it may happen that the manipulator has five d.o.f. and the gripper (i.e. the last segment) has four d.o.f. This means that one d.o.f. is lost. An extensive discussion on kinematics is necessary in order to explain the loss of some d.o.f. Such a loss can occur in some special cases of relative positions of joints axes. Such special cases are called the kinematical singularities. It should be explained why it is important to notice a difference betwen these two numbers of d.o.f. If a manipulator with n d.o.f. ($n \leq 6$) has to perform some prescribed manipulation task, it is essential that the gripper has enough d.o.f. to solve the task geometricaly. This is due to the fact that for such manipulators the tasks are usually imposed on the gripper. But, for solving, the dynamics of that task, it is necessary to solve the dynamical equations for the whole manipulator, and such equations include all n d.o.f. If we consider a manipulator with $n>6$ (redundant manipulator) then the discussion is rather complicated. Six d.o.f. of the gripper are enough for all manipulation tasks. So the tasks for manipulators with $n \geq 6$ usually contains some additional requirements which are not imposed on the gripper but on the manipulator as a whole (for instance, that a manipulator bypasses some obstacle in the working space). Hence, for solving the task geometry, not only the gripper d.o.f. are but important too are the d.o.f. of the manipulator as a whole. An extensive discussion on kinematics with the exact answers about the conditions for the disappearance of some d.o.f. will not be given here. So we shall restrict out consideration to the cases when there are no singularities. Let us be more precise. For a manipulator with $n \leq 6$ d.o.f. we shall assume that its gripper also has n d.o.f.; for a manipulator with $n>6$ d.o.f. we shall ussume that its gripper has six d.o.f., i.e., the maximal possible number. It is an important fact that this assumption holds for almost all practical problems. For instance, when imposing a task on a certain manipulator we always take care about avoiding the singularity points i.e. we chose the trajectories which can be performed. Keeping in mind the above assumption, we shall simply talk about degrees of freedom, regardless of whether the whole manipulator, or its gripper only, is concerned.

With the aim of synthesizing programmed control of functional movements in nominal dynamics, mathematical models of actuators were introduced, which, along with the mathematical models of mechanisms considered sofar, form the complete model of robot dynamics [20, 22]. The simulation algorithm is modified to allow the calculation of programmed control inputs. The influence of the actuator model order to the simulation results is also discussed. At the end of the chapter, we present for practical reasons a specific synthesis of nominal dynamics of manipulation robot functional movements on the basis of the functional subsystems of the positioning and the orientation of the last segment (gripper). Such a method of synthesizing functional motion will be considered (in the second book of the series) as the first stage of synthesis of the algorithm for the control of manipulators and robots in general. A method for calculation of the optimal velocity distribution is also presented.

3.1. Basic Ideas

The notion of the simulation of dynamics usually only involves the solution of the inverse problem of dynamics, i.e. the determination of the motion for prescribed generalized forces. In this case, the simulation is considered somewhat more liberally; so that it also includes the notion of the simulation of the direct problem. Thus, simulation of mechanism dynamics will involve the calculation of all dynamical values based on the known manipulator configuration and the input data about the manipulation task (prescribed motion). The simulation is based on the computer-aided methods for forming the mathematical model [1 - 15], working with discrete time and generalized coordinates q. Thus, if the manipulation task is prescribed in generalized coordinates, then it is possible, as in chapter II, to calculate the required driving forces and torques of the system (2.2.2), i.e.,

$$P = g(q,\ \dot{q},\ \ddot{q},\ \text{mechanism configuration}). \tag{3.1.1}$$

Thus the simulation problem is solved in principle, being reduced to the solution of the direct problem of dynamics.

However, in practice, the manipulation task is not usually given in generalized coordinates, but in terms of some so-called "external variables" (e.g.: the law of manipulator tip motion and gripper orientation in the working space). It is then also necessary to synthesize the nominal time functions of the generalized coordinates during the prescribed functional movement. In order that the task be prescribed correctly, it is necessary that the "external variables" (vector X) and the generalized coordinates (vector q) depend, and depend only, on each other. If this is not the case, for instance if a surplus of the d.o.f. is present, then a special problem arises which will only be partially considered here.

Let us designate by η a function which transforms the generalized coordinates q into the external variables X,

$$X = \eta(q), \tag{3.1.2}$$

where q and X are n-dimensional vectors.

The function η is one-place and can always be determined but calculation of q from such a system of equations (2.1.2) is difficult because not only can q not be expressed explicitly, but it cannot even be numerically approximated because of the complexity of system (3.1.2) which has to be solved.

But note the following: for the operation of the computer-aided methods of the forming the mathematical model, it is necessary to know q, $\dot{q}$ at each time instant; $\ddot{q}$ must also be known in order to calculate P. However, only $\ddot{q}$ appears as input since q, $\dot{q}$ are calculated by integration starting with the known initial state q^o, $\dot{q}^o$. So in order to realize the simulation, it is necessary to develop a procedure for calculating $\ddot{q}$ from the known state q, $\dot{q}$ and known external variables X(t).

By double differentation,

$$\dot{X} = \frac{\partial \eta}{\partial q}\dot{q}, \tag{3.1.3}$$

$$\ddot{X} = \frac{\partial \eta}{\partial q}\ddot{q} + \frac{\partial^2 \eta}{\partial q^2}\dot{q}^2. \tag{3.1.4}$$

Let $X^a = \ddot{X}$, $B = \frac{\partial \eta}{\partial q}$, $A = \frac{\partial^2 \eta}{\partial q^2}\dot{q}^2$. Equation (3.1.4) then becomes

$$X^a = B\ddot{q} + A. \tag{3.1.5}$$

B is an n×n Jacobian matrix, where n is the number of the manipulator d.o.f. and A is an n×1 matrix. The matrices B and A are functions of the state q, $\dot{q}$, which is considered known since it is calculated, as we said, by means of the algorithm. Consequently, it is necessary to prescribe X^a in a series of time instants, or calculate it starting from the manipulation task, prescribed in some way. Then methods must be found for numerically calculating the matrices B, A for a known state q, $\dot{q}$. Then, from system (3.1.5), with the assumption that B is not a singular matrix, one obtains the required generalized acceleration $\ddot{q}$:

$$\ddot{q} = B^{-1}(X^a - A). \tag{3.1.6}$$

Now it is possible to use the c.-a. methods to form and solve the mathematical model. The complete simulation algorithm can be mainly presented by the block-scheme in Fig. 3.1.

In this book, the c.-a. method of forming mathematical models based on

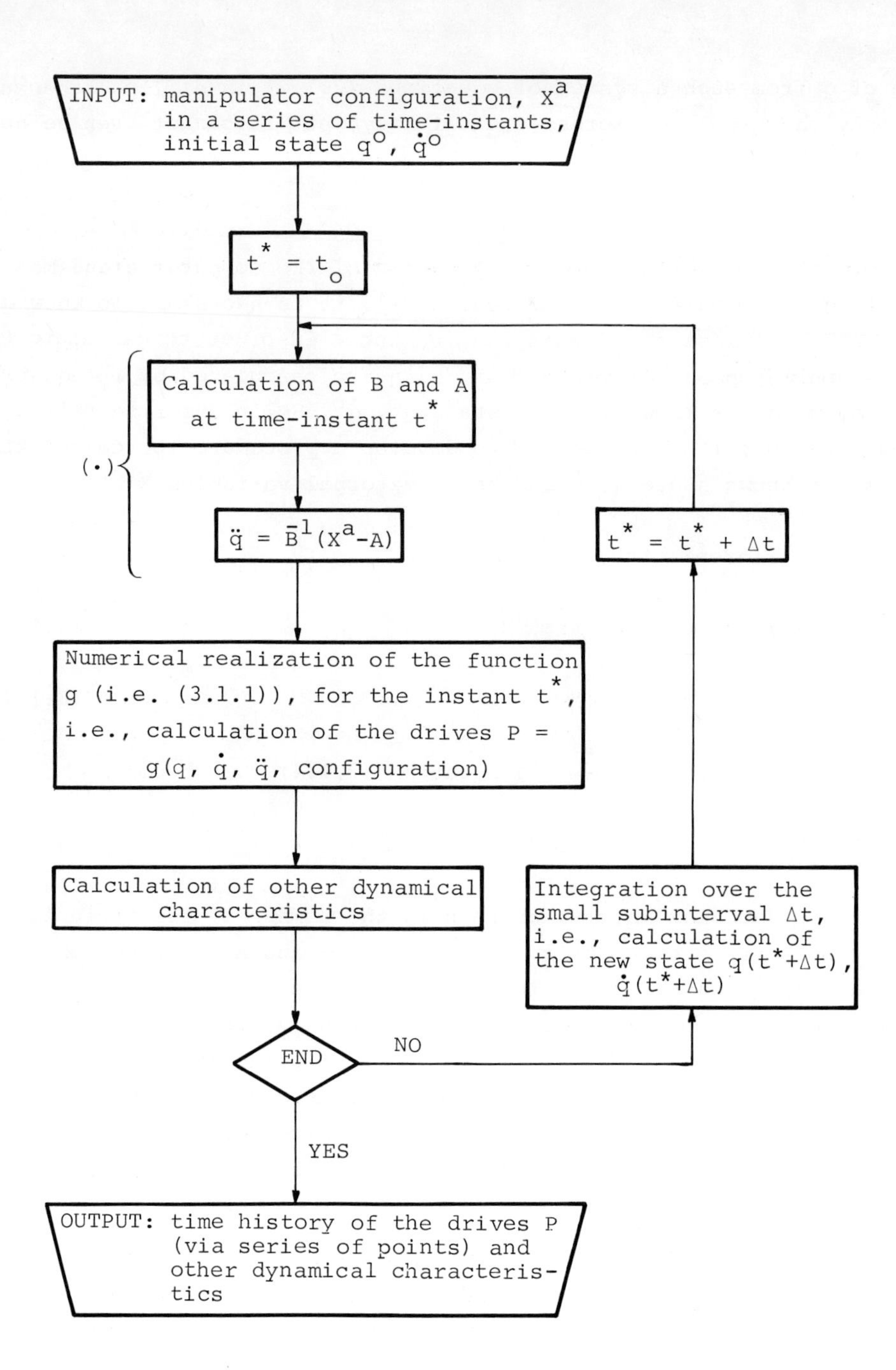

Fig. 3.1. Block sheme of the simulation algorithm

the General theorems of system dynamics has been used (2.3). The method yields the model

$$W\ddot{q} = P + U, \qquad (3.1.7)$$

i.e., it numerically calculates the matrices W, U at time instant t^*. So the function g is

$$P = g(q,\ \dot{q},\ \ddot{q},\ \text{configuration}) = W\ddot{q} - U \qquad (3.1.8)$$

which has already been described in 2.3.

Problem of the nonsingularity of matrix B will not be discussed here.

3.2. The General Simulation Algorithm

Let us consider the manipulator as an open chain of rigid bodies, without branches, as shown in Fig. 3.2. Let the manipulator have six d.o.f.

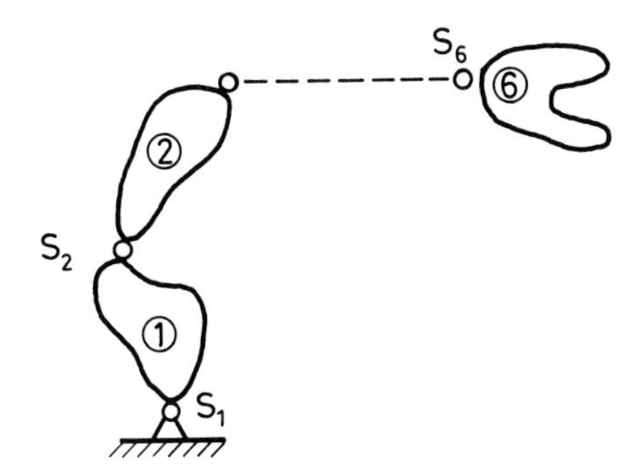

Fig. 3.2. Manipulator as a kinematic chain

The last body (segment) of the chain represents the manipulator gripper, i.e., in the phase of transfering some work object, the last segment is the gripper and object combined. Thus, the manipulation task can usually be regarded as a prescribed motion of a rigid body (the last chain segment) in space. In the development of the general simulation algorithm, one started from the fact that a rigid body motion can, in the most general way, be prescribed by means of the known initial state and the known time function of the center of gravity acceleration (or of some other point on the segment) $\vec{w}(t)$ and the known time-

-function of the angular acceleration $\vec{\varepsilon}(t)$ of the body. Such an approach is justified because for many manipulation tasks these values can easily be prescribed. For instance, gripper center of gravity motion (or its tip, or base) is usually prescribed quite easily by means of the trajectory and the velocity profile, i.e., acceleration. Thus, we will now describe the algorithm (considering the manipulation task) in terms of $\vec{w}(t)$ and $\vec{\varepsilon}(t)$, i.e. these are the input values (in discrete quantities, of course) [16, 17]. So, in matrix notation

$$X^a = \begin{bmatrix} w \\ \varepsilon \end{bmatrix}, \qquad (3.2.1)$$

where w designates a 3×1 matrix corresponding to vector $\vec{w}$, likewise for ε and all other vectors in the sequel.

In Para. 2.3. when the method of general theorems was described, it was shown how the matrices Ω, Θ, Γ, Φ were derived and calculated, so

$$w = \Omega\ddot{q} + \Theta, \qquad (3.2.2)$$

$$\varepsilon = \Gamma\ddot{q} + \Phi. \qquad (3.2.3)$$

The matrices are calculated from the recursive expressions for velocities and accelerations of segments.

The relations (3.2.2) and (3.2.3) can be combined:

$$X^a = \begin{bmatrix} w \\ \varepsilon \end{bmatrix} = \underbrace{\begin{bmatrix} \Omega \\ \Gamma \end{bmatrix}}_{B} \ddot{q} + \underbrace{\begin{bmatrix} \Theta \\ \Phi \end{bmatrix}}_{A} = B\ddot{q} + A. \qquad (3.2.4)$$

By determining the matrices B and A in this way, the problem of calculating $\ddot{q}$ is solved and simulation is performed according to the block-scheme in Fig. 3.1.

As it is necessary to prescribe the vectors $w(t)$ and $\varepsilon(t)$ in a series of time instants, it is convenient, for the purpose of programming, to introduce a separate subroutine, the input data of which are the trajectory of the manipulator tip, velocity profile, gripper orientation law and the time increments subinterval Δt. This subroutine calculates $\vec{w}$, $\vec{\varepsilon}$ in a series of time instants.

Since we shall use the body-fixed coordinate systems, let us introduce some notation: let $\vec{a}_i$ be some vector, characteristic of the i-th segment, expressed by three projections onto the axes of the immobile external coordinate system; let $\tilde{\vec{a}}_i$ be the same vector, expressed in the i-th segment b.-f. coordinate system.

3.3. Classes of Functional Tasks and Adjustment Blocks

The general algorithm for simulation which we have described can be unsuitable in some cases if the input has the form $X^a = [w\ \varepsilon]^T$. These values define the manipulation task fully but sometimes they are unsuitable for setting. Hence, adjustment of the algorithm to individual classes of tasks is necessary. Special adjustment blocks are developed corresponding to the various types of task. The blocks, incorporated into the algorithm, enable one to use the most suitable inputs, i.e., to prescribe the manipulation task in terms of the external variables X^a which can be given very simply for the types of task considered [18].

In principle, two kinds of blocks can be considered. The first calculates $\ddot{q}$ for known X^a (Fig. 3.3a) and is directly incorporated into the simulation algorithm from Fig. 3.1., in the place denoted by (•). The second calculates w, ε for known X^a and is incorporated into the general simulation algorithm described in Para. 3.2., i.e., the scheme given in Fig. 3.3b is incorporated into the algorithm in Fig. 3.1. in the place denoted by (•).

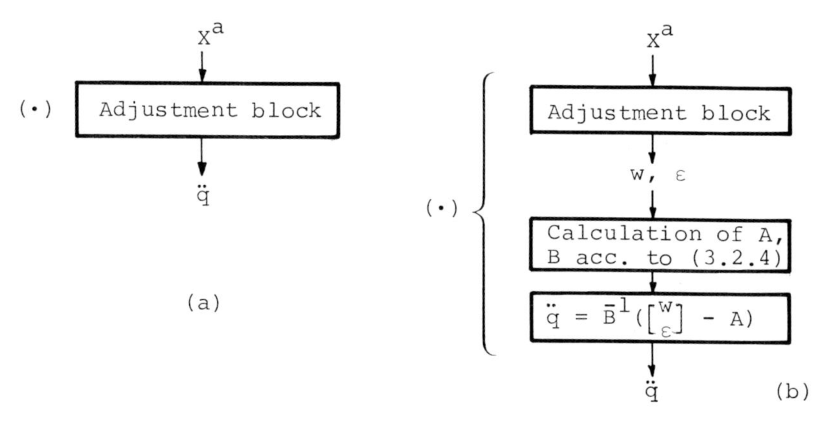

Fig. 3.3. Incorporating adjustment blocks

In order to obtain these blocks, it is necessary to analyze the need for a certain number of d.o.f. and their use in certain classes of tasks. The task classes differ essentially with respect to the use of the d.o.f. and thus also in the mathematical way in which they are treated. In displaying the basic simulation ideas, as we said there exists a dependence between the external variables X (determining the manipulation task) and the generalized coordinates q. We have to remember this because in performing many tasks, fewer d.o.f. are needed than the manipulator has. It is therefore necessary to compensate for this surplus either by holding the corresponding generalized coordinates fixed ("frozen") or using the surplus d.o.f. to fulfil some other requirements.

Let us now more precisely define some notions which will be used. By *positioning* we mean moving the center of gravity of the last segment (or some other point on the segment) to some desired point in the working space, i.e., the motion of the center of gravity (or some other point) along a prescribed trajectory according to a prescribed motion law.

Full orientation of the body in space means an exactly determined angular position of the body with respect to the external space. This can be prescribed in several ways, and it will be discussed in the sequel.

Partial orientation of the body means that the given body axis (or some arbitrary fixed direction on the body) coincides with a prescribed direction in the space.

With manipulators, the tasks of positioning and orientation refer to the last segment, i.e., the gripper. It should be added that the gripper base coincides with the tip of the minimal configuration of the manipulator (a manipulator with three d.o.f. i.e. without gripper); so positioning the gripper base is equivalent to positioning the minimal configuration tip, which term is often used.

To solve the positioning task, which is a part of every manipulation task, three d.o.f. are necessary.

To solve the positioning task along with the task of partial orientation, five d.o.f. are necessary. This class includes the tasks of transfering liquid in containers, some assembly tasks, and the like.

To solve the positioning task along with the task of full orientation, six d.o.f. are needed.

Manipulators with four, five and six d.o.f. will now be considered. Some typical classes of tasks and ways of using the d.o.f. will be analyzed.

<u>A</u> A manipulator with four d.o.f. solves the positioning task by using three d.o.f., and with the one remaining performs operations frequently sufficient for many practical tasks.

1. The task is that of positioning in the Cartesian coordinate system, i.e. $x(t)$, $y(t)$, $z(t)$ for the center of gravity of last segment (or some other point of it), and the fourth d.o.f. is prescribed directly ($q_4(t)$).

2. Same as 1., but positioning is given in cylindrical coordinates $\rho(t)$, $\theta(t)$, $z(t)$.

3. Same as 1., but positioning is given in spherical coordinates $r(t)$, $\theta(t)$, $\varphi(t)$.

<u>B</u> A manipulator with five d.o.f. solves the positioning and partial orientation task.

1. The task is given in the form of positioning (for instance in the Cartesian coordinates $x(t)$, $y(t)$, $z(t)$) and partial orientation.

<u>C</u> A manipulator with six d.o.f. solves the positioning and full orentation task, as well as all problems in which fewer d.o.f. are needed (when compensation of the d.o.f. surplus is done).

1. The task is given in the form of positioning, $x(t)$, $y(t)$, $z(t)$, and full orientation of the last segment in terms of three Euler angles.

2. Positioning is given in terms of $x(t)$, $y(t)$, $z(t)$, and full orientation in terms of one direction and the angle of rotation around it.

3. The task is given in terms of positioning as in 2. plus partial orientation. Motion along one d.o.f. is prescribed directly ($q_k(t)$).

3.4. Elaboration of the Different Adjustment Blocks

The sequence of deriving the different blocks will not be given in the same order in which they were presented in Paragraph 3.3., but it will be suited to the derivation of the mathematical apparatus.

A1. Adjustment in this case (i.e. of the corresponding block) is performed as follows. Let us introduce $w = [\ddot{x}\ \ddot{y}\ \ddot{z}]^T$. Now, with known w and $\ddot{q}_4$, calculate $\ddot{q}_1$, $\ddot{q}_2$, $\ddot{q}_3$ from the system (3.2.2), i.e., from

$$w = \Omega \begin{bmatrix} \ddot{q}_1 \\ \ddot{q}_2 \\ \ddot{q}_3 \\ \ddot{q}_4 \end{bmatrix} + \Theta. \tag{3.4.1}$$

Furthermore, the problem is solved, since $\ddot{q}_1$, $\ddot{q}_2$, $\ddot{q}_3$ have been calculated, and $\ddot{q}_4$ is known, i.e. prescribed.

An example of such a manipulator and of such use of the d.o.f. i.e. q_1, q_2, q_3 for positioning and q_4 for direct setting of supplementary actions, is the manipulator in Fig. 3.4.

Is should be said that as input into the block $\ddot{x}$, $\ddot{y}$, $\ddot{z}$, $\ddot{q}_4$ appear in a series of time instants, and not explicitly as x(t), y(t), z(t), q_4(t).

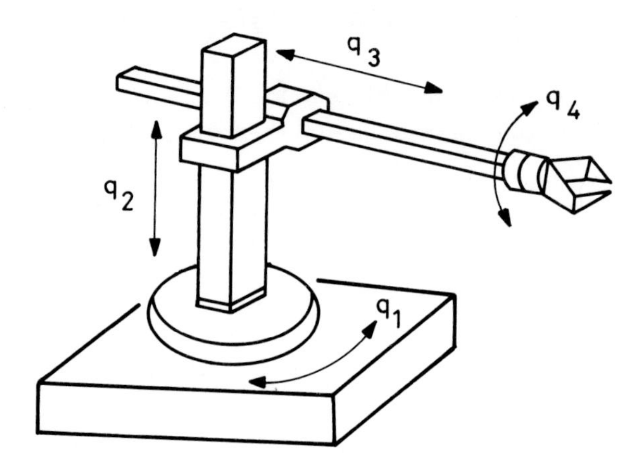

Fig. 3.4. Manipulator with four d.o.f.

A2. Depending on the form of the desired trajectory (Fig. 3.7b), in some cases it is suitable to prescribe positioning in cylindrical coordinates $\rho(t)$, $\theta(t)$, $z(t)$, (Fig. 3.5), with q_4 given separately.

Derivation of this block means calculation of the vector $\vec{w}$, i.e. its Cartesian coordinates in the absolute system, starting from prescribed positioning in cylindrical coordinates.

Projected onto the axes of the cylindrical system, the acceleration has the form

$$w_\rho = \ddot{\rho} - \rho\dot{\theta}^2, \quad w_\theta = \rho\ddot{\theta} + 2\dot{\rho}\dot{\theta}, \quad w_z = \ddot{z}, \tag{3.4.2}$$

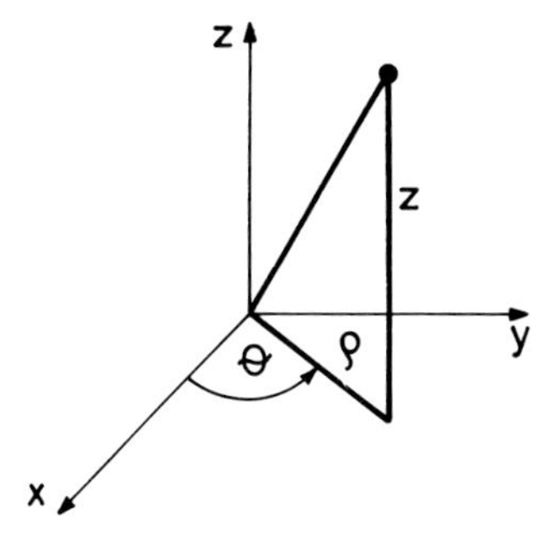

Fig. 3.5. Cylindrical coordinates

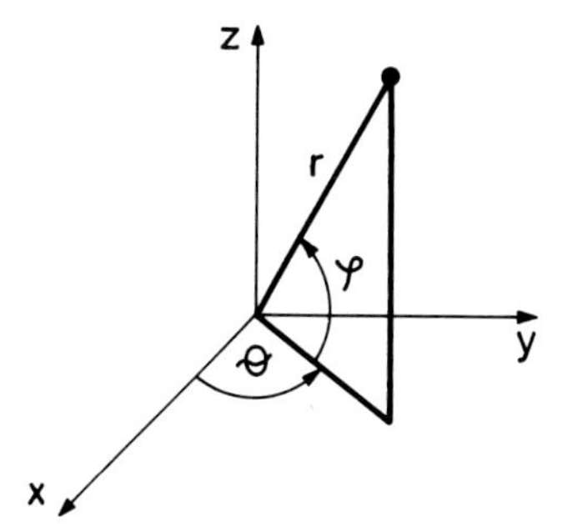

Fig. 3.6. Spherical coordinates

and further, the following Cartesian projections are obtained:

$$\begin{aligned} w_x &= w_\rho\cos\theta - w_\theta\sin\theta, \\ w_y &= w_\rho\sin\theta + w_\theta\cos\theta, \\ w_z &= \ddot{z}, \end{aligned} \tag{3.4.3}$$

i.e., the acceleration vector is

$$w = [w_x \;\; w_y \;\; w_z]^T. \tag{3.4.4}$$

Now, just as in A1., using the system (3.4.1), $\ddot{q}_1$, $\ddot{q}_2$, $\ddot{q}_3$ are calculated for w already calculated and $\ddot{q}_4$ is given in that time instant.

As input values $\ddot{\rho}$, $\ddot{\theta}$, $\ddot{z}$, $\ddot{q}_4$ appear. For the calculation of w by means of the expressions (3.4.2), (3.4.3), ρ, $\dot{\rho}$, θ, $\dot{\theta}$ are also needed; so, during the recursive calculation from one time instant to another, these required values are calculated by integration together with integration over the generalized coordinates, which is performed in the simulation algorithm.

A3. Sometimes it is more appropriate to prescribe positioning by means of spherical coordinates $r(t)$, $\theta(t)$, $\varphi(t)$ and to prescribe $q_4(t)$ separately (Figs. 3.6, 3.7c).

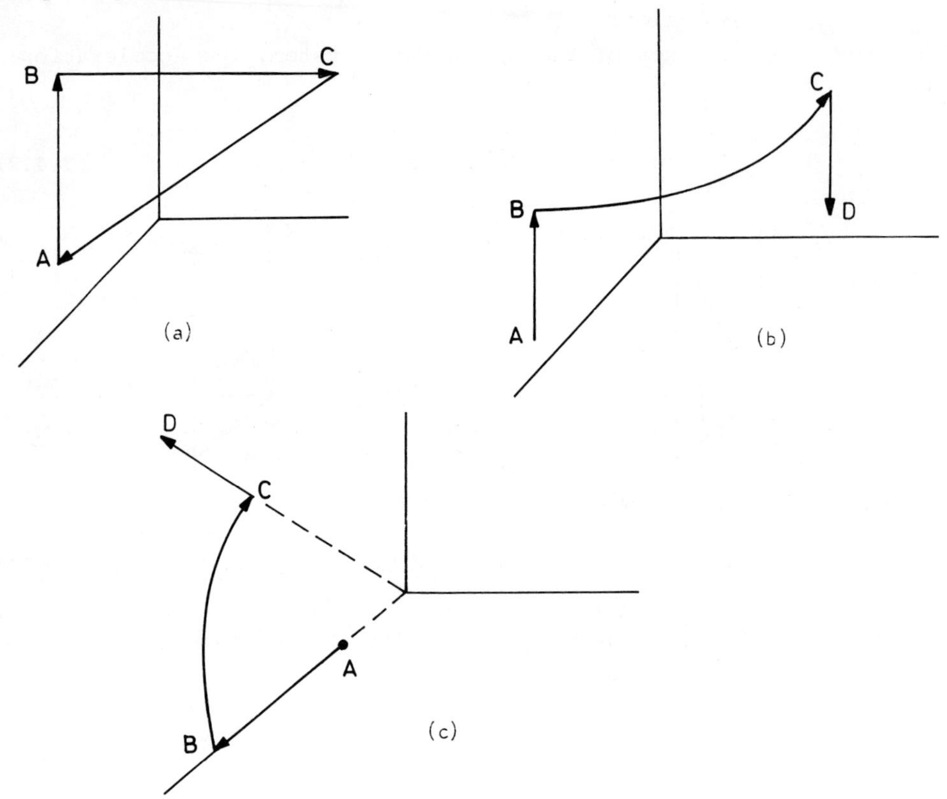

Fig. 3.7. Various trajectories of the manipulator tip

Just as in A1 and A2, it is necessary to calculate the vector w. Projecting $\vec{w}$ onto the axes of the spherical coordinate system,

$$
\begin{aligned}
w_r &= \ddot{r} - r\dot{\varphi}^2 - r\cos^2\varphi\,\dot{\theta}^2 ,\\
w_\theta &= 2\dot{r}\dot{\theta}\cos\varphi + r\ddot{\theta}\cos\varphi - 2r\dot{\varphi}\dot{\theta}\sin\varphi ,\\
w_\varphi &= 2\dot{r}\dot{\varphi} + r\ddot{\varphi} + r\sin\varphi\cos\varphi\,\dot{\theta}^2 ,
\end{aligned}
\tag{3.4.5}
$$

or, in the external Cartesian system,

$$w_x = w_r \cos\varphi\cos\theta - w_\theta \sin\theta - w_\varphi \sin\varphi\cos\theta,$$

$$w_y = w_r \cos\varphi\sin\theta + w_\theta \cos\theta - w_\varphi \sin\varphi\sin\theta, \quad (3.4.6)$$

$$w_z = w_r \sin\varphi + w_\varphi \cos\varphi;$$

so the acceleration vector is

$$w = [w_x \quad w_y \quad w_z]^T. \quad (3.4.7)$$

Now, $\ddot{q}_1$, $\ddot{q}_2$ and $\ddot{q}_3$ are calculated from (3.4.1) using w already calculated and $\ddot{q}_4$ given at that time instant.

As input data $\ddot{r}$, $\ddot{\theta}$, $\ddot{\varphi}$ appear in a series of time instants, and r, $\dot{r}$, θ, $\dot{\theta}$, φ, $\dot{\varphi}$ in expression (3.4.5) are obtained by integration within the simulation algorithm.

<u>C1.</u> Positioning in this case is prescribed in Cartesian coordinates x(t), y(t), z(t) and full orientation is given in terms of three Euler angles $\theta(t)$, $\psi(t)$, $\varphi(t)$ of the last segment (i.e. the corresponding coordinate system) relative to the external system (Fig. 3.8).

This block performs the adjustment in such way that it calculates the vectors w, ε for the time instant considered. Further calculation is carried out by the general simulation algorithm.

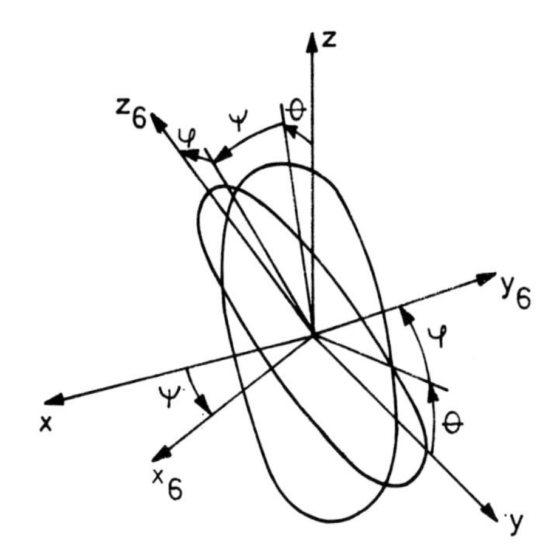

Fig. 3.8. Euler angles of the b.-f. system ($x_6y_6z_6$) relative to the external system (x y z)

The acceleration vector is

$$w = [\ddot{x}\ \ddot{y}\ \ddot{z}]^T \tag{3.4.8}$$

and angular acceleration is given by

$$\varepsilon = A_6 \left(\Pi \begin{bmatrix} \ddot{\theta} \\ \ddot{\psi} \\ \ddot{\varphi} \end{bmatrix} + {}^1\Pi \begin{bmatrix} \dot{\psi}\cdot\dot{\varphi} \\ \dot{\varphi}\cdot\dot{\theta} \\ \dot{\theta}\cdot\dot{\psi} \end{bmatrix} \right), \tag{3.4.9}$$

where

$$\Pi = \begin{bmatrix} \cos\psi & 0 & 1 \\ \sin\psi\sin\varphi & \cos\varphi & 0 \\ \sin\psi\cos\varphi & -\sin\varphi & 0 \end{bmatrix},$$

$$ {}^1\Pi = \begin{bmatrix} 0 & 0 & -\sin\psi \\ -\sin\varphi & \sin\psi\cos\varphi & \cos\psi\sin\psi \\ -\cos\varphi & -\sin\psi\sin\varphi & \cos\psi\cos\varphi \end{bmatrix}. \tag{3.4.10}$$

The transition matrix of the last segment A_6 is calculated in the method for c.-a. formation of the model, which represents a part of the simulation algorithm, or else it is calculated immediately in terms of the prescribed Euler angles. Thus,

$$A_6 = A^\theta A^\psi A^\varphi,$$

$$A^\theta = \begin{bmatrix} 1 & 0 & 0 \\ 0 & \cos\theta & -\sin\theta \\ 0 & \sin\theta & \cos\theta \end{bmatrix},$$

$$A^\psi = \begin{bmatrix} \cos\psi & 0 & \sin\psi \\ 0 & 1 & 0 \\ -\sin\psi & 0 & \cos\psi \end{bmatrix}, \tag{3.4.11}$$

$$A^\varphi = \begin{bmatrix} 1 & 0 & 0 \\ 0 & \cos\varphi & -\sin\varphi \\ 0 & \sin\varphi & \cos\varphi \end{bmatrix}.$$

The input data are $\ddot{x}$, $\ddot{y}$, $\ddot{z}$ and $\ddot{\theta}$, $\ddot{\psi}$, $\ddot{\varphi}$ in a series of time instants, and the θ, $\dot{\theta}$, ψ, $\dot{\psi}$, φ, $\dot{\varphi}$ in the expressions (3.4.9), (3.4.10) are obtained by integration within the simulation algorithm.

It should be noted that it is possible to prescribe positioning in a different way (e.g. using cylindrical or spherical coordinates). This would be done not by using (3.4.8) to calculate w but by using the expressions (3.4.2) to (3.4.4) or (3.4.5) to (3.4.7).

C2. This case considers positioning prescribed in Cartesian coordinates x(t), y(t), z(t). Full orientation is prescribed by the requirement that the arbitrarily determined fixed direction on the body always coincides with the one prescribed in space (which can be variable) and by the rotation angle around this direction. Such a method of prescribing the manipulation task is suitable in many practical cases. Let us consider for instance, the task of spraying powder along some predetermined path (Fig. 3.9). The task reduces to the need to realize the motion of an object (container) along the trajectory (a), i.e. positioning. Along that trajectory, the container should rotate around axis (b) according to the prescribed law $\psi(t)$.

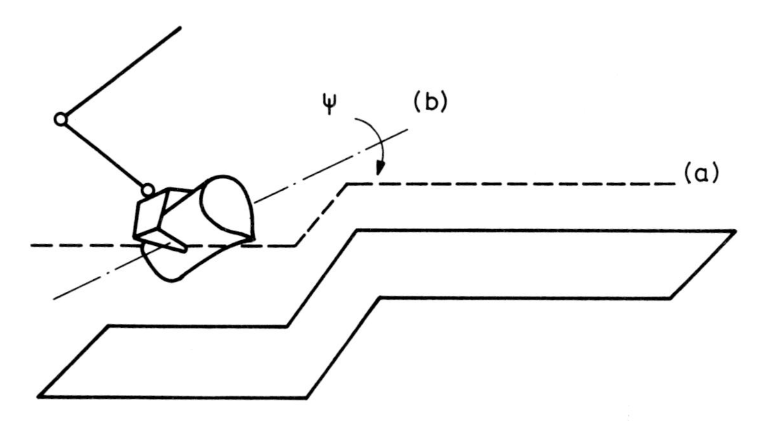

Fig. 3.9. Spraying powder along a prescribed trajectory

As a second example, let us consider the task of screwing in a bolt in an assembling operation (Fig. 3.10).

We now derive the adjustment block which calculates the vectors $\vec{w}$, $\vec{\varepsilon}$ from given inputs.

For the sake of mathematical formulation, let us introduce two parameters θ, φ determining the direction in the external space (Fig. 3.11).

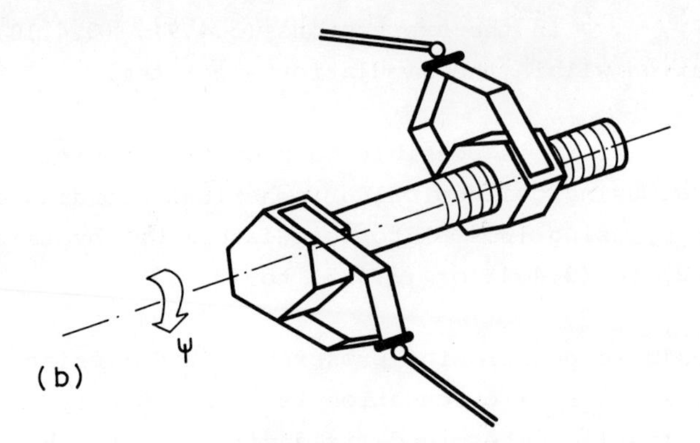

Fig. 3.10. Screwing in a bolt

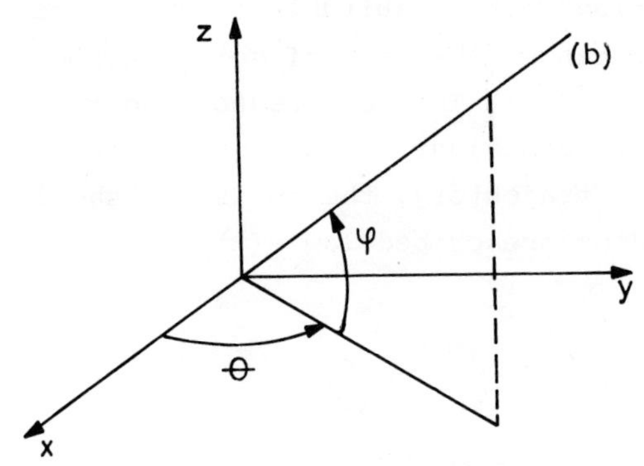

Fig. 3.11. External angles θ, φ

This is also the most suitable (the simplest) way of prescribing a direction. Now, associate with the body (the last segment) a coordinate system with x-axis coinciding with (b). This system is obtained from the external system in the following way: rotation is first made around the z-axis (angle θ), and then around the new y-axis in the negative sense (angle φ); finally, the rotation around the new x-axis represents the angle ψ. Thus, prescribing the direction and the rotation around it is reduced to prescribing $\theta(t)$, $\varphi(t)$, $\psi(t)$. The position of the b.-f. coordinate system is determined by prescribing θ^o, φ^o, ψ^o at the initial time instant.

Let us now determine w. Since positioning is given in Cartesian coordinates, w is

$$w = [\ddot{x} \quad \ddot{y} \quad \ddot{z}]^T. \tag{3.4.12}$$

The angular acceleration ε must be also determined. The expression for the angular velocity ω will be derived first and then we derive the angular acceleration $\varepsilon = \dot{\omega}$.

Let $\vec{h}$ be some vector connected to the body. Then

$$\dot{\vec{h}} = \vec{\omega} \times \vec{h}. \tag{3.4.13}$$

In addition,

$$\vec{h} = A \cdot \vec{\tilde{h}} \Rightarrow \dot{\vec{h}} = \dot{A}\vec{\tilde{h}}, \tag{3.4.14}$$

where A is the transition matrix of the b.-f. system, and $\vec{\tilde{h}}$ designates the vector $\vec{h}$, but expressed in terms of the three projections onto the axes of the b.-f. system. From (3.4.13) and (3.4.14) it follows that

$$\dot{A}\vec{\tilde{h}} = \vec{\omega} \times \vec{h} \tag{3.4.15}$$

or, in matrix notation,

$$\dot{A}\tilde{h} = \underline{\underline{\omega}}h, \tag{3.4.16}$$

where

$$\underline{\underline{\omega}} = \begin{bmatrix} 0 & -\omega_3 & \omega_2 \\ \omega_3 & 0 & -\omega_1 \\ -\omega_2 & \omega_1 & 0 \end{bmatrix}, \quad \text{for } \vec{\omega} = \{\omega_1, \omega_2, \omega_3\}. \tag{3.4.17}$$

Further,

$$\dot{A}\tilde{h} = \underline{\underline{\omega}}A\tilde{h} \tag{3.4.18}$$

or

$$(\dot{A}-\underline{\underline{\omega}}A)\tilde{h} = 0. \tag{3.4.19}$$

Since eq. (3.4.19) is satisfied for every $\tilde{h}$, it follows that

$$\dot{A} - \underline{\underline{\omega}}A = 0 \tag{3.4.20}$$

i.e.

$$\underline{\underline{\omega}} = \dot{A}A^{-1}. \tag{3.4.21}$$

Further, the angular acceleration is

$$\underline{\underline{\varepsilon}} = \dot{\underline{\underline{\omega}}} = \ddot{A}A^{-1} - \dot{A}A^{-1}\dot{A}A^{-1}, \tag{3.4.22a}$$

$$\underline{\underline{\varepsilon}} = \ddot{A}A^{-1} - \underline{\underline{\omega}}^{2}. \tag{3.4.22b}$$

In all, the problem of obtaining the angular acceleration has been reduced to determining the transition matrix A corresponding to the b.-f. system and the derivatives of A. The transition matrix has the form

$$A = A_1A_2A_3, \tag{3.4.23}$$

where

$$A_1 = \begin{bmatrix} \cos\theta & -\sin\theta & 0 \\ \sin\theta & \cos\theta & 0 \\ 0 & 0 & 1 \end{bmatrix}, \qquad A_2 = \begin{bmatrix} \cos\varphi & 0 & -\sin\varphi \\ 0 & 1 & 0 \\ \sin\varphi & 0 & \cos\varphi \end{bmatrix},$$

$$A_3 = \begin{bmatrix} 1 & 0 & 0 \\ 0 & \cos\psi & -\sin\psi \\ 0 & \sin\psi & \cos\psi \end{bmatrix} \tag{3.4.24}$$

The first and second derivatives of the transition matrix are

$$\dot{A} = \dot{A}_1A_2A_3 + A_1\dot{A}_2A_3 + A_1A_2\dot{A}_3, \tag{3.4.25}$$

$$\ddot{A} = \ddot{A}_1A_2A_3 + \dot{A}_1\dot{A}_2A_3 + \dot{A}_1A_2\dot{A}_3 + \dot{A}_1\dot{A}_2A_3 + A_1\ddot{A}_2A_3 + A_1\dot{A}_2\dot{A}_3$$
$$+ \dot{A}_1A_2\dot{A}_3 + A_1\dot{A}_2\dot{A}_3 + A_1A_2\ddot{A}_3, \tag{3.4.26}$$

So from (3.4.24),

$$\dot{A}_1 = \begin{bmatrix} -\sin\theta & -\cos\theta & 0 \\ \cos\theta & -\sin\theta & 0 \\ 0 & 0 & 0 \end{bmatrix} \dot{\theta} ,$$

$$\ddot{A}_1 = \begin{bmatrix} -\cos\theta & \sin\theta & 0 \\ -\sin\theta & -\cos\theta & 0 \\ 0 & 0 & 0 \end{bmatrix} \dot{\theta}^2, + \begin{bmatrix} -\sin\theta & -\cos\theta & 0 \\ \cos\theta & -\sin\theta & 0 \\ 0 & 0 & 0 \end{bmatrix} \ddot{\theta},$$

$$\dot{A}_2 = \begin{bmatrix} -\sin\varphi & 0 & -\cos\varphi \\ 0 & 0 & 0 \\ \cos\varphi & 0 & -\sin\varphi \end{bmatrix} \dot{\varphi},$$

$$\ddot{A}_2 = \begin{bmatrix} -\cos\varphi & 0 & \sin\varphi \\ 0 & 0 & 0 \\ -\sin\varphi & 0 & -\cos\varphi \end{bmatrix} \dot{\varphi}^2 + \begin{bmatrix} -\sin\varphi & 0 & -\cos\varphi \\ 0 & 0 & 0 \\ \cos\varphi & 0 & -\sin\varphi \end{bmatrix} \ddot{\varphi}, \qquad (3.4.27)$$

$$\dot{A}_3 = \begin{bmatrix} 0 & 0 & 0 \\ 0 & -\sin\psi & -\cos\psi \\ 0 & \cos\psi & -\sin\psi \end{bmatrix} \dot{\psi},$$

$$\ddot{A}_3 = \begin{bmatrix} 0 & 0 & 0 \\ 0 & -\cos\psi & \sin\psi \\ 0 & -\sin\psi & -\cos\psi \end{bmatrix} \dot{\psi}^2 + \begin{bmatrix} 0 & 0 & 0 \\ 0 & -\sin\psi & -\cos\psi \\ 0 & \cos\psi & -\sin\psi \end{bmatrix} \ddot{\psi}.$$

The problem of deriving the expressions for angular velocity and acceleration $\vec{\omega}$, $\vec{\varepsilon}$ when the direction and rotation around the same are given was treated in the theory of finite rotations of rigid bodies [19] by introducing the so-called "rotation vector". Here, another approach was derived by using transition matrices. This is more appropriate to the case in question and to use with a digital computer.

Thus, we have solved the problem of calculating ω, ε. Hence, the adjustment has been derived and the general simulation algorithm can flow smoothly.

As input values $\ddot{x}$, $\ddot{y}$, $\ddot{z}$ and $\ddot{\theta}$, $\ddot{\varphi}$, $\ddot{\psi}$ appear in a series of time instants, and θ^o, φ^o, ψ^o. The values θ, $\dot{\theta}$, φ, $\dot{\varphi}$, ψ, $\dot{\psi}$ in expressions (3.4.24) and (3.4.27) are obtained by integration within the simulation algorithm.

In the case of prescribing positioning in cylindrical or spherical coordinates, one proceeds just as in case C1, i.e. acceleration w is calculated from expressions (3.4.2) to (3.4.4) or (3.4.5) to (3.4.7) instead of (3.4.12).

Let us now analyze how the upper procedure is simplified in the case of the positioning and partial orientation task.

B1. In this case the manipulator with five d.o.f. should solve the task, for which positioning is prescribed in Cartesian coordinates and some fixed direction on the last segment should coincide with a given direction (which can be variable) in the space (partial orientation).

Let us designate by $\vec{h}$ the unit vector of the chosen direction on the last segment. Since the direction is fixed on the body, $\vec{\tilde{h}}$ is constant and thus the direction on the body is given. Since the direction is determined by means of the unit vector, it is unnecessary to use some new coordinate system, as in C2, but only the b.-f. system defined in the c.-a. method (2.3) is used.

Regarding the previous analysis C2, it should be noted that $\vec{h}$ is the unit vector of direction (b).

In order to compensate the five d.o.f., since much is needed for the task, five equations are needed for the calculation of the generalized accelerations $\ddot{q} = [\ddot{q}_1 \cdots \ddot{q}_5]^T$. The problem will be treated in such a way that positioning is given by $x(t)$, $y(t)$, $z(t)$ (i.e.: $\ddot{x}$, $\ddot{y}$, $\ddot{z}$ are given for a series of time instants) and the direction by $\theta(t)$, $\varphi(t)$ (i.e. $\ddot{\theta}$, $\ddot{\varphi}$ are also given for a series of time instants).

Three equations are obtained as in (3.2.2):

$$w = \Omega\ddot{q} + \Theta. \tag{3.4.28}$$

Let us now consider the first derivative of vector $\vec{h}$:

$$\dot{\vec{h}} = \vec{\omega} \times \vec{h}. \tag{3.4.29}$$

By differentiation, we obtain

$$\ddot{\vec{h}} = \dot{\vec{\omega}} \times \vec{h} + \vec{\omega} \times \dot{\vec{h}} = -\vec{h} \times \dot{\vec{\omega}} + \vec{\omega} \times (\vec{\omega}\times\vec{h}) \tag{3.4.30}$$

$$\ddot{\vec{h}} = -\vec{h} \times \vec{\varepsilon} + \vec{\alpha} \tag{3.4.31}$$

where $\vec{\alpha} = \vec{\omega} \times (\vec{\omega}\times\vec{h})$. Vector $\vec{h}$ is calculated as $\vec{h} = A\vec{\tilde{h}}$, and the transition matrix A of the last segment and the angular velocity $\vec{\omega}$ are calculated by means of the algorithm for c.-a. formation of mathematical models (2.3).

Let us introduce matrix notation:

$$\ddot{h} = -\underline{\underline{h}}\varepsilon + \alpha. \tag{3.4.32}$$

By substituting the angular acceleration (3.2.3) into (3.4.32),

$$\ddot{h} = -\underline{\underline{h}}\Gamma\ddot{q} - \underline{\underline{h}}\Phi + \alpha \tag{3.4.33}$$

and introducing

$$\Gamma' = -\underline{\underline{h}}\Gamma, \qquad \Phi' = -\underline{\underline{h}}\Phi + \alpha \tag{3.4.34}$$

it follows that

$$\ddot{h} = \Gamma'\ddot{q} + \Phi'. \tag{3.4.35}$$

The initial state and the second derivative $\ddot{h}$ fully determine the time function of vector h, i.e., of the direction (b). However, as a unit vector is in question, it is sufficient to consider two of the three equations in (3.4.35). This will be written in the form

$$\ddot{h}^p = \Gamma'^p\ddot{q} + \Phi'^p, \tag{3.4.36}$$

where the upper index "p" designates that only the upper two rows of the matrix are used.

Let us now introduce

$$X^a = \begin{bmatrix} W \\ \\ \ddot{h}^p \end{bmatrix}. \tag{3.4.37}$$

Now the three equations (3.4.28) and the two eqs. (3.4.36) can be written together:

$$X^a = \begin{bmatrix} \Omega \\ \\ \Gamma'^p \end{bmatrix} \ddot{q} + \begin{bmatrix} \Theta \\ \\ \Phi'^p \end{bmatrix} = B\ddot{q} + A. \tag{3.4.38}$$

From such a system the generalized accelerations $\ddot{q}$ are calculated and the simulation procedure is carried out as usual. The calculations of B, A necessitate calculating Ω, Θ, Γ, Φ, as in 2.3. and then calculating Γ', Φ', according to (3.4.34).

Of course it still remains to calculate w and $\ddot{h}^p$, given the prescribed values. Acceleration is direct:

$$w = [\ddot{x} \quad \ddot{y} \quad \ddot{z}]^T. \tag{3.4.39}$$

From Fig. 3.11. it can be seen that for the unit vector h of direction (b),

$$h_1 = h_x = \cos\varphi\cos\theta, \qquad h_2 = h_y = \cos\varphi\sin\theta \tag{3.4.40}$$

or, by differentiation,

$$\ddot{h}^p = \begin{bmatrix} \ddot{h}_1 \\ \\ \ddot{h}_2 \end{bmatrix}, \tag{3.4.41}$$

where

$$\begin{aligned} \ddot{h}_1 &= -\cos\varphi\dot{\varphi}^2\cos\theta - \sin\varphi\ddot{\varphi}\cos\theta + 2\sin\varphi\dot{\varphi}\sin\theta\dot{\theta} + \\ &\quad -\cos\varphi\cos\theta\dot{\theta}^2 - \cos\varphi\sin\theta\ddot{\theta}, \\ \ddot{h}_2 &= -\cos\varphi\dot{\varphi}^2\sin\theta - \sin\varphi\ddot{\varphi}\sin\theta - 2\sin\varphi\dot{\varphi}\cos\theta\dot{\theta} + \\ &\quad -\cos\varphi\sin\theta\dot{\theta}^2 + \cos\varphi\cos\theta\ddot{\theta}. \end{aligned} \tag{3.4.42}$$

We repeat, the input is $\tilde{h}$ as well as $\ddot{x}$, $\ddot{y}$, $\ddot{z}$ and $\ddot{\theta}$, $\ddot{\varphi}$ given in a series of time instants. θ, $\dot{\theta}$, φ, $\dot{\varphi}$ are obtained by integration within the simulation algorithm.

If ultimately the positioning has been prescribed differently the problem is solved as in C1.

Let us mention a practical case belonging to this class of problem. The manipulation task with five d.o.f. is well suited in practice for being prescribed by means of the position of the minimal configuration tip (three d.o.f.), i.e., positioning the gripper base, and by the direction in space which should be taken by the vector connecting the gripper base and its center of gravity [20].

C3. This case considers positioning in Cartesian coordinates x(t), y(t), z(t), and then partial orientation in terms of one direction determined by means of $\theta(t)$, $\varphi(t)$ as in B1. One d.o.f. is prescribed directly, namely $q_k(t)$.

The adjustment block is derived using expression (3.4.38). For the case in question, it represents a system of five equations in which there are six generalized accelerations, $\ddot{q} = [\ddot{q}_1 \cdots \ddot{q}_6]^T$. Since one acceleration ($\ddot{q}_k$) is prescribed, the other five unknown generalized accelerations are calculated from the system (3.4.38). With the $\ddot{q}$ thus calculated, continuation of the simulation algorithm is made possible.

We note here too that in practice it is appropriate to prescribe the positioning of the minimal configuration tip and the direction of the vector from the gripper base to its center of gravity. For such a task five degrees of freedom are used. The sixth d.o.f. is usually rotational and so designed that its axis of rotation coincides with axis of the working object. So by directly prescribing the cooresponding generalized coordinate $q_6(t)$, the rotation of the working object around its axis is also prescribed (Fig. 3.12).

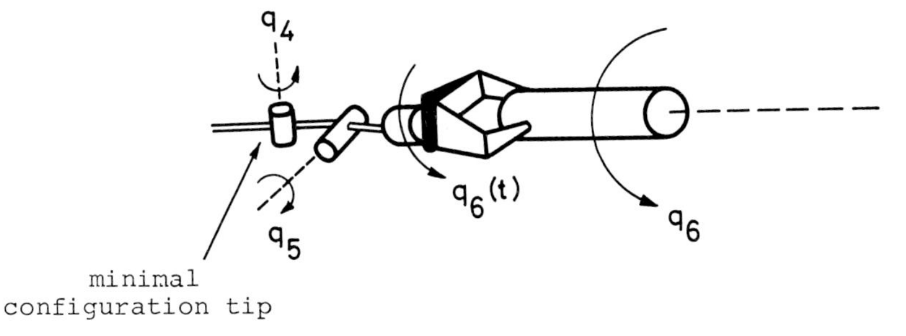

Fig. 3.12. Rotation of working object around its axis

Evidently, in the case of positioning the minimal configuration, the first three d.o.f. (q_1, q_2, q_3) perform positioning. Then the three gripper d.o.f. (q_4, q_5, q_6) decide its orientation. This will be discussed in more detail in Para. 3.11.

3.5. Calculation of Other Dynamical Characteristics and Values

$P^m - n^m$ Diagrams. Since the drives and generalized velocities $\dot{q}$ were calculated at each step of the simulation, it is possible to draw a diagram for each joint: the characteristic of the driving motor torque P^m versus the motor r.p.m. n^m, i.e. the diagram $P^m - n^m$ for each motor. Such a diagram is very useful during the synthesis and choice of the servosystems. The producer gives the $P^m_{max} - n^m$ motor characteristic in the catalogue, where P^m_{max} is the maximal motor torque at motor r.p.m.= n^m. By comparing the necessary characteristic, obtained by means of simulation, with the one from the catalogue, one can decide whether the chosen motor suits its application.

The diagram $P^m - n^m$ is obtained in such way that for each time instant and for joint "k",

$$n^m = N\dot{q}_k, \quad P^m = \eta(N)P_k/N, \tag{3.5.1}$$

where N is the reduction ratio of the subject joint and $\eta(N)$ is its mechanical efficiency. Repeating this procedure for each time instant, the desired diagram is obtained for the joint "k".

Calculation of the energy consumed. As we emphasized, the drives P in joints and the generalized coordinates q are calculated at each time instant.

The total energy is calculated in such a manner that during the simulation, summation of the energy at each step Δt_i is found:

$$E^{(i)} = E^{(i-1)} + E_{\Delta t_i}, \tag{3.5.2}$$

where $E^{(i)}$ is the total energy consumed including during the i-th time step, and $E_{\Delta t_i}$ is the energy consumed in the i-th time interval. To calculate $E_{\Delta t_i}$, we shall adopt the average drive value on the interval,

$$P^i_{med} = \frac{1}{2}(P^{i-1} + P^i), \tag{3.5.3}$$

where the upper index indicates the i-th time instant. Now,

$$E_{\Delta t_i} = \Delta q^{i^T} \cdot P^i_{med} \tag{3.5.4}$$

where $\Delta q^i = q^i - q^{i-1}$. The elements of vectors Δq^i and P^i_{med} are absolute values. Medium drive value is used to avoid more complex interpolation.

Performing summations (3.5.2) over the whole simulation, we obtain as the output the total energy consumed for the given manipulation task.

Calculation of the stresses in segments and reactions in the joints. Using the general theorems of mechanics, a procedure has been developed as in 2.3. for the calculation of the reaction forces and moments in the mechanism joints as vector and scalar values.

Let us consider the rotational joint S_k (Fig. 3.13a). In that joint, there is the driving torque $\vec{P}^M_k(||\vec{e}_k)$, reaction moment $\vec{M}_{R_k}(\perp\vec{e}_k)$, and the reaction force $\vec{F}_{R_k}$ (Fig. 3.13a).

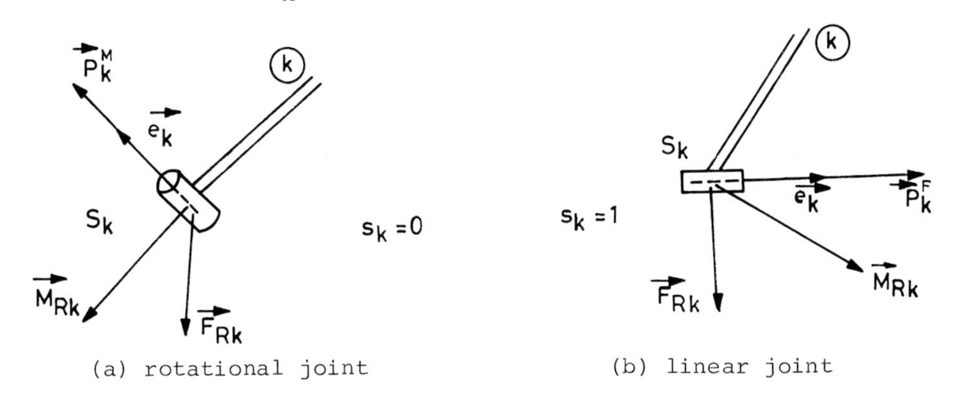

Fig. 3.13. Reactions and drives in joints

If the joint S_k is linear, there is the driving force $\vec{P}^F_k(||\vec{e}_k)$, the reaction force $\vec{F}_{R_k}(\perp\vec{e}_k)$ and the reaction moment $\vec{M}_{R_k}$ acting on the k-th segment (Fig. 3.13b).

Let us fictively disrupt the chain in joint S_k and substitute the rejected part (from the k-th joint to the base) by the reactions $\vec{F}_{R_k}$ and $\vec{M}_{R_k}$ and the drive $\vec{P}_k$. To the rest of the mechanism (from S_k to the free end) apply the theorem of the momentum moment with respect to S_k.

Thus,

$$\vec{M}_{R_k} + (1-s_k)\vec{P}_k = \sum_{i=k}^{n} \{\vec{r}_i^{(k)} \times m_i(\vec{w}_i - \vec{g}) + \vec{M}_i\}, \tag{3.5.5}$$

$$s_k = \begin{cases} 0, & \text{if the k-th joint is rotational,} \\ 1, & \text{if the k-th joint is linear,} \end{cases}$$

where $\vec{r}_i^{(k)} = \overrightarrow{S_k C_i}$ (C_i is center of gravity of the i-th segment) and the component $\vec{M}_i$ is calculated using Euler´s equations:

$$\vec{M}_i = A_i \vec{\tilde{M}}_i = A_i [\tilde{J}_i \vec{\tilde{\varepsilon}}_i - (\tilde{J}_i \vec{\tilde{\omega}}_i) \times \vec{\tilde{\omega}}_i], \tag{3.5.6}$$

where $\tilde{J}_i$ is the tensor of inertia of the i-th segment with respect to the b.-f. coordinate system.

The total moment acting on the k-th segment in the joint S_k is now

$$\vec{M}_{S_k} = \vec{M}_{R_k} + (1-s_k)\vec{P}_k = \sum_{i=k}^{n} \{\vec{r}_i^{(k)} \times m_i(\vec{w}_i - \vec{g}) + \vec{M}_i\} \tag{3.5.7}$$

and evidently does not depend on the joint type. By calculating $\vec{M}_{S_k}$ from (3.5.7) and (3.5.6), one can calculate the reaction moment in the joint. From (3.5.7) it follows that

$$\vec{M}_{R_k} = \vec{M}_{S_k} - (1-s_k)\vec{P}_k. \tag{3.5.8}$$

Let us now apply the theorem about the center of gravity motion to the part considered (from S_k to the free end), thus obtaining the total force $\vec{F}_{S_k}$ acting on the k-th segment of the joint S_k:

$$\vec{F}_{S_k} = \vec{F}_{R_k} + s_k\vec{P}_k = \sum_{i=k}^{n} m_i(\vec{w}_i - \vec{g}), \tag{3.5.9}$$

which evidently does not depend on the joint type. Now the reaction force can be calculated by

$$\vec{F}_{R_k} = \vec{F}_{S_k} - s_k\vec{P}_k. \tag{3.5.10}$$

$\vec{M}_{S_k}$ and $\vec{F}_{S_k}$, calculated in such way, are acting on the k-th segment. So when we consider the (k-1)-th segment, then $-\vec{M}_{S_k}$ and $-\vec{F}_{S_k}$ are acting on that segment because of the law of action and reaction (Fig. 3.14).

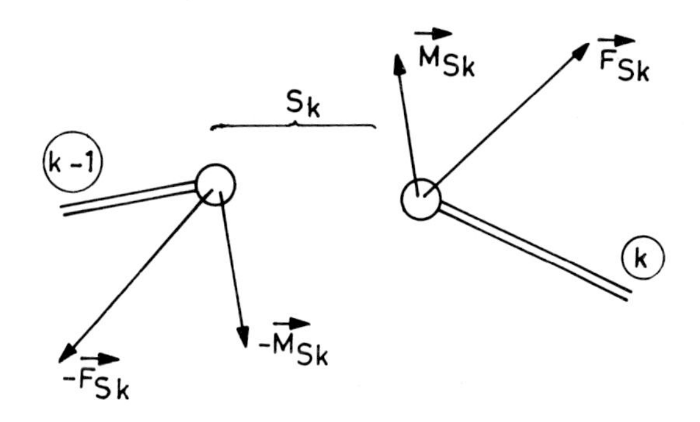

Fig. 3.14. Reactions at the fictive rupture of the chain

At the basis of equations (3.5.7) to (3.5.10), a recursive programme is derived for the calculation of reaction moments and forces in the mechanism joints.

Further the method of calculating stresses in the mechanism segments will be presented. Let us consider a segment on which the forces and moments are acting according to Fig. 3.15. and let us use some of the notation from that figure.

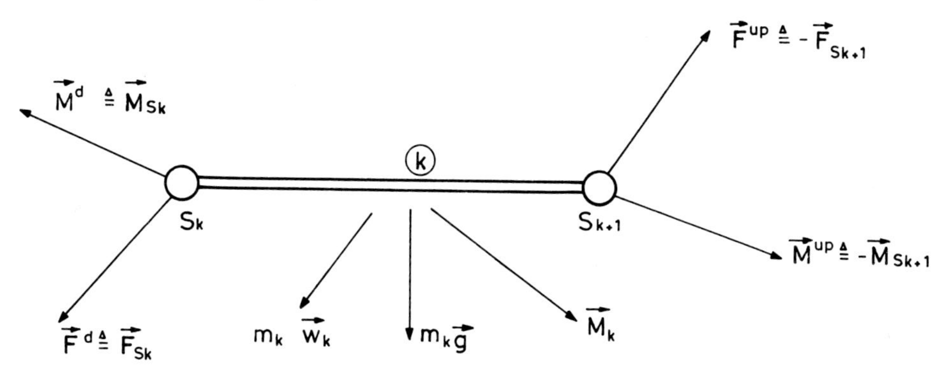

Fig. 3.15. Forces and moments acting on k-th segment

Stresses in the segments depend on the segment form so we shall show how to calculate the maximal stress in the case of a manipulator with tubular cross-section segments (Fig. 3.16).

For such a cross-section,

$$\psi = \frac{r}{R} = \frac{d}{D}, \tag{3.5.11}$$

the cross section moment of inertia,

$$I_x = I_y = \frac{D^4\pi}{64}(1-\psi^4) \tag{3.5.12}$$

and the axial section modulus,

$$W_x = W_y = \frac{D^3\pi}{32}(1-\psi^4). \tag{3.5.13}$$

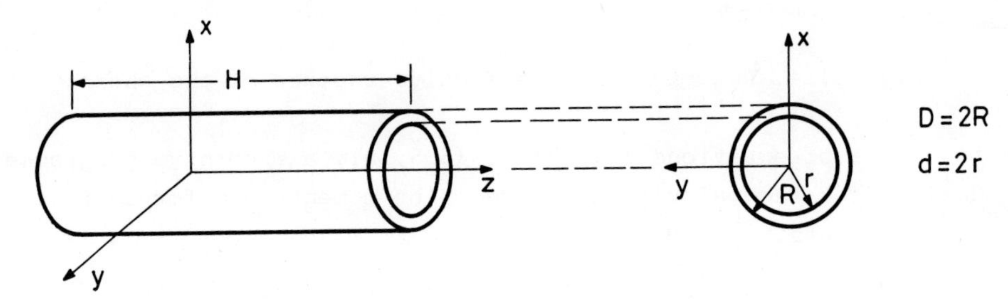

Fig. 3.16. One mechanism segment

Now the maximal bending stress is

$$\sigma_{max} = \frac{M^b_{max}}{W_x}, \tag{3.5.14}$$

where M^b_{max} is the greater one of two moments $|M^{db}|$, $|M^{upb}|$, Fig. 3.17a.

Further, the polar section modulus is

$$W_o = \frac{D^3\pi}{16}(1-\psi^4) \tag{3.5.15}$$

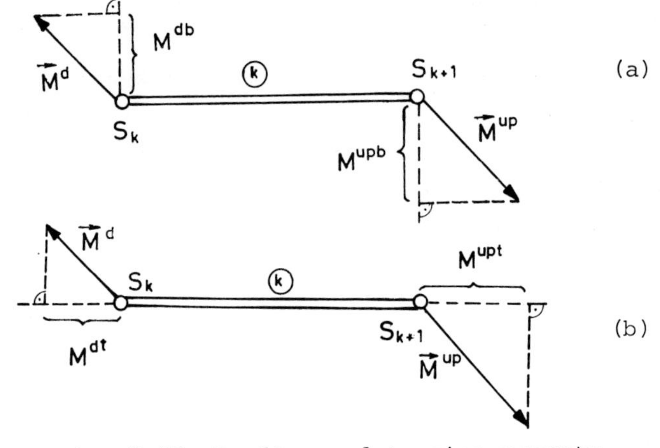

Fig. 3.17. Bending and torsion moments

and the maximal torsion stress is

$$\tau_{max} = \frac{M^t_{max}}{W_o} , \qquad (3.5.16)$$

where M^t_{max} is the greater of the two moments $|M^{dt}|$, $|M^{upt}|$, Fig. 3.17b.

When needed, the compression stress and slenderness ratio of the segment can be easily calculated. However, dominant stresses are due to bending and torsion.

Force at the interface of the gripper and the working object. In the preceding text, the influence of the working object on the manipulator dynamics has been taken into account by considering, in the phase of working object transfer, the last mechanism segment as gripper and working object together. That is, the tensor of inertia of such a last segment, and its main axes of inertia and mass, are to be given. This is not easily determined even when all the data for the gripper and working object are known separately. Hence, we shall give an approximation, which suits some gripper versions very well and yields the possibility of calculating the force and moment of the connection of the gripper and working object. Let us suppose that the contact of the gripper and working object is made in one point only (or a sufficiently small surface, compared with the object dimensions) but that the connection is still rigid (Fig. 3.18). The gripper itself will be considered as the last mechanism segment, and the object will be considered separately.

Such an approximation is realistic, for instance, in the case of manipulators equiped with vacuum grippers, such as the industrial manipulator UMS-3 [21], see (Fig. 3.19). Evidently, with the UMS-3 manipulator the contact with the object is not made in one but in several points but, considering the dimensions of the working objects, it can be safely reckoned that one point-gripper only is in question, as a substitute for the whole set.

Due to the rigid connection between the gripper and working object, both bodies move with the same accelerations w, ε, prescribed or calculated in simulation. Let us now calculate the connection force $\vec{F}_v$ and moment $\vec{M}_v$. The theorem of the center of gravity motion applied to the working object yields

$$\vec{F}_v = m_p(\vec{w}-\vec{g}) , \qquad (3.5.17)$$

where m_p is the mass of the working object. Let us now apply the theorem of the momentum moment with respect to the connecting point V:

$$\vec{M}_V = \vec{r}_p \times m_p(\vec{w}-\vec{g}) + \vec{M}_p, \tag{3.5.18}$$

where $\vec{r}_p = \overrightarrow{VC_p}$ (C_p is the object center of gravity) and $\vec{M}_p$ is the moment of momentum with respect to the center of gravity and is calculated in the b.-f. system as

$$\vec{\tilde{M}}_p = \tilde{J}_p \vec{\tilde{\varepsilon}} - (\tilde{J}_p \vec{\tilde{\omega}}) \times \vec{\tilde{\omega}}, \tag{3.5.19}$$

where $\tilde{J}_p$ is the object tensor of inertia in the b.-f. system. $\vec{r}_p$ and $\vec{M}_p$ are calculated as $\vec{r}_p = A\vec{\tilde{r}}_p$, $\vec{M}_p = A\vec{\tilde{M}}_p$, where A is the transition matrix of the object and $\vec{\tilde{r}}_p$ a given constant. In the case when the b.-f. coordinate systems of the gripper and the working object have parallel axes, the transition matrix A equals the transition matrix A_6 of the gripper calculated in the course of the c.-a. method of model formation. Then, of course, the main inertia axes of the gripper and the object must be parallel, which is often the case in practice. In the opposite case, the transition matrix between gripper and object is calculated too.

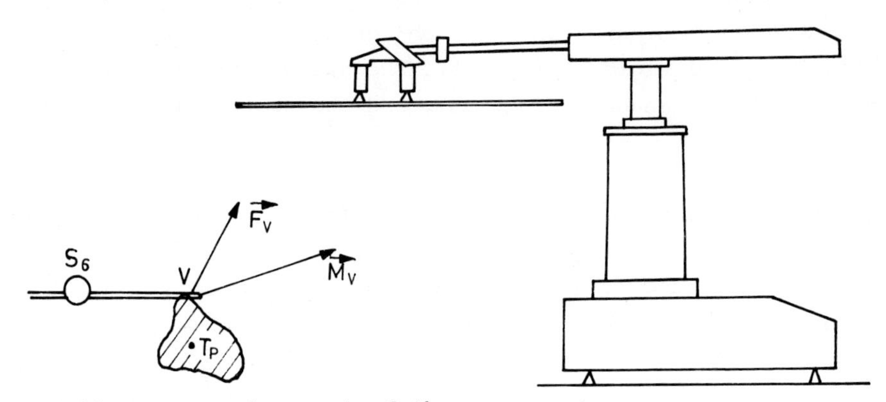

Fig. 3.18. Force and moment of the connection of the gripper and working object

Fig. 3.19. UMS-3 manipulator

The connection force and moment ($\vec{F}_V$, $\vec{M}_V$), calculated thus numerically, are considered in the method of c.-a. formation of the mathematical model to be the known external forces and they appear the model in that way.

The practical benefits of thus considering the gripper and object lies in the fact that the calculated force and moment represent useful data for dimensioning the vacuum cups.

We can likewise consider the inverse problem of dynamics, i.e. determine motion for known drives, but in that case the basic system (3.1.7) acquires a more complicated form.

At the end of this presentation of the simulation of manipulators dynamics, it would be suitable to conclude that we have described an algorithm which has the following input: the manipulator configuration, initial state and the manipulation task in the form which is most suited for prescribing (according to the class of functional movements). As the output we obtain time history of the generalized coordinates, velocities and drives, the reactions in the joints, stresses in the segments, connection force and moment between the gripper and working object (in some cases), the torque - r.p.m. diagram for each actuator, and the total energy consumed. In such an algorithm we can also have some special-purpose blocks, such as the blocks for calculating the elastic oscillations, nominal control or other desirable characteristics.

An algorithm which calculates all dynamical characteristics during the manipulation task clearly represents a useful means for the process of manipulator design, notably because a simple change in the manipulator configuration and task is possible.

3.6. Examples

First we considered the anthropomorphic manipulator with 6 d.o.f. and the kinematic scheme given in Fig. 3.20. The last segment, i.e. the manipulator gripper, is connected to the minimal configuration by means of a 3-d.o.f. joint. In the course of simulation, the joint is partitioned into series-connected simple joints, according to the real axes of rotation in the complex joint. Segments are adopted in the form of cylindrical tubes of $R = 0.02$ m outer and $r = 0.015$ m inner radius. As material of the segments, the light alloy AℓMg3 was adopted. Segment lengths are given in Fig. 3.20.

After that, a cylindrical manipulator with 6 d.o.f. (Fig. 3.21) was consider. Segments were also in the form of cylindrical tubes of radii: $R = 0.022$ and $r = 0.0165$ m. The material was again AℓMg3. Lengths are given in Fig. 3.21.

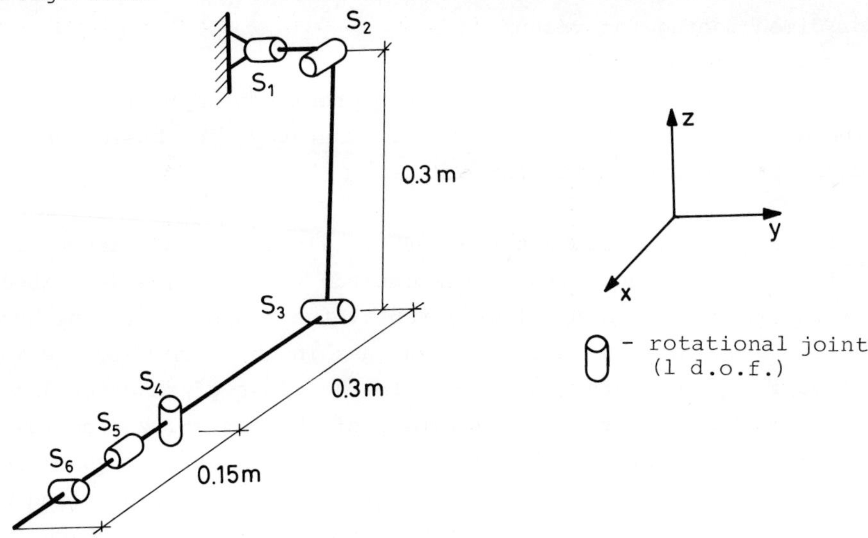

Fig. 3.20. Anthropomorphic manipulator

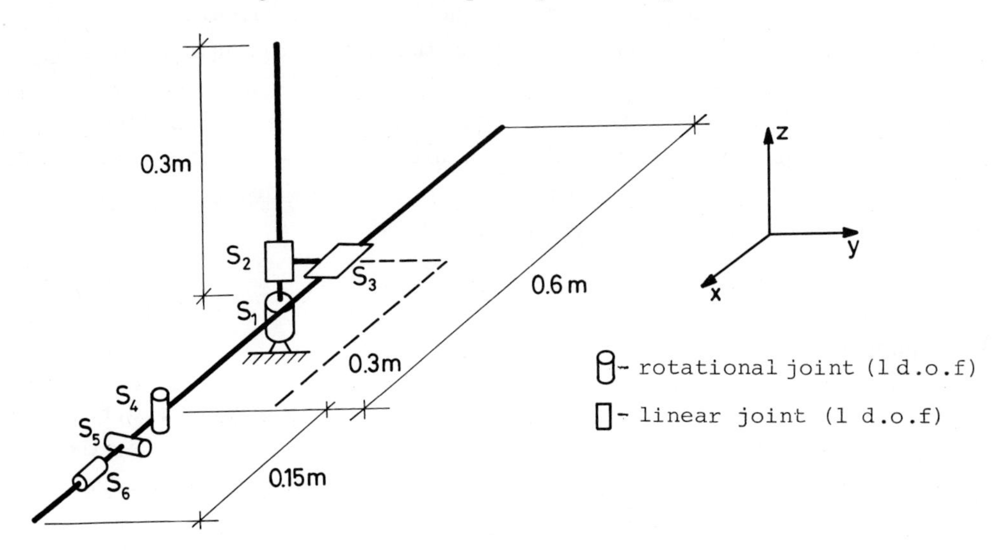

Fig. 3.21. Cylindrical manipulator

The manipulation task is to transfer a 5 kg mass object along the trajectory ABCA in Fig. 3.22 so that all during the motion the gripper maintains a constant orientation in space. On the rectilinear parts of the trajectory the velocity profile is triangular. For instance, until half way along AB, the manipulator accelerates constantly. It then decelerates constantly so that it reaches the point B with zero velocity. The total execution time was chosen to be T = 3 s. Each of the parts: AB, BC, CD should be passed in equal time T/3. Manipulator initial positions are given in Fig. 3.20, 3.21. and manipulators start from a stand still.

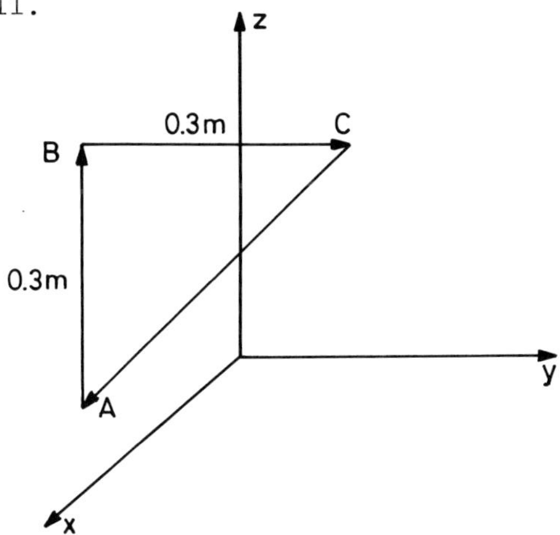

Fig. 3.22. Trajectory of object transfer

For the simulation of dynamics in this manipulation task, the general simulation algorithm (3.2) was used. Further, the simulation results are presented. Fig. 3.23 shows the time history of the drives in joints and Fig. 3.24. shows the same for the generalized coordinates. Both apply to the anthropomorphic manipulator. Figs. 3.25. and 3.26. show the analogues for the cylindrical manipulator.

In performing the task the anthropomorphic manipulator consummed the 76.03 J of energy. The cylindrical manipulator consumed 44.23 J.

Then, we considered the arthropoid manipulator whose kinematical scheme is shown in Fig. 3.27. The material used is the same as in the preceding example, the light alloy AℓMg3. The segments are also in the form of cylindrical tybes, of outer radius R = 0.02 meters and inner radius r = 0.015 m. The segment lengths are given in Fig. 3.27.

The manipulation task consists of the following. In its initial position (Fig. 3.27), the manipulator holds in its gripper a screw of 5 kg mass (Fig. 3.28). In the first time interval T_1=1 s, the manipulator tip (screw head) should pass the part AB (Fig. 3.29) and during this phase the screw turns around its vertical axis through an angle of $\pi/2$ rad = 90° (Fig. 3.29). The translational and rotational velocity profile is trangular. In time period T_2 = 1s, the screw is transfered along the path BC (Fig. 3.29) whereby the gripper and the screw do not

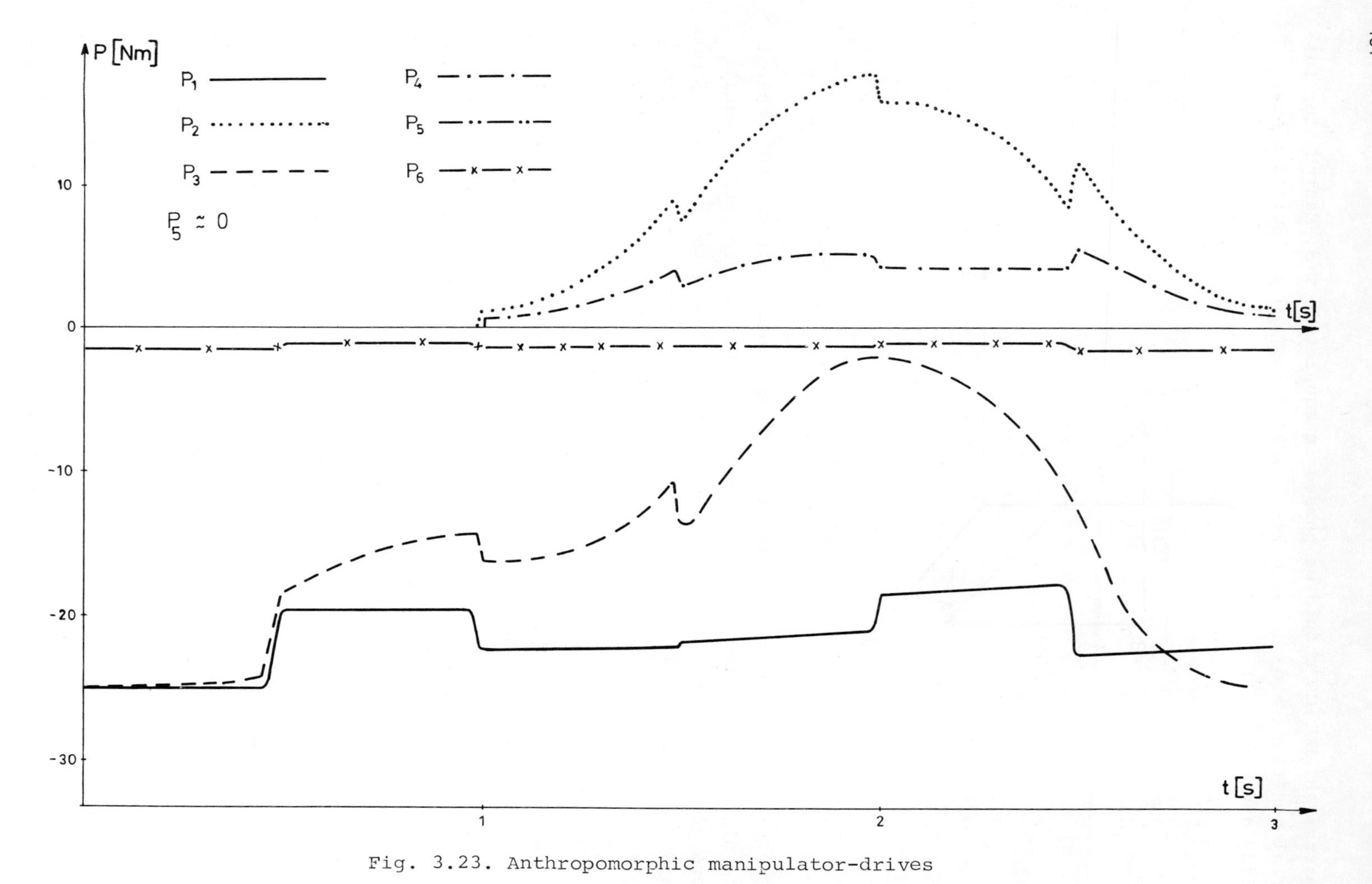

Fig. 3.23. Anthropomorphic manipulator-drives

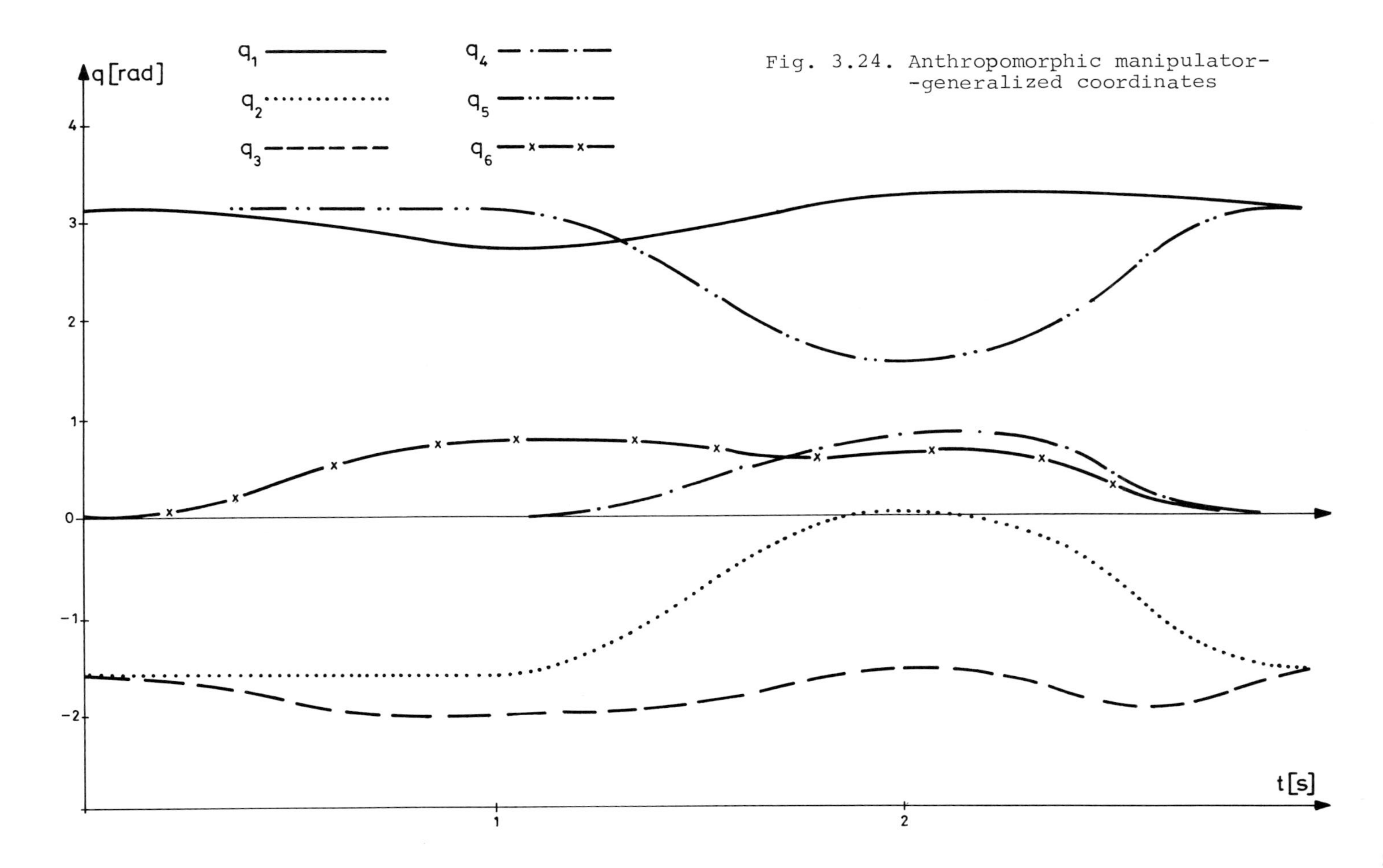

Fig. 3.24. Anthropomorphic manipulator-
-generalized coordinates

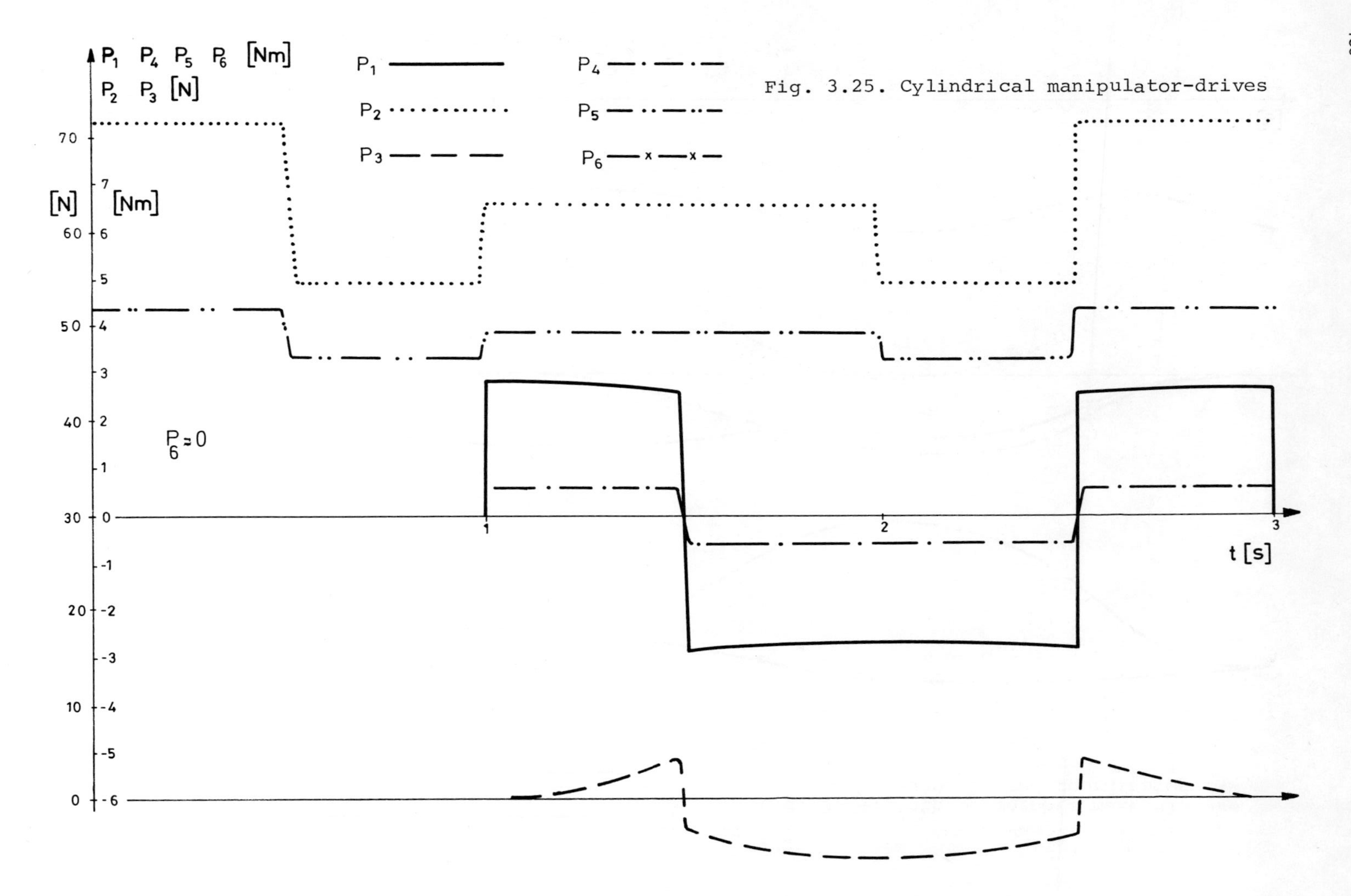

Fig. 3.25. Cylindrical manipulator-drives

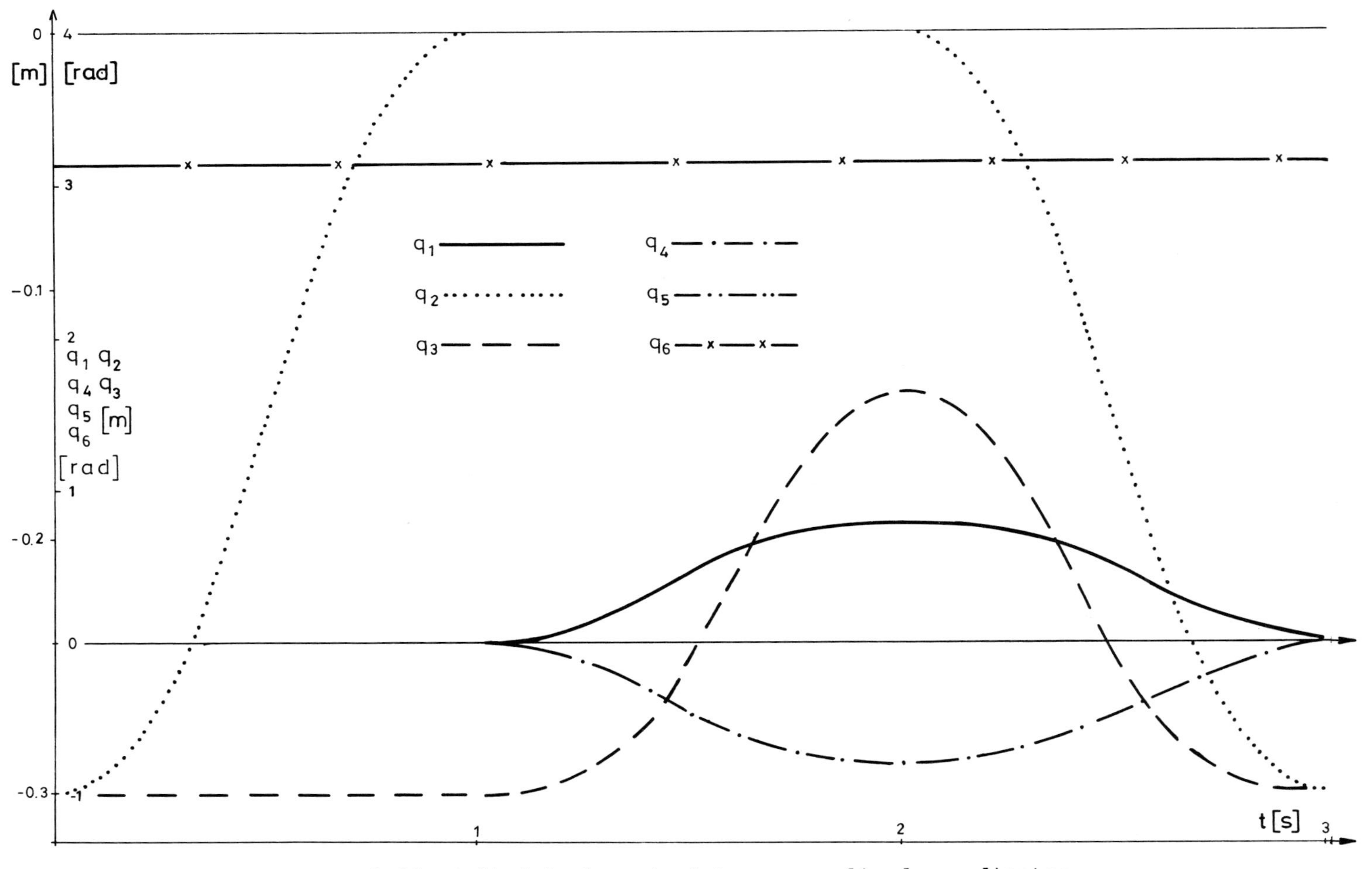

Fig. 3.26. Cylindrical manipulator-generalized coordinates

shange their orientation. The velocity profile is trangular. Finally, in the interval T_3 = 1 s, the screw is screwed-in in position C (Fig. 3.29), i.e. it rotates around its longitudinal axis for an angle of 4πrad = 720° = 2 full turns. Angular velocity profile is also traingular.

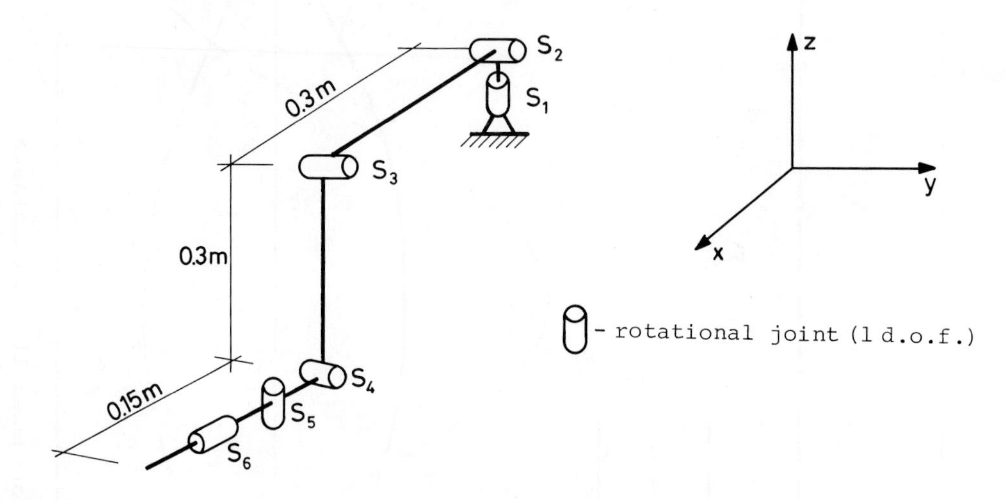

Fig. 3.27. Arthropoid manipulator

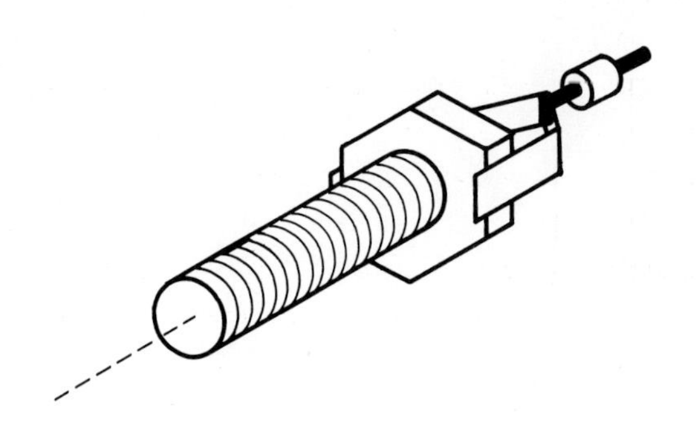

Fig. 3.28. Way of grasping the screw

For the simulation of dynamics of this task, the algorithm, supplemented by the adapting blocks C2 and C3 (3.4), was used. Fig. 3.30 shows the time history of the drives in joints and Fig. 3.31 shows the generalized coordinates. The values obtained for the drives possess a qualitative character only because the screw mass was considered to be concentric and the external friction forces during screwing-in were not taken into account.

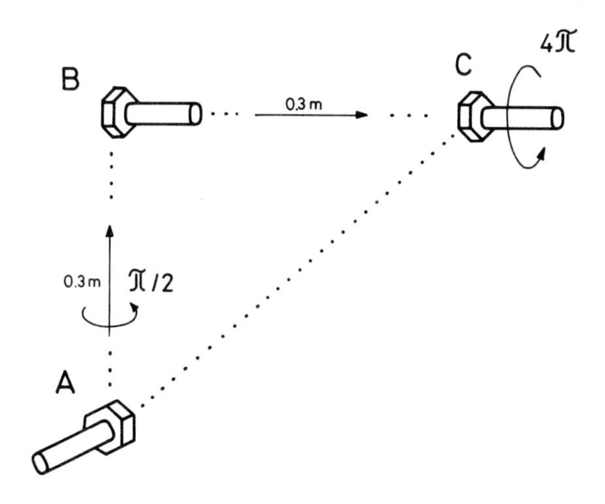

Fig. 3.29. Scheme of the manipulation task

3.7. Synthesis of Nominal Dynamics of Manipulation Movements

In the general case, the control task can be defined as a task of transfering the system state from any initial state into a defined point in the state space during a finite time interval. The initial state can usually only belong to a bounded zone in the system state space X^I. It is not necessary to transfer the system state into a point but a bounded zone in the state space around the desired point X^F. The system is observed during a determined time interval τ and it is required that transfer of the state from the zone of initial conditions X^I into the zone X^F is performed during a defined time interval τ_s, $\tau_s \leq \tau$. It is also required that during transfer from zone X^I to zone X^F the system state belongs to a certain bounded zone X^t. At the stage of nominal dynamics the task of control synthesis is the following: programmed control $u^o(t)$ should be synthesized for $t \varepsilon T$, which should transfer the system S state from a defined initial state $x(0) \varepsilon X^I$ into a desired state $x(\tau_s) \varepsilon X^F$ during the time interval $\tau_s \leq \tau$, where the nominal trajectories of the system state coordinates should satisfy the conditions $x(t, x(0)) \varepsilon X^t$ for $t \varepsilon T$.

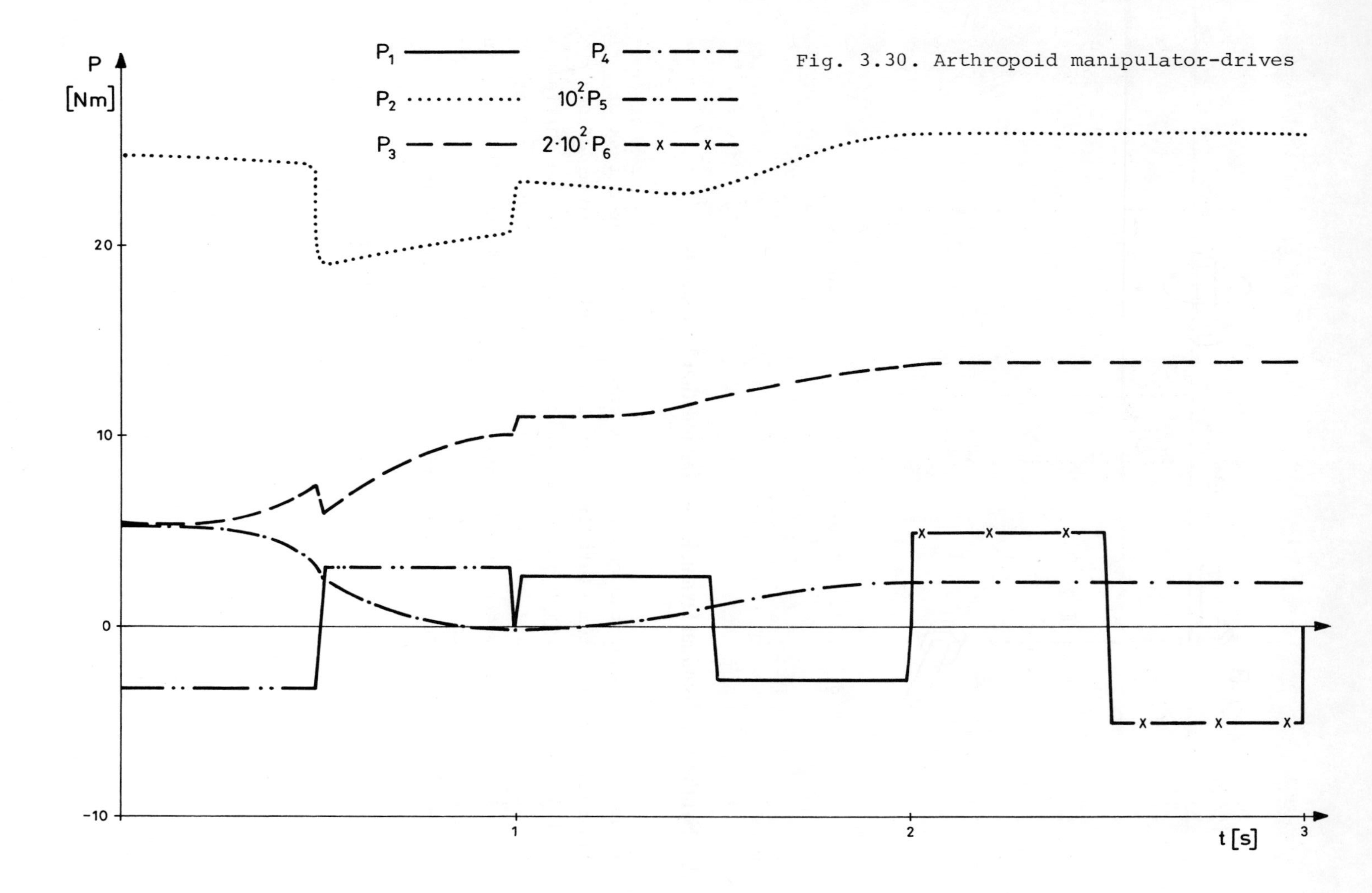

Fig. 3.30. Arthropoid manipulator-drives

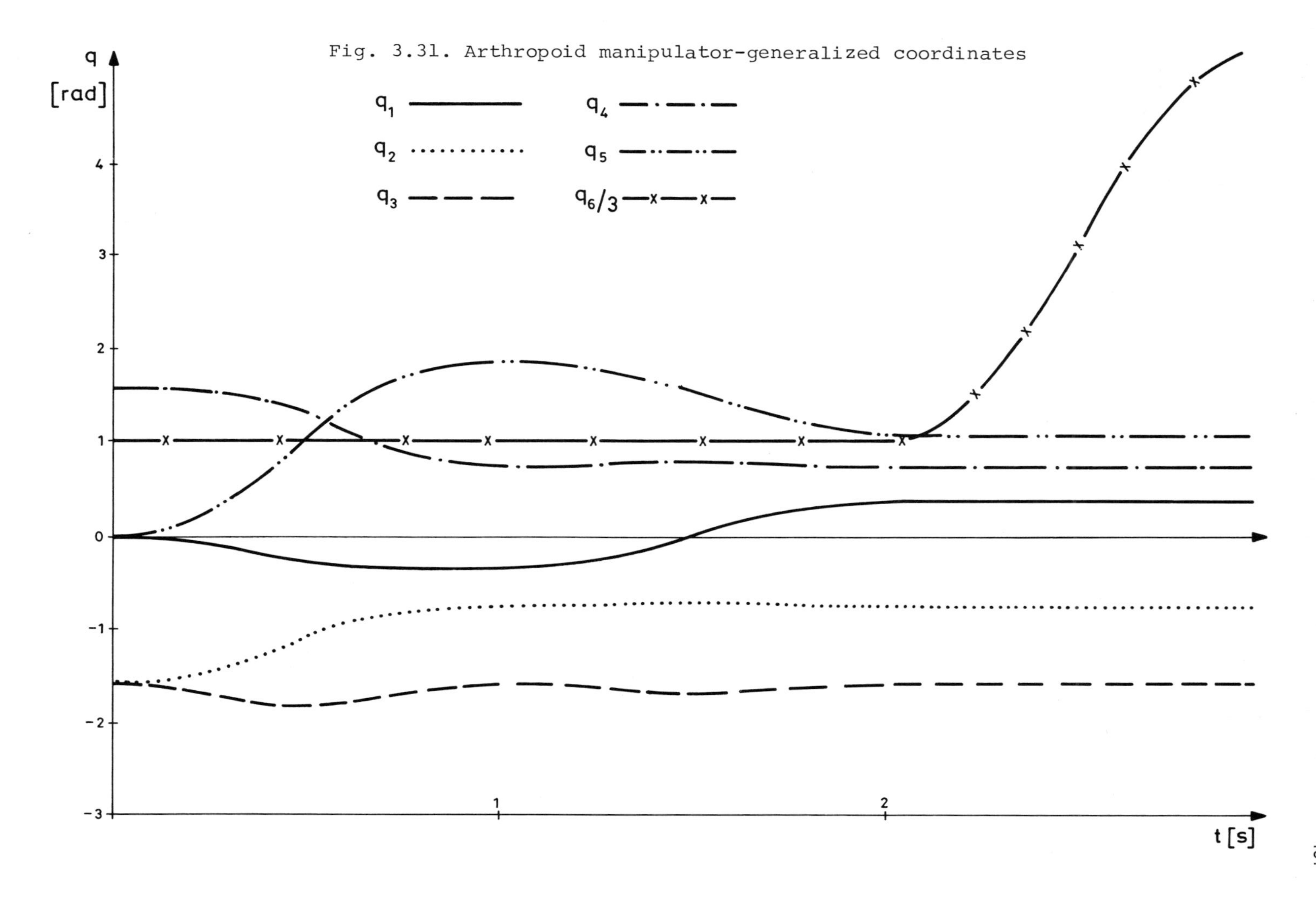

Fig. 3.31. Arthropoid manipulator-generalized coordinates

Having in mind the fact that for most robots and especially manipulators in industrial practice the working conditions are known in advance and that the functional tasks, performed by the system, are repeated in cycles, so that they can be foreseen, it has been proposed that the control be synthesized in two stages [22, 23]. At the first stage of control synthesis, the nominal programmed control is synthesized, which produces the prescribed system motion from a certain chosen initial state under the supposition that no perturbations are acting on the system.

At the second stage of control synthesis, control of the tracking of the nominal trajectories is synthesized when the initial state deviates from the nominal initial state (but belongs to a bounded zone of initial states) and perturbations of the initial conditions type act on the system. At this stage, decentralized control is applied.

As this monograph only treats the active mechanisms dynamics, this chapter will only consider the problem of the synthesis of nominal dynamics, which is the first stage in the synthesis of the control system.

The complete dynamical model of the manipulation robot will therefore be constructed, including both the mechanism and the actuator system. The simulation algorithm described earlier is then supplemented with a control block in order to permit the calculation of nominal control inputs for the given motion. The example of synthesis at the basis of functional subsystems for positioning and orientation is given. The influence of the complexity of the actuators models is also analyzed. Finally, a method for optimal synthesis of functional movements is presented, offering the possibility of dynamically programming the exact nonlinear model of the manipulation mechanism.

3.8. The Complete Dynamical Model

In this paragraph, the complete mathematical model of the manipulator dynamics will be derived. This model includes the model of the manipulator mechanical part as well as the model of the driving actuators.

The manipulator is considered as an active mechanism of the open chain type, with n d.o.f. (Fig. 3.2). It is assumed that the manipulator joints have one d.o.f. each, which can be rotational or linear. There is a driving actuator acting at each joint.

Manipulator mechanical part. This term refers to the manipulator mechanism as a mechanical system. If we introduce the generalized coordinates $q = [q_1 \dots q_n]^T$ as in previous chapters, then the mechanical part dynamics can be described by means of a system of n second-order differential equations in matrix form:

$$S_M : W(q)\ddot{q} = P + U(q, \dot{q}), \quad q(t_o) = q^o, \quad \dot{q}(t_o) = \dot{q}^o, \tag{3.8.1}$$

where $P = [P_1 \dots P_n]^T$ is the n-dimensional vector of the driving forces and torques in the joints. P_i is a force if S_i is a linear joint, or a torque if S_i is a rotational joint.

By introducing a 2n-dimensional state vector

$$\xi = \begin{bmatrix} \xi_1 \\ \xi_2 \end{bmatrix}; \qquad \xi_1 = q, \quad \xi_2 = \dot{q}, \tag{3.8.2}$$

the system (3.8.1) can be reduced to canonical form

$$\dot{\xi} = \begin{bmatrix} \xi_2 \\ W(\xi_1)^{-1}U(\xi_1, \xi_2) \end{bmatrix} + \begin{bmatrix} 0 \\ W^{-1}(\xi_1) \end{bmatrix} P, \quad \xi(t_o) = \xi^o. \tag{3.8.3}$$

Introducing

$$K_M(\xi) = \begin{bmatrix} \xi_2 \\ W(\xi_1)^{-1}U(\xi_1, \xi_2) \end{bmatrix}, \qquad D_M(\xi) = \begin{bmatrix} 0 \\ W^{-1}(\xi_1) \end{bmatrix} \tag{3.8.4}$$

the model (3.8.3) of the mechanical part becomes

$$S_M : \dot{\xi} = K_M(\xi) + D_M(\xi)P, \quad \xi(t_o) = \xi^o \tag{3.8.5}$$

Mathematical model of the driving actuators. Driving actuators will be considered such that the model of the i-th actuator can be written in the form

$$S_A^i : \dot{x}_i = C_i x_i + f_i P_i + d_i N(u_i), \quad x_i(t_o) = x_i^o, \tag{3.8.6}$$

where x_i is an n_i-dimensional state vector of the i-th subsystem S_A^i, C_i, f_i and d_i are constant matrices of the model, P_i is the driving torque (or force) of the i-th actuator (scalar value), u_i is the control input of the i-th actuator (scalar value) and $N(u_i)$ is a nonlinearity of the saturation type.

Further, let k_i elements of vector x_i coincide with the elements of vector ξ, i.e., let the k_i state coordinate of the i-th actuator S_A^i be already contained in the state vector of the mechanical part S_M. For instance, the generalized coordinate q_i and the generalized velocity $\dot{q}_i$ are usually included in the state vector x_i of the i-th joint actuator. So, usually $k_i = 2$ and $\sum_{i=1}^{n} k_i = 2n$ = the dimension of the vector ξ. In general, q_i and $\dot{q}_i$ need not be included in x_i, but a nonlinear dependence exists.

The subsystems S_A^i, i=1,...,n can be united into a system of dimension $N = \sum_{i=1}^{n} n_i$, i.e.,

$$S_A : \dot{x} = Cx + FP + Du, \quad x(t_o) = x^o, \tag{3.8.7}$$

where $x=[x_1^T \ldots x_n^T]^T$, $P=[P_1 \ldots P_n]^T$, $u=[u_1 \ldots u_n]^T$, $C=\mathrm{diag}[C_1 \ldots C_n]$, $F=\mathrm{diag}[f_1 \ldots f_n]$, $D=\mathrm{diag}[d_1 \ldots d_n]$. P is the vector of the drives and u is the control vector. Thus, the model of driving actuators is written in the form (3.8.7). In the model (3.8.7), care should be taken with the nonlinearity of the saturation type.

Complete model. The models S_M: (3.8.1) and S_A: (3.8.7) can now be united into a complete dynamical model. Let it be assumed that q_i and $\dot{q}_i$ are included in x_i, i.e., $k_i = 2$, i=1,...,n, and let us introduce the transformation matrices T_i, i=1,...,n (dimension $1\times n_i$) such that $\ddot{q}_i = T_i\dot{x}_i$. Now from (3.8.1) it follows that

$$P = W\ddot{q} - U = WT\dot{x} - U, \tag{3.8.8}$$

where $T = \mathrm{diag}[T_1 \ldots T_n]$ is an $n\times N$ matrix. Substituting $\dot{x}$ in (3.8.7) into (3.8.8), one obtains

$$P = (E_n - WTF)^{-1}[WT(Cx + Du) - U], \tag{3.8.9}$$

where E_n is an $n\times n$ unit matrix.

Now, substituting P in (3.8.9) into (3.8.7), the complete system is obtained in the form

$$\dot{x} = \hat{C}(x) + \hat{D}(x)u, \tag{3.8.10}$$

where the N×N matrix $\hat{C}$ and the N×n matrix $\hat{D}$ are

$$\hat{C} = Cx + F(E_n - WTF)^{-1}(WTCx - U), \quad \hat{D} = D + (E_n - WTF)^{-1}WTD. \tag{3.8.11}$$

This form of manipulator mathematical model is used in the second book of this series which discusses the control problems of industrial manipulation.

3.9. Mathematical Models of the Actuator Systems

Permanent magnet D.C. motors are widely used as the actuators for industrial manipulators. The scheme of such a motor is shown in Fig. 3.32. Let i_r be the rotor current. Then the 3-dimensional state vector for such a system is

$$x = [q_i \ \dot{q}_i \ i_{r_i}]^T \tag{3.9.1}$$

if the i-th joint actuator is considered. In this case $k_i = 2$. The third-order mathematical model ($n_i = 3$) can be written in the form

$$S_A^i : \dot{x}_i = C_i x_i + f_i P_i + d_i N(u_i), \tag{3.9.2}$$

where P_i is the motor torque and u_i is the control voltage. The amplitude of the control is constrained:

$$N(u_i) = \begin{cases} -u_{imax} & \text{for} \quad u_i \leq -u_{imax} \\ u_i & \text{for} \quad -u_{imax} < u_i < u_{imax} \\ u_{imax} & \text{for} \quad u_i \geq u_{imax} \end{cases} \tag{3.9.3}$$

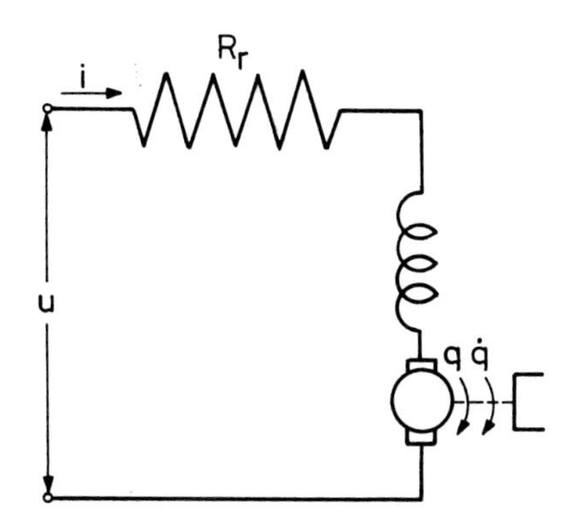

Fig. 3.32. Scheme of a D.C. motor

The system matrices are

$$C_i = \begin{bmatrix} 0 & 1 & 0 \\ 0 & -\frac{B_c}{J_r} & \frac{C_M}{J_r} \\ 0 & -\frac{C_E}{L_r} & -\frac{R_r}{L_r} \end{bmatrix}; \quad f_i = \begin{bmatrix} 0 \\ -\frac{1}{J_r} \\ 0 \end{bmatrix}; \quad d_i = \begin{bmatrix} 0 \\ 0 \\ \frac{1}{L_r} \end{bmatrix}, \qquad (3.9.4)$$

where C_M and C_E are the constants of moment and electromotor force, L_r and R_r are the rotor inductivity and resistence, J_r is the rotor moment of inertia, and B_c is the viscous friction coefficient. Index "i" is omitted.

If the rotor inductivity is neglected, then the actuator model reduces to the second-order form ($n_i = 2$). The state vector is $x_i = [q_i \; \dot{q}_i]^T$ and the system matrices of the second-order model are

$$C_i = \begin{bmatrix} 0 & 1 \\ 0 & \frac{C_E C_M}{J_r R_r} \end{bmatrix}; \quad f_i = \begin{bmatrix} 0 \\ -\frac{1}{J_r} \end{bmatrix}; \quad d_i = \begin{bmatrix} 0 \\ \frac{C_M}{J_r R_r} \end{bmatrix}. \qquad (3.9.5)$$

The rotor current is then

$$i_{r_i} = (u_i - C_{E_i} \dot{q}_i)/R_{r_i}. \qquad (3.9.6)$$

The viscous friction is also neglected. For this model $n_i = k_i = 2$ i.e. all state variables of the actuator are included in the state vector ξ of the manipulator mechanical part.

The electrohydraulic actuator consists of a servovalve and a cylinder. The scheme of such an actuator is given if Fig. 3.33.

The 5-dimensional state vector is

$$x_i = [\ell_i \; \dot{\ell}_i \; p_i \; Q_i' \; \dot{Q}_i'] \qquad (3.9.7)$$

if the i-th joint actuator is considered. ℓ is the stroke of the cylinder piston. $p = p^1 - p^2$, where p^1 and p^2 are pressures on the piston sides. Q' is the flow due to servovalve piston motion (theoretic flow). Index "i" is omited. If the actuator drives a linear joint, then assuming $\ell_i = q_i$, $\dot{\ell}_i = \dot{q}_i$, it follows that $k_i = 2$. If such an actuator drives a rotational joint, then there is usually nonlinear dependence between the actuator state variables ℓ_i, $\dot{\ell}_i$ and the joint variables q_i, $\dot{q}_i$.

Now, the fifth order mathematical model can be written in the form (3.9.2). P_i represents the actuator force and u_i represents the control current. The system matrices are

$$C_i = \begin{bmatrix} 0 & 1 & 0 & 0 & 0 \\ 0 & -\frac{B_c}{m} & \frac{A}{m} & 0 & 0 \\ 0 & -\frac{4\beta A}{V} & -\frac{4\beta(k_c+C_\ell)}{V} & \frac{4\beta}{V} & 0 \\ 0 & 0 & 0 & 0 & 1 \\ 0 & 0 & 0 & -c_1 & -c_2 \end{bmatrix}; \quad f_i = \begin{bmatrix} 0 \\ -\frac{1}{m} \\ 0 \\ 0 \\ 0 \end{bmatrix}; \quad d_i = \begin{bmatrix} 0 \\ 0 \\ 0 \\ 0 \\ k_q \end{bmatrix} \tag{3.9.8}$$

where B_c is the viscous friction coefficient, m is the mass of the cylinder piston and the corresponding load, A is the piston area, β is the compressibility coefficient of the fluid depending on the percentage of air in the oil, V is the total volume including the volume of the valve, the cylinder and the pipes, k_c is the slope of the servovalve flow-pressure characteristic in the working point, $C_\ell = C_{i\ell} + C_{e\ell}/2$, where $C_{i\ell}$ and $C_{e\ell}$ are the coefficients of internal and external leakage, c_1 and c_2 are the coefficients depending on the servovalve frequency characteristic and k_q is a servovalve coefficient.

If the servovalve bandwidth is enough large, we can assume that its dynamics do not influence the behaviour of the whole system. The actuator model can then be reduced to the third order form ($n_i = 3$) with the state vector $x_i = [\ell_i \ \dot{\ell}_i \ p_i]^T$ and the system matrices

$$C_i = \begin{bmatrix} 0 & 1 & 0 \\ 0 & -\frac{B_c}{m} & \frac{A}{m} \\ 0 & \frac{4\beta}{V}A & -\frac{4\beta}{V}(k_c+C_\ell) \end{bmatrix}; \quad f_i = \begin{bmatrix} 0 \\ -\frac{1}{m} \\ 0 \end{bmatrix}; \quad d_i = \begin{bmatrix} 0 \\ 0 \\ \frac{4\beta}{V}\frac{k_q}{C_\ell} \end{bmatrix} \tag{3.9.9}$$

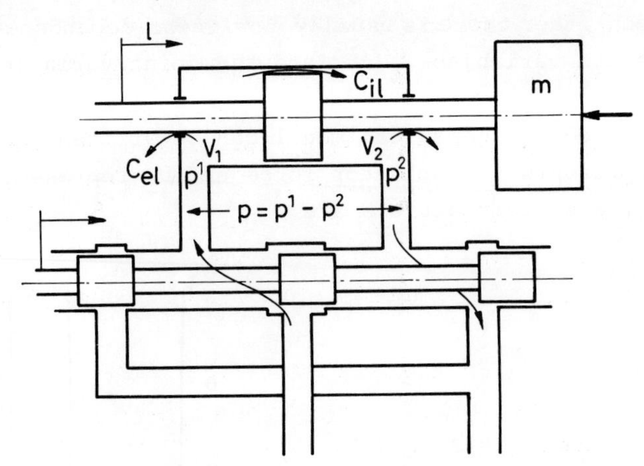

Fig. 3.33. Scheme of an electrohydraulic actuator

3.10. Simulation Algorithm for the Synthesis of Nominal Dynamics

The notion of simulation will now be broadned a little. The complete model will be considered and the simulation algorithm will include the calculation of the control inputs which have to produce the prescribed motion of the manipulator.

We shall now describe an algorithm for the simulation of manipulators with d.c. motors. In a similar way, the algorithm for manipulators with hydraulic actuators can be derived. The third order model of D.C. motor will be used. It is assumed that the system starts with nominal initial conditions.

Let us consider a time instant t^* when the state $(q, \dot{q})$ of the mechanical part is known. The accelerations $\ddot{q}$ can be computed from (3.1.6). The driving forces and torques P are then computed from the S_M model (3.8.1), i.e.,

$$P = W\ddot{q} - U. \tag{3.10.1}$$

The control inputs should be derived from the actuator models S_A^i, i = 1,...,n i.e. (3.9.2). The matrix equation (3.9.2), (3.9.4) consists of three scalar equations which will be denoted by (3.9.2a), (3.9.2b) and (3.9.2c), respectively. The rotor current i_{r_i} is computed from (3.9.2b).

The derivative $\frac{d}{dt} i_{r_i}$ can be found in the form $\frac{d}{dt} i_{r_i} = (i_{r_i}(t^*) - i_{r_i}(t^*-\Delta t))/\Delta t$. The control input u_i is now somputed from (3.9.2c).

If a second-order actuator model S_A^i is used, then the procedure is simplified. The second-order matrix model (3.9.2), (3.9.5) consists of two scalar equations, which will be denoted by (3.9.2A), (3.9.2B). For known $(q, \dot{q})$ and computed $\ddot{q}$ and P, the control input u_i can be obtained from (3.9.2B). If i_{r_i} is needed, it can be computed from (3.9.6).

The block-scheme of the whole simulation algorithm is given in Fig. 3.34. The scheme is similar to the scheme in Fig. 3.1. but with the addition of the control block.

3.11. Example of the Synthesis

A manipulator with 6 d.o.f. will be considered with a task of positionning and partial orientation. As we said in 3.4., for such a task, 5 d.o.f. are necessary. Hence, in this case, motion along one d.o.f. is directly prescribed by the requirement that the corresponding generalized coordinate always equals zero ($q_k(t) = 0$). Such a case was considered in 3.4., in class C3. The approach will be somewhat different here because, from the standpoint of practical control realization, it is appropriate to divide the degrees of freedom into two groups. Thus, let us consider the minimal manipulator configuration (manipulator without gripper). It possesses 3 d.o.f., $q^m = [q_1\ q_2\ q_3]^T$, which perform the positioning of the minimal configuration tip. The second group of 3 d.o.f., $q^g = [q_4\ q_5\ q_6]^T$, represents the d.o.f. by means of which the gripper is connected to the minimal configuration. These 3 d.o.f. perform the gripper orientation. When we use the term "gripper", we mean the last manipulator segment. In the phase of working object transfer, we mean the gripper and working object together.

Let us now consider how to perform the simulation. Let the manipulation task be given in the form of positioning the minimal configuration and the partial orientation in terms of the position of vector $\vec{h}$ connecting the gripper base (minimal configuration tip) to its center of gravity (or gripper tip). Finally, one generalized coordinate is prescribed directly (Fig. 3.35). Let the positioning be given in Cartesian coordinates $x(t)$, $y(t)$, $z(t)$ and let $q_k(t)$ $k \in \{4, 5, 6\}$ be prescribed directly.

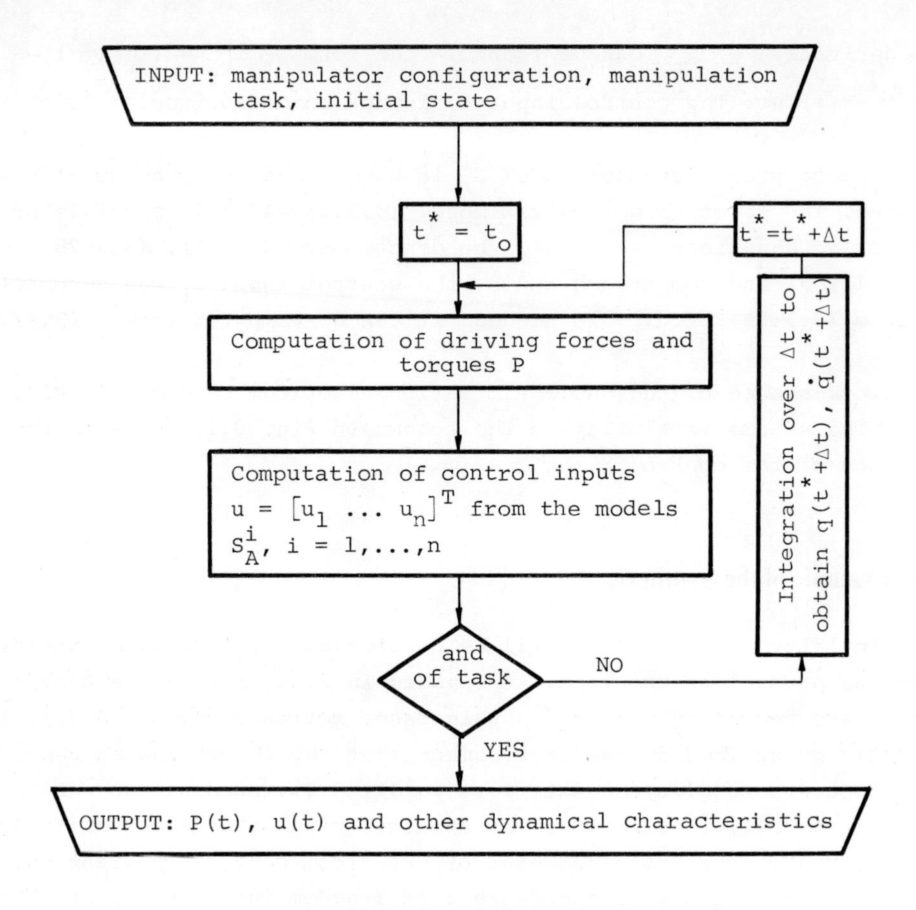

Fig. 3.34. Block-scheme of the simulation algorithm for synthesis of nominal dynamics

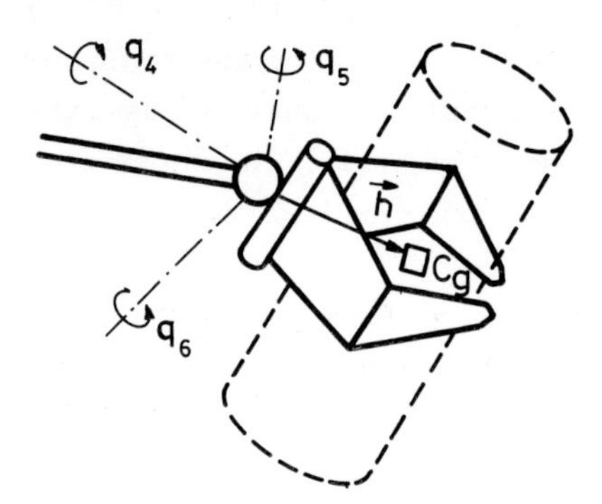

Fig. 3.35. Determination of partial orientation

D.C. motors are adopted and second-order mathematical models of such actuators are used.

Let us suppose a time instant t^* when q_i, $\dot{q}_i$, i=1,...,6 are known. Consider the derivation of case B1 in 3.4. The same procedure holds for C3 and consequently also for the case considered now. However, the vector q now has six elements: $q = [q_1 \; \dots \; q_6]^T$. In the case of positioning the minimal configuration tip, eq. (3.4.28) becomes

$$w = \begin{bmatrix} \ddot{x} \\ \ddot{y} \\ \ddot{z} \end{bmatrix} = \Omega^m \ddot{q}^m + \Theta, \tag{3.11.1}$$

where Ω^m is a 3×3 matrix containing the first three columns of the matrix Ω. This is used because the other three columns of the matrix Ω equal zero. Now, from (3.11.1),

$$\ddot{q}^m = \begin{bmatrix} \ddot{q}_1 \\ \ddot{q}_2 \\ \ddot{q}_3 \end{bmatrix} = \Omega^{m^{-1}} (w - \Theta) \tag{3.11.2}$$

With $\ddot{q}_1$, $\ddot{q}_2$, $\ddot{q}_3$ calculated (ensuring positioning) and with prescribed $\ddot{q}_k$, the remaining two unknown generalized accelerations (by which the orientation is ensured) are calculated from the system (3.4.36). The vector $\ddot{q}$ has now been calculated, so the drives are obtained from the system (3.10.1). The control inputs $u_1, \dots, u_6$ are derived from the actuators models (3.9.2), (3.9.5).

In the numerical example, the semianthropomorphic manipulator UMS-1 (Fig. 3.36) was considered.

The manipulation task: The initial manipulator position A is $q(t_o) = [0.1 \;\; 1 \;\; 0.5 \;\; 0 \;\; 0 \;\; 0]^T$, $\dot{q}(t_o) = 0$. For the minimal configuration tip, $[x \;\; y \;\; z]^T = [0.425 \;\; 0.167 \;\; 0.571]^T$. The gripper center of gravity C_g has the coordinates $[0.455 \;\; 0.203 \;\; 0.630]^T$. At the end position B, $[x \;\; y \;\; z]^T = [0.368 \;\; 0.365 \;\; 0.355]^T$ and the coordinates of C_g are $[0.391 \;\; 0.410 \;\; 0.410]$. Between the positions A and B, the minimal configuration tip and center of gravity C_g should move along a straight line with a triangular velocity profile. The movement should last 0.9 sec. The coordinate q_4, representing gripper rotation around its proper longitudinal axis, is directly prescribed as $q_4(t) = 0$, i.e., it is kept constant.

D.C. servomotors GLOBE type 102A200-8 were used for the drives and data were taken from the corresponding catalogue.

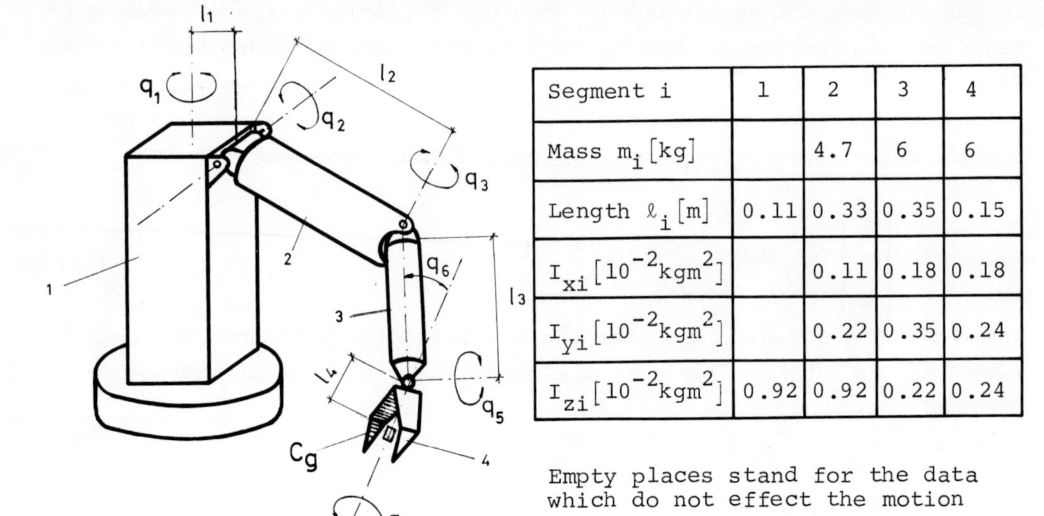

Segment i	1	2	3	4
Mass m_i [kg]		4.7	6	6
Length ℓ_i [m]	0.11	0.33	0.35	0.15
I_{xi} [10^{-2}kgm^2]		0.11	0.18	0.18
I_{yi} [10^{-2}kgm^2]		0.22	0.35	0.24
I_{zi} [10^{-2}kgm^2]	0.92	0.92	0.22	0.24

Empty places stand for the data which do not effect the motion

Fig. 3.36. UMS-1 manipulator

Figs. 3.37. and 3.38. present the simulation results.

3.12. Influence of Actuator Models Complexity

In paragraph 3.9, the mathematical models of the most frequent driving actuators were derived. It has been shown that the actuators can be described by models of different complexity (second-order models, third-order models, etc.). Let us now discuss how the complexity of the actuator model influences the simulation results. The discussion will be presented in the form of an example [24].

Configuration. Let as consider the minimal configuration of the manipulator UMS-1 (Fig. 3.36) having 3 rotational d.o.f. driven by permanent magnet D.C. motors. The motor parameters are: B_C = 1.5 Nm/rad/s, J_r = 1.52 kg m^2 C_M = 4.31 Nm/A, C_E = 7. V/rad/s, R_r = 2.45 Ω.

The manipulator tip is loaded with a concentrated mass of 4kg, which stands for the gripper.

Manipulation task. The manipulator tip should follow a straight line between the points A and B defined by the corresponding generalized

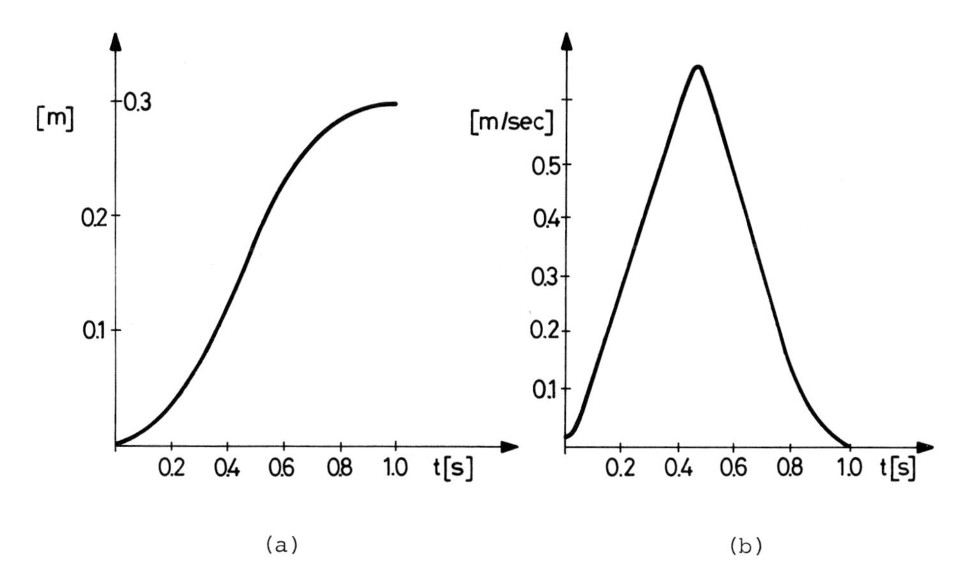

Time history of minimal configuration tip motion along the straight line

Velocity profile of minimal configuration tip

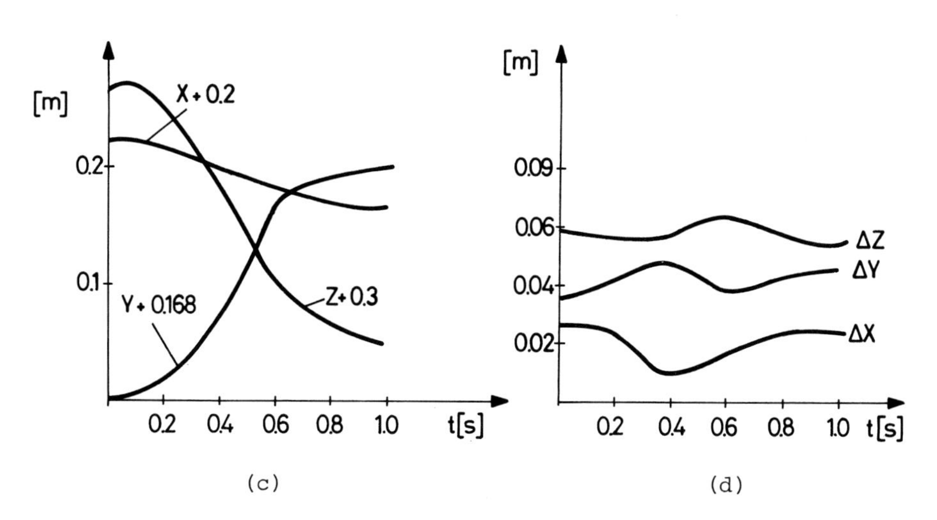

Coordinates of minimal configuration tip

Differences of the manipulator tip coordinates and the coordinates of the gripper center of gravity i.e. projections of the vector h

Fig. 3.37. Simulation results

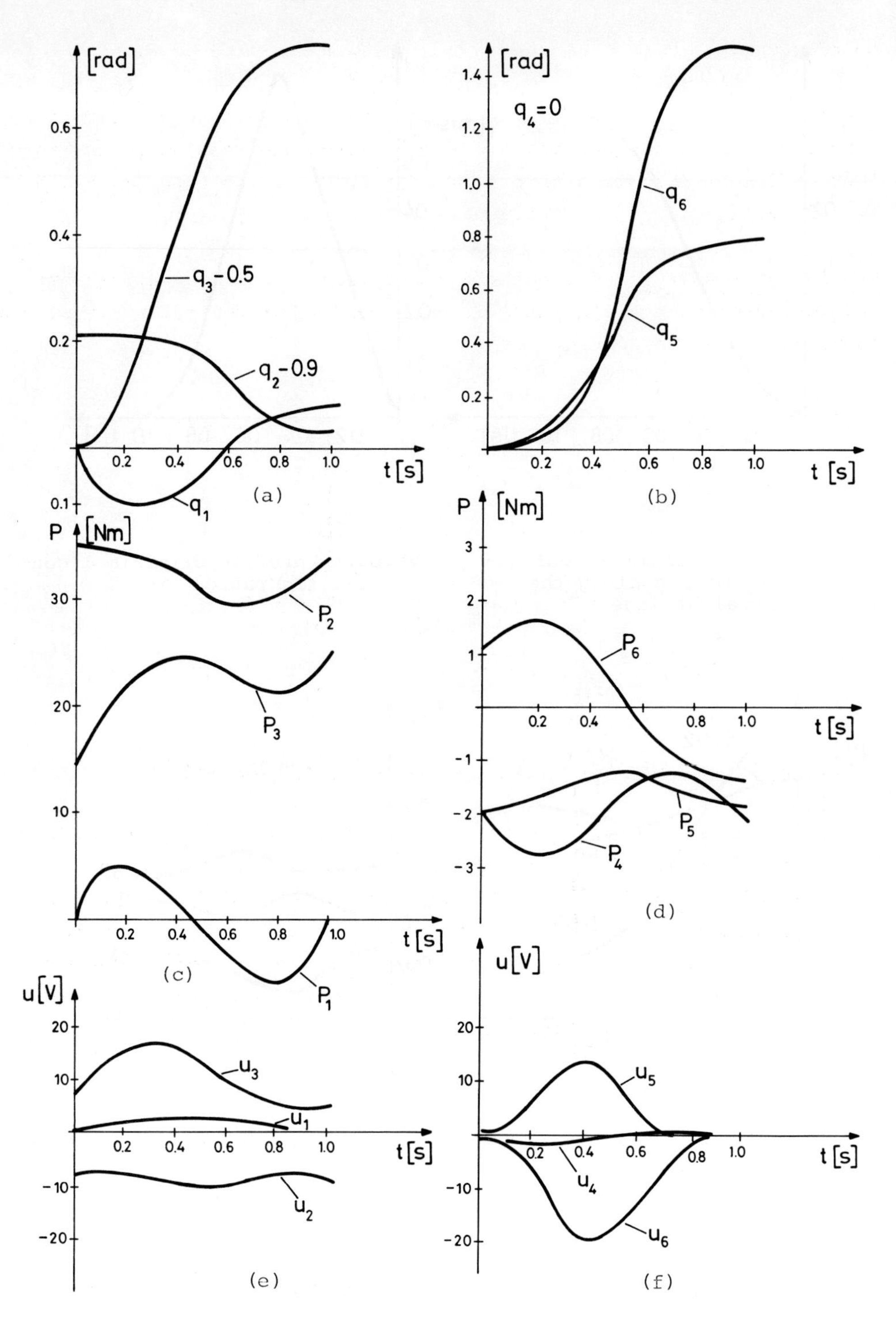

Fig. 3.38. Nominal dynamics; angle trajectories, moments and control

coordinates: $q_A = [0 \quad 1.1 \quad 0.5]^T$, $q_B = [0.08 \quad 0.94 \quad 1.25]^T$.

The quasi-triangular velocity profile (Fig. 3.39) is adopted. Four different values of the maximal tip acceleration are considered and thus, four different execution times. The tip accelerations are: $w_1 = 1.5$, $w_2 = 2.13$, $w_3 = 3.33$, $w_4 = 4.9$ m/s^2.

Simulation results. The simulations are performed using third-order and second-order actuator models. The results for control voltages are compared and shown in Fig. 3.40 (a) - (d).

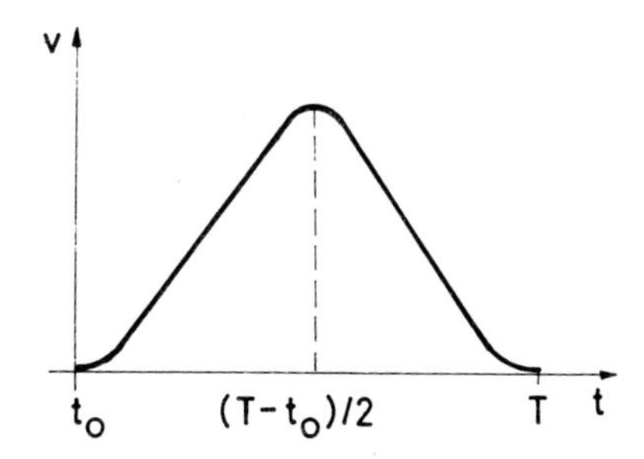

Fig. 3.39. Quasi-triangular velocity profile

3.13. One Method for the Optimal Synthesis of Functional Movements

Synthesis of the manipulator functional movements can also in principle be performed by applying optimal systems theory. It is supposed that the functions $\hat{C}$ and $\hat{D}$ in (3.8.10) are continuous and differentiable in the region of the state space in which optimization is performed. It is also supposed that explicit contraints on the control vector are not present but that there are constraints on one part of the state vector in the terminal instant in the form

$$x_i(T) = x_i^F \quad i=1,\ldots,i_1 \tag{3.13.1}$$

Let us restrict the discussion to the quadratic optimality criterion:

$$J(u, x^o) = \int_{t_o}^{T} (\tfrac{1}{2} u^T R u + x^T S u)\, dt, \tag{3.13.2}$$

where $R>0$, $S \geq 0$ are symmetric constant matrices. Optimal control $u^*(t)$, the optimal trajectory $x^*(t)$ and the adjoined vector $p^*(t)$ must satisfy the following equations with boundary conditions [25, 26].

$$\dot{x}^*(t) = \frac{\partial H^T}{\partial p}(x^*(t), p^*(t), u^*(t)), \; x^*(t_o)=x^o, \; x_i^*(T)=x_i^F, \; i=1,\ldots,i_1$$

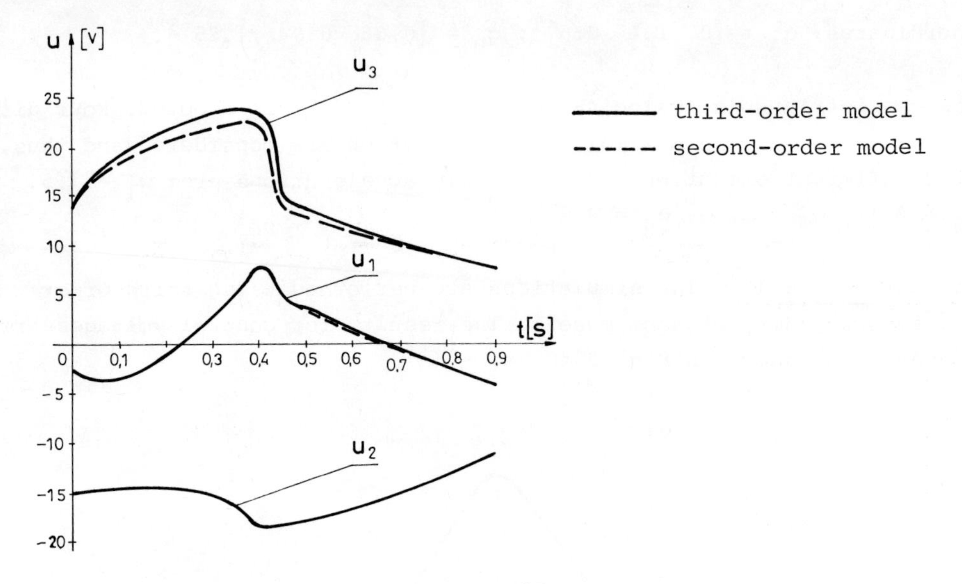

(a) Tip acceleration $w_1 = 1.5\ m/s^2$

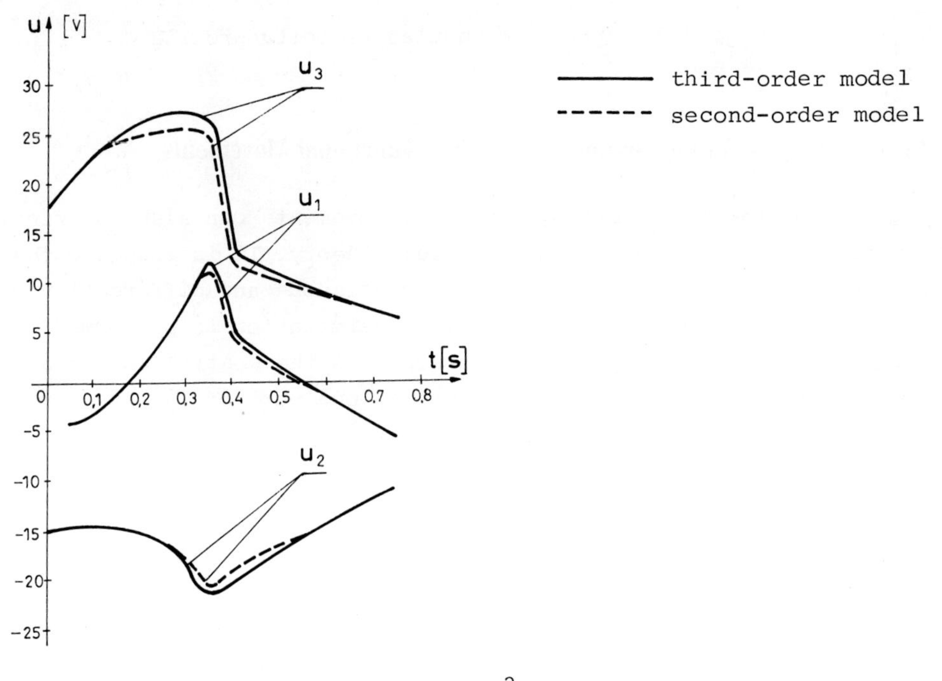

(b) Tip acceleration $w_2 = 2.13\ m/s^2$

Fig. 3.40. Comparison of simulation results

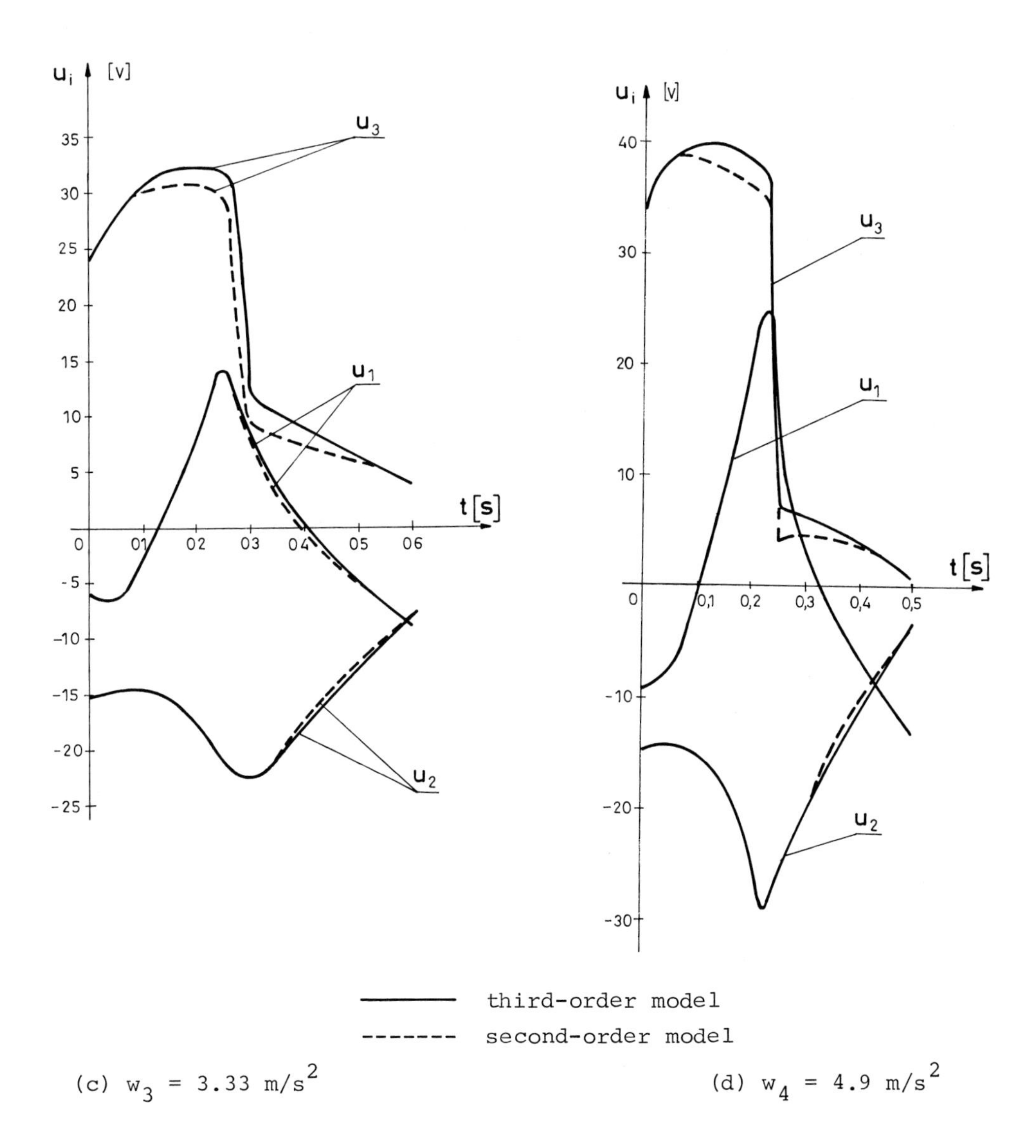

(c) $w_3 = 3.33\ m/s^2$

(d) $w_4 = 4.9\ m/s^2$

Fig. 3.40. Comparison of simulation results

$$\dot{p}^*(t) = -Su^*(t) - \left(\frac{\partial \hat{C}^T(x^*)}{\partial x} + u^{*T}(t)\frac{\partial \hat{D}^T(x^*)}{\partial x}\right)p^*(t), \quad p_i^*(T)=0,$$

$$i=i_1+1,\ldots,N$$

$$Ru^*(t) + S^T x^*(t) + \hat{D}^T(x^*)p^*(t) = 0, \qquad (3.13.3)$$

where H is the system Hamiltonian.

Solving this system of equations, apart from determining the nonlinear functions $\hat{C}(x)$ and $\hat{D}(x)$ in each integration step, requires knowing the matrices of the partial derivatives of these functions $\frac{\partial \hat{C}(x)}{\partial x}$ and $\frac{\partial \hat{D}(x)}{\partial x}$.

The system (3.13.3) can be solved in two ways using first-order gradient procedures. The first of these algorithms is based on the elimination of the control vector from the first two equations of the system (3.13.3) with the aid of the third equation. The system can be solved in such way that some initial vector $p(t_o)$ is supposed. Then for t_o, $p(t_o)$ and prescribed $x(t_o)$, the system (3.13.3) is integrated over $\dot{x}^*(t)$ and $\dot{p}^*(t)$. It is then checked to what extent the boundary conditions are satisfied, i.e. the error vector $e(T) = [e_1(T) \ldots$ $\ldots e_N(T)]^T$ is calculated using the equations

$$e_i(T) = \begin{cases} x_i(T) - x_i^F & i = 1,2,\ldots,i_1 \\ \Delta p_i(T) & i = i_1+1,\ldots,N \end{cases} \qquad (3.13.4)$$

The procedure should be repeated, i.e. the corrections Δp_o which diminish the error $e(T)$ should be determined, until the error is reduced to below some value, this value being given in advance. This procedure, however, is difficult to realize because the adjoined system p behaves very unstably.

The second algorithm is based on an exact solution of the first two equations, while the error in the third equation in (3.13.3) is reduced by an iterative procedure. This algorithm does not create problems in the control space during the integration of equations. However, the problems which arise in the choice of $p_i(T)$, $i=1,\ldots,i_1$, which ensure the satisfaction of the terminal conditions of the state vector, are rather great. Consequently, it can be concluded that the solution of the canonical system of equations for the case of manipulator motion between two given points, represents a very complex problem, i.e., that the synthesis of the optimal nominal trajectories using the "two-point-boundary value" problem is an extremely complex task. In addition, this method does not take into account the constraints on the control, which exist in a real manipulation system, or the possibility of supplementary constraints on the manipulator tip trajectory, which are occur frequently in industrial manipulation.

Thus, the problems arising during the synthesis of the trajectories at the basis of the complete dynamic model of manipulation mechanisms, are extermely difficult to solve because of the complexity of the nonlinear model. The impossibility of a simple introduction of constraints on the trajectory of the state and control vector complicates its application still more.

It is possible to form the nominal trajectories suboptimally in such a way that some optimization criterion (speed, energy consumption) is imposed and such that programmed control is synthesized at the basis of the simplified manipulator dynamic model. In this way, one solved the problem of time-optimal motion [27], as well as the problem of the synthesis of suboptimal trajectories at the basis of the simplified, decentralized dynamical model and the approximate energy criterion [28].

Velocity profile optimization. With the aim of partially surmounting these problems, an algorithm was formed [29] for the synthesis of optimal nominal trajectory and the corresponding control at the basis of the exact dynamical model. This algorithm is presented in more detail in the second book of this series.

According to this procedure, the manipulator tip velocity along the prescribed path is optimized using the methods of dynamical programming. Although we have abandoned the "optimal" synthesis of the nominal manipulator trajectories under the supposition that the trajectory of the tip is prescribed by the functional task requirements, it is very useful to optimize some other parameters of the manipulator motion such as the velocity profile (velocity distribution).

The complete mathematical model (3.8.10) is used to describe the manipulator dynamics. The second-order models of actuators systems are adopted, so the state vector is $x = [q^T \; \dot{q}^T]^T$. The manipulator motion is considered in terms of so-called "external variables" X. This was discussed in 3.1. The vector X defines the position and orientation of the manipulator gripper. According to 3.1, we introduce a function

$$X = \eta(q) \tag{3.13.5}$$

for computing X when the generalized coordinates q are known. Such a function-algorithm can easily be performed for an arbitrary manipulator configuration. The relation (3.13.5) is called the position model of a manipulator. The solution of the inverse problem (evaluation of joint coordinates for a given position vector X) is difficult but can be carried out using various numerical procedures, [17, 18]. We shall consider only nonredundant manipulators, i.e. manipulators whose number of degrees of freedom is equal to the dimension of the vector X(dim q = dim X). For this class of mechanisms it is possible to obtain the solution of the inverse problem by the use of the Newton-Raphson algorithm. For any given position vector X (on a manipulator trajectory) it is thus possible to determine the corresponding vector of joint coordinates q so that $\eta(q) = X$.

Optimization procedure. Let us consider the following optimization task. It is necessary to determine the optimal trajectory of the control vector u(t) which realizes optimal velocity distribution along a prescribed path in the work-space. The performance index is chosen as

$$J = \int_{t_o}^{T} L(x(t),\ u(t))dt + g[x(T)], \tag{3.13.6}$$

where $\tau = T - t_o$ represents the given movement execution time and L and g are limited, arbitrary functions. The manipulation system is modelled by a complete dynamic model (3.8.10) and a position model (3.13.5).

The manipulator tip moves between two end-points X^o and X^F along a prescribed path, which might be represented in parameter form

$$X = r(\lambda), \tag{3.13.7}$$

where $\lambda \varepsilon [0,\ 1]$ is a scalar parameter, r is a given vectorial function, $r(0) = X^o$, $r(1) = X^F$. For example, equation (3.13.7) may represent a staight line or a parabolic curve in the work-space. We also assume that the initial and terminal conditions are known:

$$x(t_o) = x^o = [q^{oT}\ \dot{q}^{oT}]^T,\ x(T) = x^F = [q^{FT}\ \dot{q}^{FT}]^T, \tag{3.13.8}$$

where q^o and q^F correspond to the vectors X^o and X^F : $X^o = \eta(q^o)$ and $X^F = \eta(q^F)$ and where x is the state vector.

The problem of determining optimal velocity distribution is solved by the use of dynamic programming [30]. The time interval τ is devided into N_D small subintervals $\Delta t = \frac{\tau}{N_D}$. The dynamic model (3.8.10) is considered as a discrete time system:

$$x(k+1) = x(k) + \Delta t\hat{C}(x(k)) + \Delta t\hat{D}(x(k))u(k), \quad k=0,\dots,N_D-1, \qquad (3.13.9)$$

where x(k) and u(k) are values of the state-space vector and the input control vector in the time instant $t_k = t_o + k\Delta t$. The performance index (3.13.6) is also discrete.

$$J_o = \sum_{k=0}^{N_D-1} \Delta t \cdot L(x(k), u(k)) + g[x(N_D)] \qquad (3.13.10)$$

with the following set of boundary conditions and constraints:

$$x(0) = x^o, \quad x(N_D) = x^F$$

$$X(k) = r(\lambda(k)), \qquad k=1,\dots,N_D-1 \qquad (3.14.11)$$

$$u_i(k) \le U_m^i, \qquad i=1,\dots,n \quad k=0,\dots,N_D-1 \qquad (3.13.12)$$

For any given value of $\lambda(k)$, the position of the manipulator tip X(k) is determined from (3.13.11). The vector of joint coordinates q(k) is now numerically obtained from the equation (3.13.5). We thus obtain the complete state-space vector $x(k) = [q^T(k) \ \dot{q}^T(k)]^T$, where $\dot{q}(k)$ is approximated by $(q(k+1) - q(k))/\Delta t$.

The optimization task is now reduced to the problem of evaluating the control vectors u(k), $k=0,\dots,N_D-1$ and the state-space vectors x(k) (related to u(k) by (3.13.9)) which minimize the performance index (3.13.10), constrained by (3.13.11) and (3.13.12). The dynamic programming approach reduces this problem to a series of successive minimizations with respect to u(k), $k = 0,\dots,N_D-1$.

The optimization procedure starts at the last time interval $[N_D-1, N_D]$, and finishes at the first. Let us assume that the set of optimal trajectories (control and state vectors) in the time interval $[N_D-j+1, N_D]$ are allready determined. In order to extend the optimization to the

$[N_D-j]$-th interval, we choose a set of parameters $\lambda^i(N_D-j)$, $i=1,\ldots,M$, which determine the positions of the manipulator tip. For any pair of parameters $\lambda^i(N_D-j)$ and $\lambda^k(N_D-j+1)$, $i, k=1,\ldots,M$, it is possible to evaluate the corresponding state vector $x^m(N_D-j)$, $m = (i-1)M + k$. The control vector $u^{m\ell}(N_D-j)$ which transfers the system from the state $x^m(N_D-j)$ to the state $x^\ell(N_D-j+1)$ is now determined from (3.13.9).

$$u^{m\ell}(N_D-j) = [\hat{D}^T(x^m(N_D-j))\hat{D}(x^m(N_D-j))]^{-1}\hat{D}^T(x^m(N_D-j))$$

$$\cdot\ [\frac{x^\ell(N_D-j+1) - x^m(N_D-j)}{\Delta t} - \hat{C}(x^m(N_D-j))] \qquad (3.13.13)$$

$$m=1,\ldots,M^2,\ \ell=(m^*-1)M+1,\ldots,m^*M,\ \text{where } m^*=m - \left[\frac{m-1}{M}\right]M.$$

Following the dynamic programming approach the performance index (3.13.10) on the time interval $[N_D-j,\ N_D]$ is obtained in the form

$$J^{m\ell}_{N_D-j}(x^m(N_D-j),\ u^{m\ell}(N_D-j)) = L(x^m(N_D-j),$$

$$u^{m\ell}(N_D-j)) + S_{N_D-j+1}(x^\ell(N_D-j+1)),$$

$$S_o = 0,\quad \ell = (m^*-1)M + 1,\ldots,m^*M \qquad (3.13.14)$$

We now define the optimal control $u^m(N_D-j)$ to be that which satisfies (3.13.12) and

$$S^m_{N_D-j}(x^m(N_D-j)) = \min_{u^{m\ell}(N_D-j)} J^{m\ell}_{N_D-j}(x^m(N_D-j),\ u^{m\ell}(N_D-j)),\ m=1,\ldots,M^2 \qquad (3.13.15)$$

The whole procedure is repeated for the indices $j=1,\ldots,N_D$. In the first and terminal iterations the state vectors are already determined by (3.13.11).

The flow chart of this algorithm is presented in Fig. 3.41.

Applying the algorithm to a concrete manipulation system we obtian the optimal distribution of the parameters $\lambda(k)$, $k=0,\ldots,N_D$, the corresponding optimal control signals $u(k)$ and the optimal trajectories $x(k)$ and $X(k)$.

Example. We apply the procedure to the minimal configuration of the

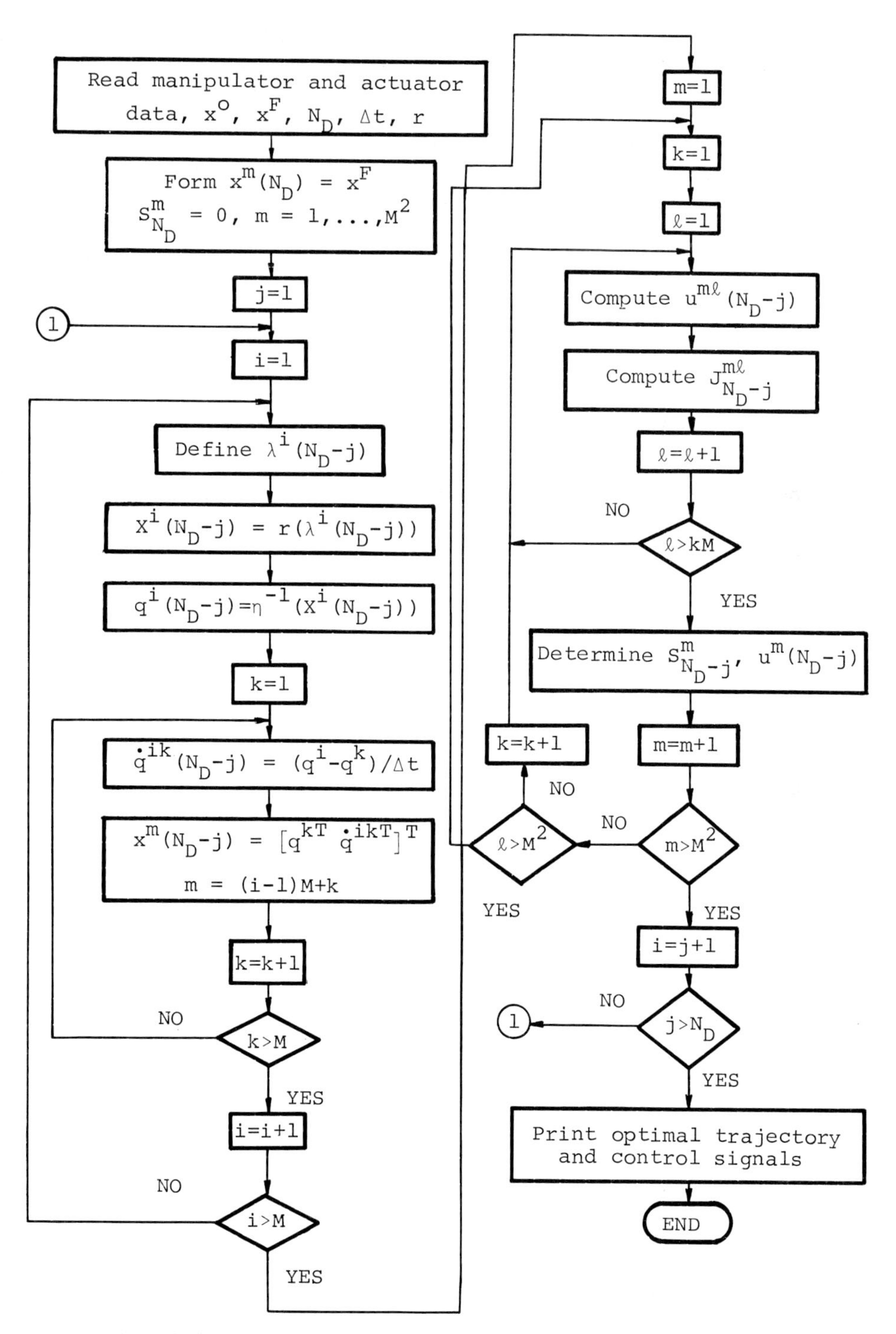

Fig. 3.41. Flow chart of the algorithm for determining optimal velocity distribution

semi-anthropomorphic manipulator UMS-1 (Fig. 3.36) driven by DC motors [20]. In the case of DC motors the actuator matrices become

$$C_i = \begin{bmatrix} 0 & 1 \\ 0 & \alpha_i \end{bmatrix}, \quad d_i = \begin{bmatrix} 0 \\ \beta_i \end{bmatrix}, \quad f_i = \begin{bmatrix} 0 \\ \gamma_i \end{bmatrix} \tag{3.13.16}$$

For the manipulator UMS-1,

Joint	α_i	β_i	γ_i
1,2	-10.8	1.42	-0.66
3	-15.2	10.7	-30

Let us choose energy consumption to be the performance index

$$J = \int_{t_o}^{T} (u^T Q_1 u + x^T Q_2 u) dt. \tag{3.13.17}$$

The matrix Q_1 is diagonal: $Q_1 = \text{diag}[1/R_{r1}, 1/R_{r2}, 1/R_{r3}]$. All elements of matrix Q_2 are equal to zero, except for $Q_{2_{3+i,i}} = -C_{Mi}/2R_{ri}$.

We chose the straight line manipulator tip trajectory between two terminal points X^o and X^F:

$$X = r(\lambda) = X^o + \lambda(X^F - X^o), \tag{3.13.18}$$

with $X^o = [0.49 \quad 0.36 \quad 0.]^T$, $X^F = [0.44 \quad 0.47 \quad 0.101]^T$. The initial and terminal velocities of the manipulator tip are taken to be zero.

The resultant optimal velocity distribution $\dot{\lambda}(t)$ is presented in Fig. 3.42. The corresponding distribution of manipulator generalized velocities $\dot{q}_i(t)$ is illustrated in Fig. 3.43. Optimal control inputs are presented in Fig. 3.44.

The proposed algorithm has been illustrated by several examples. Optimization results point to the fact that the optimal velocity distributions are most like parabolic functions and that this profile should be applied in practical manipulator control in order to minimize actuator consumptions.

The proposed algorithm for manipulator trajectory synthesis permits the evaluation of optimal trajectories using the complete dynamic model of the active mechanism. This procedure has the advantage of being very convenient for optimal synthesis when state vector trajectories and controls are subjected to various types of constraints (design constraints, presence of obstacles in the work-space, velocity and acceleration constraints, limited input signals for the actuators, etc.). However, the deficiencies of this procedure arise, in principle, from the dynamic programming algorithm itself. They are: a dimensionality problem, relatively long optimization time and large computer memory requirement. On the other hand, this procedure is especially suitable for optimal velocity distribution synthesis for nonredundant manipu-

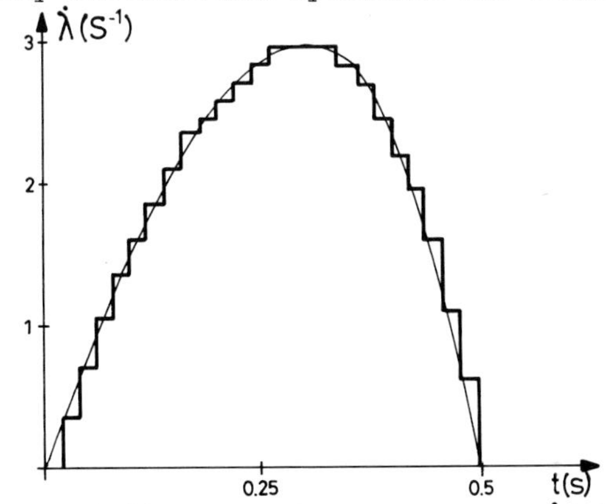

Fig. 3.42. Optimal distribution $\dot{\lambda}(t)$

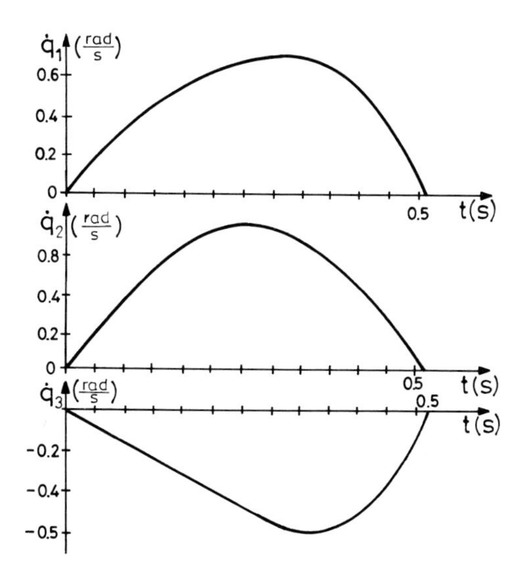

Fig. 3.43. Optimal distributions $\dot{q}(t)$

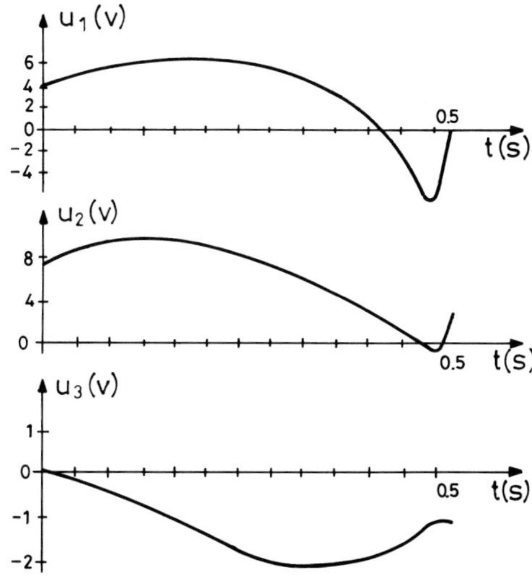

Fig. 3.44. Optimal control signals $u(t)$

lators. In this case, the problem reduces to optimization with respect to one parameter only. The long optimization time and large memory capacities are, in the case of optimal velocity distribution synthesis, due to the complexity of the nonlinear dynamic model and also depend on the desired optimization accuracy and the duration of the movement. On the other hand, dynamic programming is the only optimization technique which could be applied to such a complex, nonlinear and multiconstrained system.

The proposed procedure is assigned to optimal nominal trajectory synthesis in the off-line dynamics. This is not a great drawback for automatic industrial robots, considering that industrial tasks are very often completely defined in advance.

References

[1] Stepanenko Yu., "Method of the Analysis of Spatial Articulated Mechanisms" (in Russian), Mekhanika mashin, Vol. 23, Moscow, 1970.

[2] Juričić D., Vukobratović M., "Mathematical Modelling of a Biped Walking System", ASME Publ. 72-WA/BHF-13.

[3] Vukobratović M., Stepanenko Yu., "Mathematical Models of General Anthropomorphic Systems", Math. Biosciences, Vol. 17, pp. 191-242, 1973.

[4] Stepanenko Yu., Vukobratović M., "Dynamics of Articulated Open-Chain Active Mechanisms", Math. Biosciences, Vol. 28, No 1/2, 1976.

[5] Vukobratović M., "Computer Method for Mathematical Modelling of Active Kinematic Chains via Generalized Coordinates", Journal of IFToMM Mechanisms and Machine Theory, Vol. 13, No 1, 1978.

[6] Vukobratović M., Legged Locomotion Robots and Anthropomorphic Mechanisms, Monograph, Mihailo Pupin Institute, 1975, P.O.B. 15, Beograd, Yugoslavia.

[7] Orin D., Vukobratović M., R.B. Mc Ghee., G.Hartoch, "Kinematic and Kinetic Analysis of Open-Chain Linkages Utilizing Newton-Euler Methods", Math. Biosciences, Vol. 42, pp. 107-130, 1978.

[8] Luh J.Y.S., Walker M.W., Paul R.P.C., "On Line Computational Scheme for Mechanical Manipulators", Trans. of ASME, Journal of Dynamic Systems, Measurement and Control, June 1980, Vol. 102/09.

[9] Vukobratović M., Potkonjak V., "Contribution to the Forming of Computer Methods for Automatic Modelling of Active Spatial Mechanisms Motion", PART1: "Method of Basic Theoremes of Mechanics", Journal of Mechanisms and Machine Theory, Vol. 14, No 3, 1979.

[10] Vukobratović M., Potkonjak V., "Contribution to the Computer Methods for Generation of Active Mechanisms via Basic Theoremes of Mechanics", Teknitcheskaya kibernetika ANUSSR, No 2, 1979.

[11] Vukobratović M., "Computer Method for Mathematical Modelling of Active Kinematic Chains via Euler´s Angles", Journal of IFToMM Mechanisms and Machine Theory, Vol. 13, No 1, 1978.

[12] Vukobratović M., Potkonjak V., "Contribution to Automatic Forming of Active Chain Models via Lagrangian Form", Journal of Applied Mechanics, No 1, 1979.

[13] Hollerbach J.M., "A Recursive Formulation of Lagrangian Manipulator Dynamics", Proc. JACC, June, 1980, San Francisco.

[14] Potkonjak V., Vukobratović M., "Two Methods for Computer Forming of Dynamic Equations of Active Mechanisms", Journal of Mechanism and Machine Theory, Vol. 14, No 3, 1979.

[15] Paul R.C., Modeling, Trajectory Calculation and Servoing of a Computer Controlled Arm, A.1 Memo 177, Sept. 1972, Stanford Artificial Intelligence Laboratory, Stanford University, Sept. 1972.

[16] Vukobratović M., Potkonjak V., Hristić D., "Dynamic Method for the Evaluation and Choice of Industrial Manipulators", Proc 9th International Symposium on Industrial Robots, Washington, 1979.

[17] Vukobratović M., Potkonjak V., "Contribution to Computer-Aided Design of Industrial Manipulators Using Their Dynamic Properties", Journal of IFToMM, Mechanisms and Machine Theory, Vol. 16, No 2, 1982.

[18] Vukobratović M., Potkonjak V., "Transformaiton Blocks in the Dynamic Simulation of Typical Manipulator Configurations", IFToMM Journal of Mechanism and Machine Theory, No 3, 1982.

[19] Lurie A.I., Analytical Mechanics (in Russian), Gostehizdat, Moscow, 1961.

[20] Vukobratović M., Stokić D., "Engineering Approach to Dynamic Control of Industrial Manipulators", PART 1: Synthesis of Nominal Regimes", (in Russian), Journal of Mashinovedenia ANUSSR, No 3, 1981.

[21] Vukobratović M., Hristić D., Stokić D., "Synthesis and Design of Anthropomorphic Manipulator for Industrial Application, Journal of Industrial Robot, Vol. 5, No 2, 1978.

[22] Vukobratović M., Stokić D., "Contribution to the Decoupled Control of Large-Scale Mechanical Systems", IFAC Automatica, Jan. 1980.

[23] Vukobratović M., Stokić D., "One Engineering Concept of Dynamical Control of Manipulators", Trans. of ASME, Journal of Dynamic Systems, Measurement and Control, Special issue, June 1981.

[24] Vukobratović M., Borovac B., Stokić D., "Influence Analysis of the Actuator Mathematical Model Complexity on the Manipulator Control Synthesis", IFToMM Journal of Mechanisms and Machine Theory, (in press).

[25] Athans M., Falb P., Optimal Control, Mc Graw-Hill Book Company, 1966.

[26] Bryson A., Yu-Chi Ho, Applied Optimal Control, New York, 1975.

[27] Kahn M.E., Roth B., "The Near-Minimum-Time Control of Open-Loop Articulated Kinematic Chains", Jour. of Dynamic Systems, Measurement and Control, Trans. of the ASME, Sept. 1971.

[28] Cvetković V., Vukobratović M., "Contribution to Controlling Non-Redundant Manipulators", Proc. of III CISM-IFToMM Symp., Udine, 1978.

[29] Kirćanski M., "Contribution to Synthesis of Nominal Trajectories of Manipulators via Dynamic Programming", International Conf. on Systems Engineering, Coventry, Sept. 1980.

[30] Bellman R., Dynamic Programming, Princeton University Press, Princeton N.J., 1957.

Chapter 4
Dynamics of Manipulators with Elastic Segments

4.1. Introduction

In earlier research into the dynamics of active mechanisms in robotics it was always assumed that the mechanism consisted of rigid segments. This was justified because the dimensions of cross-sections of segments in practice were such that their elastic properties were insignificant. Only recently have papers appeared which consider the problems of manipulators with certain elastic properties [22 - 28]. Among all the methods dealing with elastic manipulators, the most general is the one presented in [23, 24]. A manipulator with rigid and elastic segments is properly modelled by a "hybrid multibody system". The method uses the generalized rigid-body coordinates ($q_i(t)$, $i=1,\ldots,n$) to describe the "transport" motion of manipulator segments, and the distributed elastic coordinates ($v_i(x_i, t)$, $w_i(x_i, t)$, $\alpha_i(x_i, t)$, $i = 1,\ldots,m$) to characterize the bending and torsion deformations. By means of shape functions and a (Ritz-Kantorovitch) series expansion of the distributed deformation coordinates, the equations of motion are ordinary differential equations. They are, in general, nonlinear; a linearization with respect to a prescribed reference motion yields linear equations. This method is derived for using in the elastic manipulator control problems. The main disadvantage of the method is its high dimensionality. For a manipulator with six one-d.o.f. joints and two elastic segments, the model order is

$$N = 6 + \sum_{i=1}^{2} (N_{vi}+N_{wi}+N_{\alpha i})$$

where N_{vi}, N_{wi}, $N_{\alpha i}$ are the numbers of terms (i.e. shape functions) in the (Ritz-Kantorovitch) series expansion. If we use four shape functions for bending coordinates ($N_{vi}=N_{wi}=4$) and two shape functions for torsional coordinates ($N_{\alpha i}=2$) then the model order is N=26.

In this chapter, another method is presented. The approach its different because of a different aim of the method. The method is not intended for control problems but for computer-aided design of a manipulator mechanism where the upper bounds of deformations are essential. The method results in a model of considerably lower dimensionality. This chapter deals with such approach.

The development of computer methods for formulating and solving mathematical models of active mechanisms [1 - 18] and the development of corresponding algorithms for the simulation of manipulator dynamics [19, 20, 21], have made possible a systematic choice of dimensions and other manipulator parameters, which are optimal for the intended application of the device. Such a systematic choice leads to a reduction of overlarge dimensions up to the point where elastic properties of the manipulator segments must also be taken into account. This has enhanced the study of elastic manipulator dynamics.

The trend towards the choice of optimal manipulator parameters has led, as we said, to a reduction of oversize dimensions. When it was felt that the elastic properties of the segments should be considered, it was necessary to introduce constraints so that deviations, as a result of elastic deformations, do not surpass the permitted values. In order to achieve this, it was necessary to simulate elastic manipulator dynamics. Such an algorithm is described in this chapter. Thus, in the investigations which are the subject of this book, elastic manipulator dynamics has become a very practical problem. Taking into account its role in the method for manipulator evaluation and choice, it was necessary to develop an algorithm which would provide the models and solve the elastic manipulator dynamics in a sufficiently simple way, suitable for programming and use in the general simulation method. Hence, in the course of algorithm derivation, certain approximations were introduced. They produce certain errors but the resulting design calculations are conservative, i.e., the results obtained are always on the safe side. Hence this procedure corresponds to its purpose.

The proposed algorithm calculates the time function of the elastic deviations, i.e., the small oscillations around the nominal manipulator motion.

4.2. Basic Ideas and Postulates

The computer-aided (c.-a.) methods for formation and solving mathematical models of active mechanisms (Chapter II of this book and [1 - 18]) made possible the development of the algorithms for the simulation of manipulator dynamics (Chapter III), [19 - 21]. This simulation algorithm considers the manipulator as an open chain of rigid bodies (Fig. 4.1a) linked by one rotational or linear d.o.f. Here we shall consider a special case, where segments can be in the form of canes (Fig. 4.1b). The motion of such a chain of n rigid segments will be called the nominal motion of the mechanism. The simulation algorithm described earlier makes it possible to compute all characteristics of nominal dynamics: driving forces and torques in joints, reactions in joints, positions, velocities and accelerations of all mechanism characteristic points, etc. All these values are obtained as functions of time (a sequence of time instants).

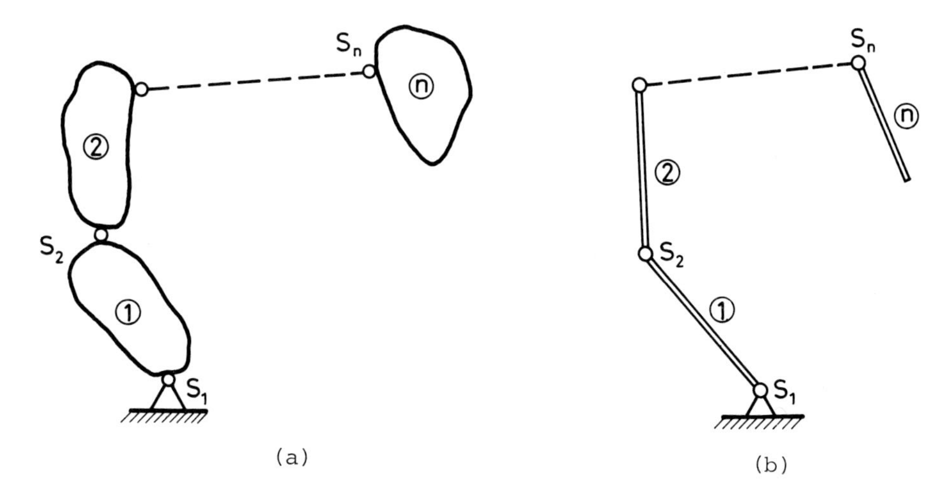

Fig. 4.1. Open kinematic chain

The basic idea of the approach to the elastic manipulator is to consider the manipulator as an open chain of elastic canes and to consider weights, inertial forces of nominal motion, nominal driving forces and torques and nominal reactions in joints as known external forces and moments. These values are computed in the block of nominal dynamics.

In this treatment of elastic manipulators we consider small elastic deformations, i.e., small oscillations around nominal motion.

Let us introduce the values $\vec{u}_i$, i=1,...,n and $\vec{\varphi}_i$, i=1,...,n, which will distinguish elastic manipulator deviation from the nominal motion of the rigid case. $\vec{u}_i$ is the linear deviation vector of joint S_{i+1}

from its nominal motion, Fig. 4.2a. $\vec{\varphi}_i$ is the rotation vector, i.e., the angular deviation (change of the orientation in space) of the point S_{i+1} with respect to the nominal orientation (Fig. 4.2b). Taking into account the assumption that the deviations were small, the angles can be treated as vectors. Thus, $\vec{u}_i$, $\vec{\varphi}_i$, $i=1,\ldots,n$ represent the characteristic values of the elastic oscillations, which will later be called micro-motion. Total motion is consequently regarded as superposition of nominal and micro motion. Let us introduce matrix notation: u_i denotes a 3×1 matrix corresponding to the vector $\vec{u}_i$; likewise for other vectors in the text. Let us also introduce the $3n\times1$ matrix

$$u = \begin{bmatrix} u_1 \\ \vdots \\ u_n \end{bmatrix} \tag{4.2.1}$$

The way is now set to develop the c.-a. method for the formation of the mathematical model of micro-motion for known nominal motion. Thus a procedure will be derived by which, for known characteristics and values of nominal dynamics in some time instant t^*, the matrices $D(3n\times3n)$ and $c(3n\times1)$ can be calculated so that

$$D\ddot{u} + u = c \tag{4.2.2}$$

and by means of which the model of micro-motion for time instant t^* would be formed.

It is clear that there is no amortization in this model. The elastic oscillations are not considered as amortized due to the aim of the method i.e. we are especially interested in the maximal elastic deviation. It should be mentioned that the amortization may easily be included in the calculation.

One of the important assumptions which will be adopted is that micro-motion due to the segment elasticity does not influence the generalized coordinates in the joints, the time function of which remains the same as in the nominal motion. In this case deviations are due to elastic segment deformations only. Such an assumption would permit the incorporation of the block of micro-dynamics into the existing algorithm for the simulation of nominal motion (Chapter III). This incorporation is performed according to the block-scheme in Fig. 4.3. The integration of deviations u can be calculated as for the generalized coordinates q; for instance consider $\ddot{u}$ constant over the subinterval Δt so that

$$u(t^*+\Delta t) = \frac{1}{2}\,\ddot{u}(t^*)\Delta t^2 + \dot{u}(t^*)\Delta t + u(t^*),$$
$$\dot{u}(t^*+\Delta t) = \ddot{u}(t^*)\Delta t + \dot{u}(t^*). \qquad (4.2.3)$$

Applications of other numerical integration methods are also possible.

The algorithm from Fig. 4.3. can be modified a little in order to use different time increments for the integration of nominal dynamics and of microdynamics. It means that the integrations of q and u may be separated. It is suitable because the microdynamics is must faster than the nominal dynamics.

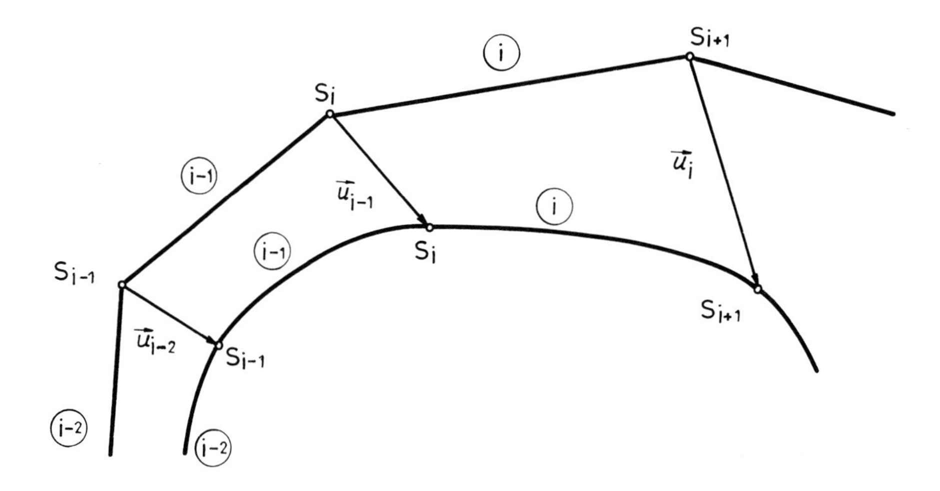

Fig. 4.2a. Linear elastic deviation-deflexion

In the input of such a supplemented algorithm, in addition to the former input values, the initial deviation values $u(t_o)$ and $\dot{u}(t_o)$ are also included. ($\dot{u}(t_o) = 0$ when the manipulator starts from rest).

In the output, besides the nominal dynamical values, the time functions of the linear deviation u and the angular φ (orientation), as well as the other micro-dynamical values, are obtained. More will be said about these output values and their calculation later in the text.

It still remains to derive the matrices D and c determining the mathematrical model of micro-dynamics in one time instant. Something more should be said about the dimensionality of the micro dynamics model (4.2.2). If a general case is considered i.e. a chain with n elastic segments, then the model dimension is 3n. But, with the real manipulator configurations there are usually not more then two segments which should be considered as elastic. Hence, the model dimension is usually less or equal to six.

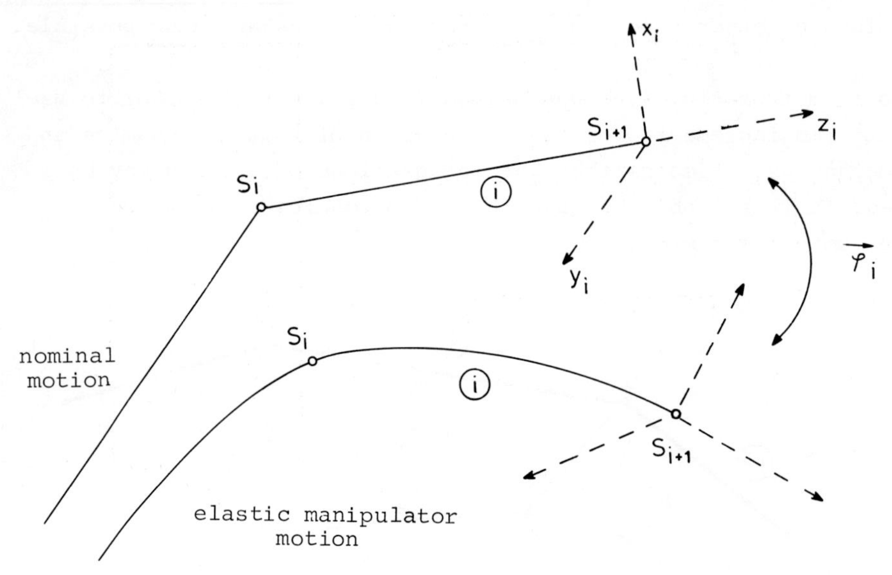

Fig. 4.2b. Angular elastic deviation-slope (orientation change in space)

4.3. Nominal Dynamical Characteristic

As we said, the simulation algorithm, completely solves the manipulator nominal dynamics. At each time instant the following are calculated: driving forces and torques in the joints, $\vec{P}_i$, i=1,...,n, then the reaction forces and moments in the joints $\vec{F}_{R_i}$, $\vec{M}_{R_i}$, i=1,...,n, as well the positions, velocities and accelerations of all mechanism points. Let us introduce some notation: let $\vec{F}^n_{S_i}$ be the total force acting in nominal motion on the i-th segment in the i-th joint, i.e.,

$$\vec{F}^n_{S_i} = \vec{F}^n_{R_i} + s_i \vec{P}^n_i, \tag{4.3.1}$$

and $\vec{M}_{S_i}$ be the total nominal moment in that joint:

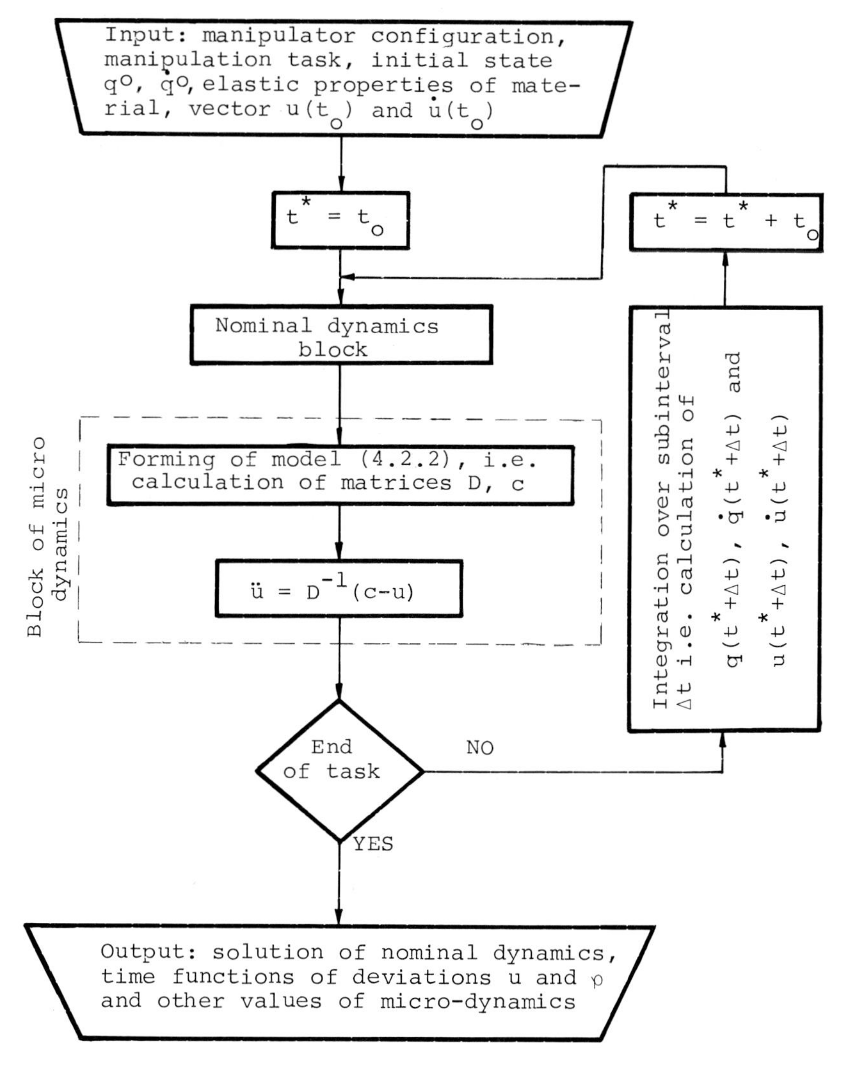

Fig. 4.3. Block-scheme of the simulation algorithm with the micro-dynamics block implemented

$$\vec{M}^n_{S_i} = \vec{M}^n_{R_i} + (1-s_i)\vec{P}^n_i. \tag{4.3.2}$$

The upper index n indicates that nominal values are in question, and the index s_i determined the joints type:

$$s_i = \begin{cases} 0, & \text{if joint } S_i \text{ is rotational,} \\ 1, & \text{if joint } S_i \text{ is linear.} \end{cases} \tag{4.3.3}$$

Further, let us denote by $\vec{w}_i^n$ the acceleration of the point S_{i+1} in nominal motion, which will be needed in calculations to follow.

Finally, knowing the nominal mechanism position means knowing also, or rather calcualting, the transition matrices of the nominal motion. Let A_i, $i=1,\ldots,n$ be the transition matrices of the segments, i.e., their body-fixed systems, relative to the external system.

Some geometrical characteristics. As already stated in 4.2., the mechanism consists of segments in the form of canes. Let us consider a linear joint and note two cases as in Figs. 4.4a and 4.4b.

Let us introduce the indicator p_i, which determines to which of the two cases the linear joint S_i belongs:

$$p_i = \begin{cases} 0, & \text{if the joint is of type (a),} \\ 1, & \text{if the joint is of type (b).} \end{cases} \tag{4.3.4}$$

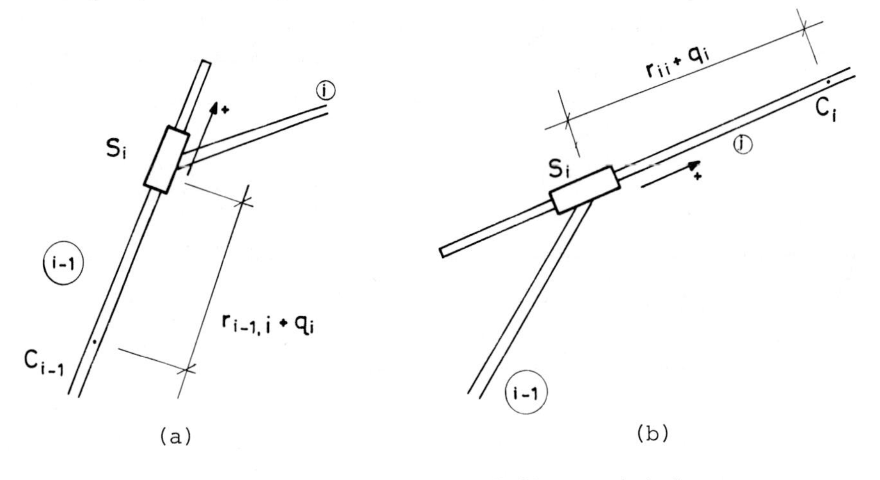

Fig. 4.4. Two sorts of linear joint

Let us now define the length $\vec{\ell}_i$ of the segment (cane) between two joints (Fig. 4.5b):

$$\vec{\ell}_i = \overrightarrow{S_{i-1}S_i} \tag{4.3.5}$$

The magnitude of this vector is determined by

$$|\ell_i| = |r_{ii}| + s_i p_i q_i + (|r_{i,i+1}| + s_{i+1}(1-p_{i+1})q_{i+1}) \tag{4.3.6}$$

The lengths $|r_{ii}|$ and $|r_{i,i+1}|$ represent the magnitudes of vectors $\vec{r}_{ii}$ and $\vec{r}_{i,i+1}$ (Fig. 4.5a), which are used in the method for c.-a. formation of the mathematical model (Paragraphs 2.3. and 2.9).

In the case of segments in the form of canes, these vectors are parallel (Fig. 4.5b) so their magnitudes can be added as in relation (4.3.6). Thus, the magnitudes $|r_{ii}|$ and $|r_{i,i+1}|$ determine the positions from which the generalized coordinates q_i and q_{i+1} are measured. Of course, if some of these joints are rotational (for instance S_i), then the corresponding length $|r_{ii}|$ determines the real position of joint S_i. It should be stressed that the $|r_{ii}|$ and $|r_{i,i+1}|$ are set values, i.e., segment characteristics.

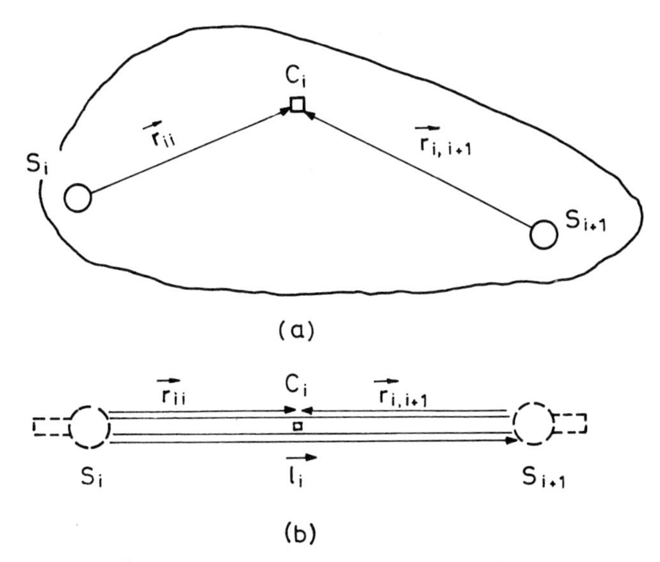

Fig. 4.5. Position of joints with respect to the center of gravity

Let us introduce body-fixed (b.-f.) coordinate system as in Fig. 4.6. as well as the immobile external system with the z-axis vertical.

Further, let us adopt the notation that $\vec{a}_i$ denotes some vector, characteristic of the i-th segment, expressed in terms of three projections onto the axes of the external system, while $\vec{\tilde{a}}_i$ denotes the same vector in the b.-f. system.

Now

$$\vec{\tilde{\ell}}_i = \{0, \ 0, \ |\ell_i|\} \tag{4.3.7}$$

and in the external system

$$\vec{\ell}_i = A_i \tilde{\vec{\ell}}_i . \tag{4.3.8}$$

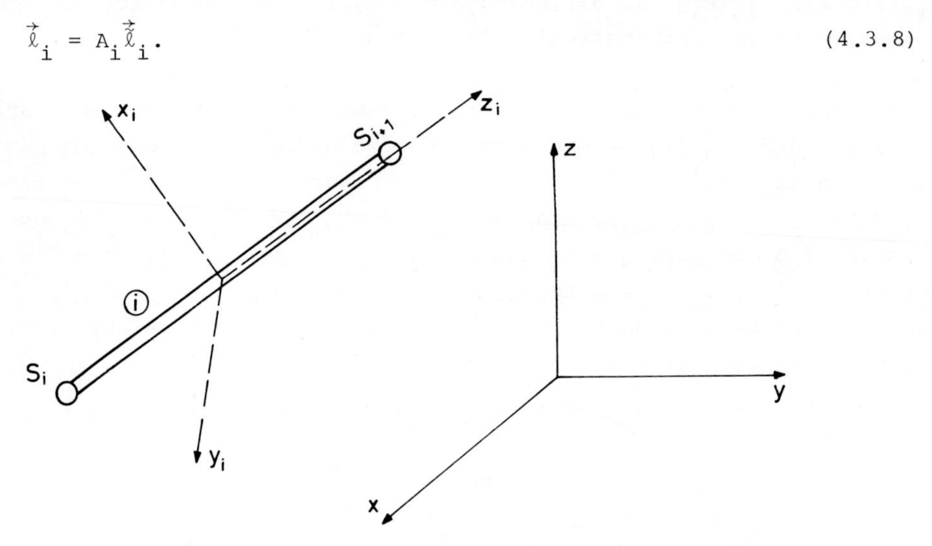

Fig. 4.6. Body-fixed and external coordinate systems

In this way, the vectors $\vec{\ell}_i$, i=1,...,n are calculated for the nominal mechanism motion.

4.4. Deriving the Model of Elastic Oscillations

Right at the start let us introduce two basic suppositions:

(1) Small elastic deformations, i.e., small deviations from nominal motion, will be considered.

(2) Superposition of the nominal and micro motion will be considered, assuming that the micro motion, i.e., elastic oscillations, do not influence the nominal dynamics.

These assumptions enable us to write the mathematical model of micro--dynamics in the form (4.2.2), and then incorporate it as a separate block into the algorithm for the simulation of nominal dynamics, as shown in Fig. 4.3.

Kinematical and dynamical connections. For the sake of brevity, we shall for the moment refer to the linear elastic deviation $\vec{u}_i$ as deflection and the angular deviation $\vec{\varphi}_i$ as tilt. In addition, if $\vec{a}_i$ is some vector, then a_i denotes a 3×1 matrix corresponding to it. This notation will be used for all the vectors in the sequel.

The deflection $\vec{u}_i$ consists of three components (Fig. 4.7):

$$\vec{u}_i = \vec{u}_{i-1} + \vec{u}_i^{e\ell} + \vec{\varphi}_{i-1} \times \vec{\ell}_i, \tag{4.4.1}$$

where $\vec{u}_i^{e\ell}$ represents the elastic deflection of segment "i" due to its elastic deformation under the action of forces and moments. $\vec{\varphi}_{i-1} \times \vec{\ell}_i$ represents the component of deflexion $\vec{u}_i$ due to the tilt of segment i-1.

One thus obtains the recursive formula for deflexion (4.4.1) or, in matrix notation,

$$u_i = u_{i-1} + u_i^{e\ell} - \underline{\underline{\ell}}_i \varphi_{i-1}, \tag{4.4.2}$$

where $\underline{\underline{\ell}}_i$ is the 3×3 matrix

$$\underline{\underline{\ell}}_i = \begin{bmatrix} 0 & -\ell_{i_z} & \ell_{i_y} \\ \ell_{i_z} & 0 & -\ell_{i_x} \\ -\ell_{i_y} & \ell_{i_x} & 0 \end{bmatrix} \tag{4.4.3}$$

corresponding to the vector $\vec{\ell}_i = \{\ell_{i_x}, \ell_{i_y}, \ell_{i_z}\}$ and facilitates vector multiplication in matrix calculus. The boundary condition for the recursive formula is $u_o = 0$.

For the angular deviation,

$$\vec{\varphi}_i = \vec{\varphi}_{i-1} + \vec{\varphi}_i^{e\ell}, \tag{4.4.4}$$

where $\vec{\varphi}_i^{e\ell}$ is the angular elastic deviation of segment "i" due to its deformation. In matrix notation,

$$\varphi_i = \varphi_{i-1} + \varphi_i^{e\ell}. \tag{4.4.5}$$

Let us now find expressions for the elastic deflexion $\vec{u}_i^{e\ell}$ and tilt $\vec{\varphi}_i^{e\ell}$. Let us consider the segment "i". Fig. 4.8. In order to avoid partial equations, the mass of each cane will be considered as a collection of discrete masses in such a way that the mass m_i of cane "i" is concentrated in only two points S_i and S_{i+1} (at the cane ends). Let us denote the mass at the point S_i by $\mu_i^d = \frac{m_i}{2}$ and in S_{i+1} by $\mu_i^{up} = \frac{m_i}{2}$. Such an approximation permits the simple inclusion of the motor masses in the joints by adding the motor mass in the joint S_i to the mass μ_i^d or μ_{i-1}^{up}. The motor mass is thereby also considered to be concentrated at one point. It is clear that inclusion of the motor masses reduces the error which appears because of mass division.

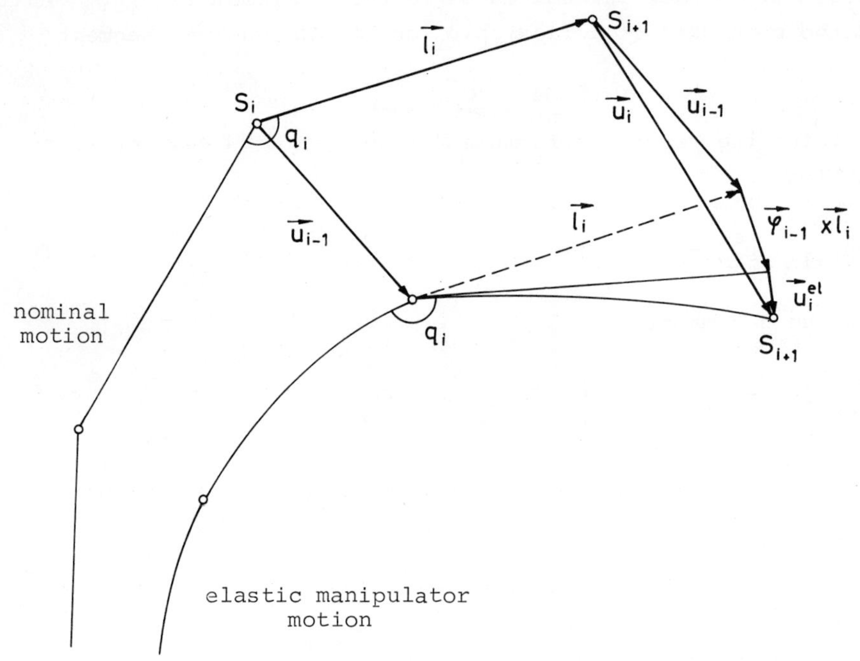

Fig. 4.7. Components of deflexion

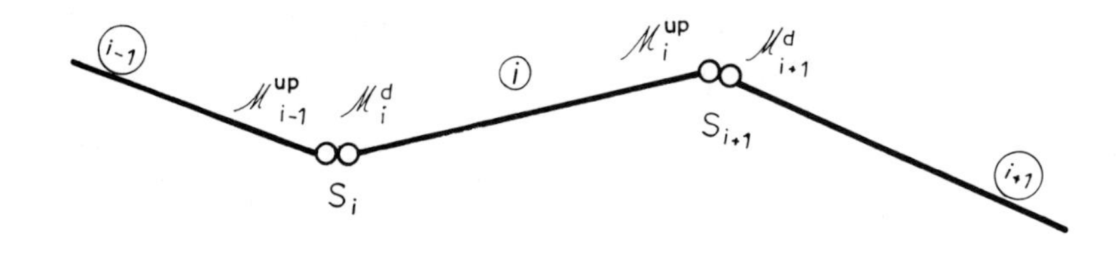

Fig. 4.8. Concentrated masses in joints

For a more exact calculation one could assume that the cane mass is distributed in k concentrated masses arranged along the cane length. However, this would result in much more complex equations, and we consider that the approximation with two concentrated masses satisfies all our needs.

The segment "i" will be considered to have its lower end S_i fixed and the upper end S_{i+1} will be considered free, replacing the action of the next segment by reactions. Thus, the following forces and moments (Fig. 4.9) act on the free end:

- $\vec{F}^n_{S_{i+1}}$ is the nominal force in the joint S_{i+1} determined by (4.3.1).
- $\vec{F}^m_{R_{i+1}}$ is the micro reaction force in joint S_{i+1} due to micro-motion. It was shown earlier, that the nominal reaction $\vec{F}^n_{R_{i+1}}$ is perpendicular to the translation axis if S_{i+1} is a linear joint. But, since the generalized coordinate, i.e., displacement in the joint, does not depend on micro motion and does not permit variation, the micro reaction $\vec{F}^m_{R_{i+1}}$ is not, in general, perpendicular to the translation axis.
- $\vec{M}^n_{S_{i+1}}$ is the nominal moment in S_{i+1} determined by (4.3.2).
- $\vec{M}^m_{R_{i+1}}$ is the micro-moment of the reaction in joint S_{i+1}; by reasoning as in the case of the micro reaction force, one concludes that $\vec{M}^m_{R_{i+1}}$, in general, is not perpendicular to the rotation axis (if the joint S_{i+1} is rotational), which is the case for the nominal reaction moment $\vec{M}^n_{R_{i+1}}$.
- $\mu_i^{up}\vec{g}$ is the gravity force ($\vec{g}$ is the gravitational acceleration of earth $\vec{g} = \{0, \ 0, \ -9.81\}$).
- $\mu_i^{up}\vec{w}_i^n$ is the nominal inertial force ($\vec{w}_i^n$ is the nominal acceleration of the point S_{i+1}).
- $\mu_i^{up}\ddot{\vec{u}}_i$ is the micro inertial force.

In order to make a formalism, let us assume that the reactions $\vec{F}^n_{S_{i+1}}$, $\vec{M}^n_{S_{i+1}}$, $\vec{F}^m_{R_{i+1}}$ and $\vec{M}^m_{R_{i+1}}$ in the joint S_{i+1} act on the next segment i.e. the segment "i+1". So, in this joint, $-\vec{F}^n_{S_{i+1}}$, $-\vec{M}^n_{S_{i+1}}$, $-\vec{F}^m_{R_{i+1}}$ and $-\vec{M}^m_{R_{i+1}}$ act on the segment "i".

The elastic deflexion $u_i^{e\ell}$ is now in matrix form:

$$u_i^{e\ell} = \alpha_i(-F^n_{S_{i+1}} - F^m_{R_{i+1}} + \mu_i^{up}g - \mu_i^{up}w_i^n - \mu_i^{up}\ddot{u}_i) + \beta_i(-M^n_{S_{i+1}} - M^m_{R_{i+1}}) \quad (4.4.6)$$

and the elastic tilt $\varphi_i^{e\ell}$:

$$\varphi_i^{e\ell} = \gamma_i(-F^n_{S_{i+1}} - F^m_{R_{i+1}} + \mu_i^{up}g - \mu_i^{up}w_i^n - \mu_i^{up}\ddot{u}_i) + \delta_i(-M^n_{S_{i+1}} - M^m_{R_{i+1}}). \quad (4.4.7)$$

α_i, β_i, γ_i, δ_i are matrix influence coefficients (3×3). They will be discussed later. In matrix notation μ_i^{up} and μ_i^d are diagonal 3×3 matrices, with the masses along the diagonal:

$$\mu_i^{up} = \begin{bmatrix} \mu_i^{up} & & \\ & \mu_i^{up} & \\ & & \mu_i^{up} \end{bmatrix}, \qquad \mu_i^d = \begin{bmatrix} \mu_i^d & & \\ & \mu_i^d & \\ & & \mu_i^d \end{bmatrix}$$

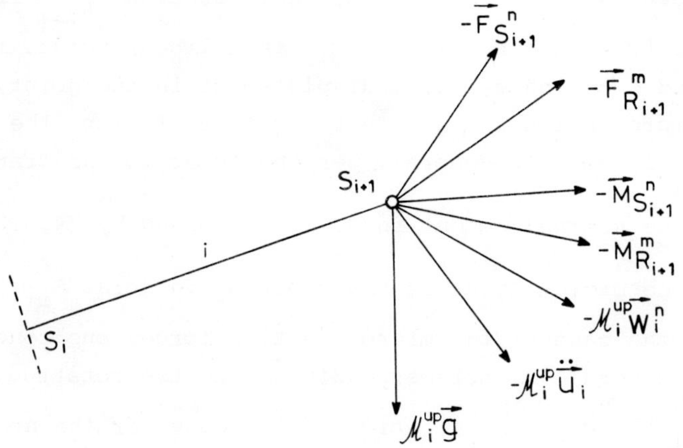

Fig. 4.9. Forces and moments at the "free" end of a segment

Let us now consider the separated segment "i", with the forces and moments acting on it (Fig. 4.10).

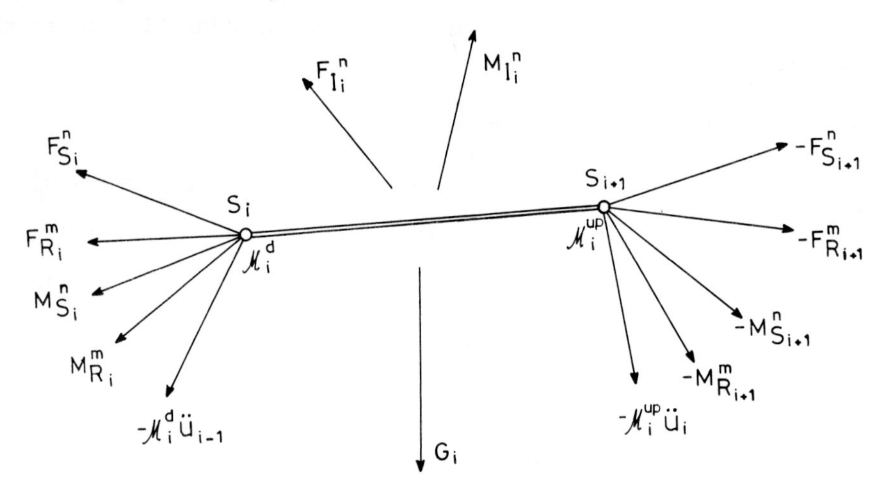

Fig. 4.10. Forces and moments acting on the i-th segment

Let us apply to the segment D´Alambert´s principle of the equilibrium of forces:

$$F^n_{S_i} + F^m_{R_i} + G_i + F^n_{I_i} - \mu^d_i \ddot{u}_{i-1} - \mu^{up}_i \ddot{u}_i - F^n_{S_{i+1}} - F^m_{R_{i+1}} = 0, \tag{4.4.8}$$

where G_i is the total gravity force and $F^n_{I_i}$ is the total inertial force of nominal motion for the whole segment. Dynamic equilibrium of the nominal motion yields

$$F^n_{S_i} + G_i + F^n_{I_i} - F^n_{S_{i+1}} = 0 \tag{4.4.9}$$

so eq. (4.4.8) becomes

$$F^m_{R_i} = \mu^d_i \ddot{u}_{i-1} + \mu^{up}_i \ddot{u}_i + F^m_{R_{i+1}}, \tag{4.4.10}$$

a recursive formula for the reaction micro-forces. Boundary condition for the recursion is $F^m_{R_{n+1}} = 0$.

Further, let us apply to the i-th segment D´Alambert´s principle of equilibrium of moments for the point S_i:

$$M^n_{S_i} + M^m_{R_i} + M_{G_i} + M^n_{IS_i} - \underline{\underline{\ell}}_i \mu^{up}_i \ddot{u}_i - \underline{\underline{\ell}}_i F^n_{S_{i+1}} - \underline{\underline{\ell}}_i F^m_{R_{i+1}} - M^n_{S_{i+1}} - M^m_{R_{i+1}} = 0, \tag{4.4.11}$$

where M_{G_i} is the moment of gravitational forces and $M^n_{IS_i}$ is the nominal moment of the inertial forces relative to the point S_i.

Dynamic equilibrium of nominal motion gives

$$M^n_{S_i} + M_{G_i} + M^n_{IS_i} - \underline{\underline{\ell}}_i F^n_{S_{i+1}} - M^n_{S_{i+1}} = 0, \tag{4.4.12}$$

so (4.4.11) becomes

$$M^m_{R_i} = \underline{\underline{\ell}}_i F^m_{R_{i+1}} + \underline{\underline{\ell}}_i \mu^{up}_i \ddot{u}_i + M^m_{R_{i+1}}. \tag{4.4.13}$$

One thus obtains the recursive formula for micro-moments of reactions. The boundary condition is $M^m_{R_{n+1}} = 0$.

The equations (4.4.2), (4.4.5), (4.4.6), (4.4.7) and (4.4.10), (4.4.13), $i=1,\ldots,n$ determine the mathematical model of micro-dynamics and permit the calculation of $\ddot{u}_1,\ldots,\ddot{u}_n$. It will later be shown how this set of equations can be written in matrix notation and reduced to n equations. The reduced system will also be written in matrix form (4.2.2).

Deriving the model in matrix form. We will now show how, by introducing block-matrices, the mathematical model of micro motion can be written in a suitable matrix form (4.2.2).

From the recursive relation (4.4.2) it follows that

$$u_i = \sum_{k=1}^{i} u_k^{e\ell} - \underline{\underline{\ell}}_k \varphi_{k-1}; \quad \varphi_o = 0. \tag{4.4.14}$$

By substituting (4.4.6) in (4.4.14) one finds

$$u_i = \sum_{k=1}^{i} -\alpha_k (F_{S_{k+1}}^n + F_{R_{k+1}}^m) + \alpha_k \mu_k^{up}(g - w_k^n - \ddot{u}_k) + \beta_k (-M_{S_{k+1}}^n - M_{R_{k+1}}^m) - \underline{\underline{\ell}}_k \varphi_{k-1} \tag{4.4.15}$$

Likewise, from the recursive relations for tilt (4.4.5):

$$\varphi_i = \sum_{k=1}^{i} \varphi_k^{e\ell} \tag{4.4.16}$$

and by substituting from (4.4.7),

$$\varphi_i = \sum_{k=1}^{i} -\gamma_k (F_{S_{k+1}}^n + F_{R_{k+1}}^m) + \gamma_k \mu_k^{up}(g - w_k^n - \ddot{u}_k) + \delta_k (-M_{S_{k+1}}^n - M_{R_{k+1}}^m). \tag{4.4.17}$$

Let us now introduce the (3n×1) block-vectors. Let a_i, i=1,...,n be a set of 3×1 vectors. Now introduce the block-vector

$$a = \begin{bmatrix} a_1 \\ \vdots \\ a_n \end{bmatrix}. \tag{4.4.18}$$

Likewise, introduce the block vectors F_S^n, F_R^m, w^n, M_S^n, M_R^m, φ, u.

$$g = \begin{bmatrix} g \\ \vdots \\ g \end{bmatrix} \tag{4.4.19}$$

Further, let us introduce diagonal block-matrices of dimensions 3n×3n. Let $b_1,\ldots,b_n$ be a set of 3×3 matrices. Then the diagonal block-matrix is

$$b = \begin{bmatrix} b_1 & & 0 \\ & \ddots & \\ 0 & & b_n \end{bmatrix} \tag{4.4.20}$$

We shall also use the following notation:

$$b_{(dt)} = \begin{bmatrix} b_1 & & & & & & & \\ b_1 & b_2 & & & 0 & & & \\ \cdot & \cdot & \cdot & & & & & \\ b_1 & b_2 & \cdots & b_i & & & & \\ \cdot & \cdot & \cdot & & & & & \\ b_1 & b_2 & \cdots & b_i & \cdots & b_n \end{bmatrix} \quad (a) \qquad (4.4.21)$$

$$b_{(upt)} = \begin{bmatrix} b_1 & b_2 & \cdots & b_i & \cdots & b_n \\ & b_2 & \cdots & b_i & \cdots & b_n \\ & & & & & \cdots \\ & & & b_i & \cdots & b_n \\ & & & & & \cdots \\ & 0 & & & & b_n \end{bmatrix} \quad (b)$$

$$b^{(r)}_{(dt)} = \begin{bmatrix} 0 & b_1 & & & & & & & \\ 0 & b_1 & b_2 & & & & 0 & & \\ \cdot & \cdot & \cdot & & & & & & \\ 0 & b_1 & b_2 & \cdots & b_{i-1} & b_i & & & \\ \cdot & \cdot & \cdot & & & & & & \\ 0 & b_1 & b_2 & \cdots & b_{i-1} & b_i & \cdots & b_{n-1} \\ 0 & b_1 & b_2 & & b_{i-1} & b_i & \cdots & b_{n-1} \end{bmatrix} \quad (c)$$

$$b^{(r)}_{(upt)} = \begin{bmatrix} 0 & b_1 & b_2 & \cdots & b_{i-1} & b_i & \cdots & b_{n-1} \\ & & b_2 & \cdots & b_{i-1} & b_i & \cdots & b_{n-1} \\ & & & & & & & \cdots \\ & & & & & b_i & \cdots & b_{n-1} \\ & & & & & & & \cdots \\ & & 0 & & & & & b_{n-1} \\ & & & & & & & 0 \end{bmatrix} \quad (d)$$

$$
b^{(\ell)}_{(dt)} = \begin{bmatrix}
0 & & & & & & & \\
b_2 & & & & & 0 & & \\
b_2 & b_3 & & & & & & \\
\cdots & & & & & & & \\
b_2 & b_3 & \cdots & b_i & & & & \\
\cdots & & & & & & & \\
b_2 & b_3 & \cdots & b_i & & \cdots & b_n & 0
\end{bmatrix} \qquad (e)
$$

$$
b^{(\ell)}_{(upt)} = \begin{bmatrix}
b_2 & b_3 & \cdots & b_i & b_{i+1} & \cdots & b_n & 0 \\
b_2 & b_3 & \cdots & b_i & b_{i+1} & \cdots & b_n & 0 \\
 & b_3 & \cdots & b_i & b_{i+1} & \cdots & b_n & 0 \\
 & & & & & & & \\
 & & & b_i & b_{i+1} & \cdots & b_n & 0 \\
 & 0 & & & & & \cdots & \\
 & & & & & & b_n & 0
\end{bmatrix} \qquad (f)
$$

We also use the following block-matrices for index shifting ($3n\times 3n$):

$$
\sigma^{(\ell)} = \begin{bmatrix}
0 & \cdots & \cdots & 0 & 0 \\
I & \ddots & 0 & & 0 \\
 & & \ddots & & \vdots \\
 & 0 & & I & 0
\end{bmatrix}
\qquad
\sigma^{(r)} = \begin{bmatrix}
0 & I & \ddots & 0 & \\
\vdots & & & \ddots & \\
0 & & 0 & & I \\
0 & 0 & \cdots & \cdots & 0
\end{bmatrix}
\qquad (4.4.22)
$$

and the summation block-matrices ($3n\times 3n$)

$$
\Sigma_{(dt)} = \begin{bmatrix}
I & \ddots & & 0 \\
\vdots & & \ddots & \\
\vdots & & & \ddots \\
I & \cdots & \cdots & I
\end{bmatrix},
\qquad
\Sigma_{(upt)} = \begin{bmatrix}
I & \cdots & \cdots & I \\
 & \ddots & & \vdots \\
 & 0 & \ddots & \vdots \\
 & & & I
\end{bmatrix}
\qquad (4.4.23)
$$

where I is a (3×3) unit matrix.

Now

$$b^{(\ell)}_{(dt)} = b_{(dt)}\sigma^{(\ell)}, \quad b^{(r)}_{(dt)} = b_{(dt)}\sigma^{(r)},$$
$$b^{(\ell)}_{(upt)} = b_{(upt)}\sigma^{(\ell)}, \quad b^{(r)}_{(upt)} = b_{(upt)}\sigma^{(r)}. \qquad (4.4.24a)$$

and

$$b_{(dt)} = \sum_{(dt)} b, \qquad b_{(upt)} = \sum_{(upt)} b. \qquad (4.4.24b)$$

Let us introduce the following block-diagonal matrices: α, β, γ, δ, μ^{up}, μ^{d}, $\underline{\underline{\ell}}$.

For the sake of clarity, block-matrices will be introduced gradually. Let us consider the expression (4.4.15). We first introduce the block--vectors. Then (4.4.15) can be written in the form

$$\begin{aligned} u_i = & -[0 \;\; \alpha_1 \cdots\cdots \alpha_i \;\; 0 \cdots 0](F^n_S + F^m_R) + \\ & +[\alpha_1 \;\; \alpha_2 \cdots \alpha_i 0 \cdots\cdots 0]\mu^{up}(g-w^n-\ddot{u}) + \\ & -[0 \;\; \beta_1 \cdots\cdots \beta_i \;\; 0 \cdots 0](M^n_S + M^m_R) + \\ & -[\underline{\underline{\ell}}_2 \;\; \underline{\underline{\ell}}_3 \cdots \underline{\underline{\ell}}_i \;\; 0 \cdots\cdots 0]\varphi, \quad i=1,\ldots,n. \end{aligned} \qquad (4.4.25)$$

By using block-matrices expresssions (4.4.25), i=1,...,n can be written together:

$$u = -\alpha^{(r)}_{(dt)}(F^n_S+F^m_R)+\alpha_{(dt)}\mu^{up}(g-w^n-\ddot{u})-\beta^{(r)}_{(dt)}(M^n_S+M^m_R)-\underline{\underline{\ell}}^{(\ell)}_{(dt)}\varphi \qquad (4.4.26)$$

or, using (4.4.24),

$$\begin{aligned} u = \sum_{(dt)} [& \alpha(-\sigma^{(r)}F^n_S - \sigma^{(r)}F^m_R + \mu^{up}g - \mu^{up}w^n - \mu^{up}\ddot{u}) + \\ & \beta(-\sigma^{(r)}M^n_S - \sigma^{(r)}M^m_R) - \underline{\underline{\ell}}\sigma^{(\ell)}\varphi]. \end{aligned} \qquad (4.4.27)$$

Likewise, the expressions (4.4.17) for i=1,...,n can be written together in the form:

$$\begin{aligned} \varphi = \sum_{(dt)} [& \gamma(-\sigma^{(r)}F^n_S - \sigma^{(r)}F^m_R + \mu^{up}g - \mu^{up}w^n - \mu^{up}\ddot{u}) + \\ & + \delta(-\sigma^{(r)}M^n_S - \sigma^{(r)}M^m_R)]. \end{aligned} \qquad (4.4.28)$$

From the recursive expression for micro reaction forces (4.4.10) it follows that

$$F^m_{R_i} = \sum_{k=i}^{n} \mu^d_k \ddot{u}_{k-1} + \mu^{up}_k \ddot{u}_k \tag{4.4.29}$$

and from the recursive expression for micro reaction moments (4.4.13),

$$M^m_{R_i} = \sum_{k=i}^{n} \underline{\underline{\ell}}_k F^m_{R_{k+1}} + \underline{\underline{\ell}}_k \mu^{up}_k \ddot{u}_k \tag{4.4.30}$$

By introducing the block-matrices, the relations (4.4.29) for i=1,...,n can be written together in the form

$$F^m_R = \mu^{d(\ell)}_{(upt)} \ddot{u} + \mu^{up}_{(upt)} \ddot{u} \tag{4.4.31}$$

and from (4.4.24),

$$F^m_R = \sum_{(upt)} (\mu^d \sigma^{(\ell)} + \mu^{up}) \ddot{u} \tag{4.4.32}$$

Analogously, the expressions (4.4.30), for i=1,...,n can be written together:

$$M^m_R = \sum_{(upt)} (\underline{\underline{\ell}} \sigma^{(r)} F^m_R + \underline{\underline{\ell}} \mu^{up} \ddot{u}) = \sum_{(upt)} \underline{\underline{\ell}} (\sigma^{(r)} F^m_R + \mu^{up} \ddot{u}) . \tag{4.4.33}$$

The matrix equations (4.4.27), (4.4.28), (4.4.32) and (4.4.33) determine the micro motion mathematical model. By elimination, these four matrix equations can be reduced to

$$D\ddot{u} + u = c \tag{4.4.34}$$

where

$$D = \sum_{(dt)} (f + h\sigma^{(r)} \sum_{(upt)} \underline{\underline{\ell}}) [\mu^{up} + \sigma^{(r)} \sum_{(upt)} (\mu^d \sigma^{(\ell)} + \mu^{up})] \tag{4.4.35a}$$

and

$$f = \alpha - \underline{\underline{\ell}} \sigma^{(\ell)} \sum_{(dt)} \gamma, \quad h = \beta - \underline{\underline{\ell}} \sigma^{(\ell)} \sum_{(dt)} \delta, \tag{4.4.35b}$$

$$c = \sum_{(dt)} [f(-\sigma^{(r)} F^n_S + \mu^{up}(g - w^n)) - h\sigma^{(r)} M^n_S] . \tag{4.4.36}$$

The matrices D and c, determine the mathematical model of micro-motion, for known nominal dynamics. These matrices are calculated from expressions (4.4.35) and (4.4.36). The calculation is carried out for each time instant. Thus the micro-dynamics block, or the block of dynamic analysis of micro-motion, has been completed. The block for calculating D and c is incorporated into the simulation algorithm according to Fig. 4.3. After calculating $\ddot{u}$ from (4.4.34), other micro-dynamic values, F_R^m, M_R^m, φ, are calculated from (4.4.32), (4.4.33) and (4.4.28).

We should also mention that the quasi-static deflexion is determined by $\ddot{u} = 0$, i.e.,

$$u^{qs} = c. \tag{4.4.37}$$

This deflexion results from static and nominal inertial forces.

4.5. Influence Coefficients

As we said earlier, each segment "i" is considered to be fixed at its lower end S_i. The reduced mass of the cane μ_i^{up} is concentrated at its free end S_{i+1} where there are forces and moments (Fig. 4.9). We now derive the matrix influence coefficients of that segment.

Let us consider the cane AB with its fixed end at the point A and with reduced mass μ^B concentrated in point B (Fig. 4.11).

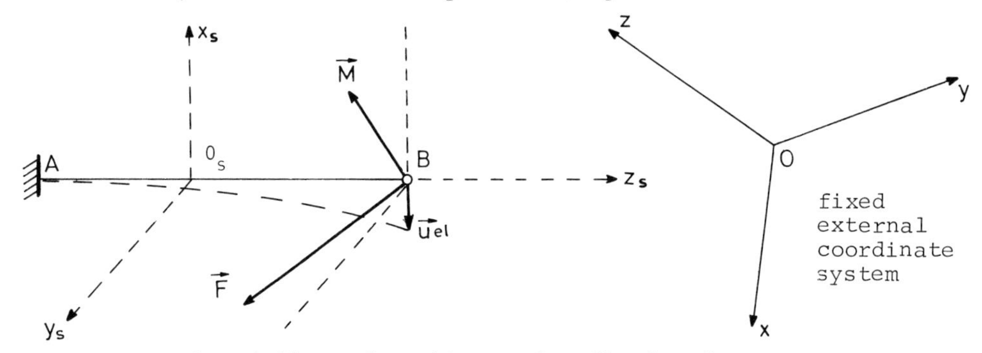

Fig. 4.11. Deformations of a fixed-end cane

Let us note the coordinate systems. System $O_s x_s y_s z_s$ is the coordinate system of the cane and Oxyz is the absolute, or external, coordinate system. Suppose that at the point B there is an arbitrary force $\vec{F}$ and moment $\vec{M}$. Let the cane be deformed elastically so that at the point B three is a linear deviation $\vec{u}^{e\ell}$ and angular deviation $\vec{\varphi}^{e\ell}$. For $\vec{u}^{e\ell}$ in the cane coordinate system the equations of elastic displacements can be written

$$u^{e\ell}_{x_s} = \alpha_{x_s}(F_{x_s} - \mu^B w^B_{x_s}) + \beta_{x_s} M_{y_s},$$

$$u^{e\ell}_{y_s} = \alpha_{y_s}(F_{y_s} - \mu^B w^B_{y_s}) - \beta_{y_s} M_{x_s}, \tag{4.5.1}$$

$$u^{e\ell}_{z_s} = \alpha_{z_s}(F_{z_s} - \mu^B w^B_{z_s}) + 0 \cdot M_{z_s},$$

where

(i) $u^{e\ell}_{x_s}$ is the projection (component) of the linear deviation (deflexion) onto the x_s-axis, F_{x_s} and M_{x_s} are projections onto the same axis of the force and moment, and $w^B_{x_s}$ is the projection of the acceleration of point B. Likewise for the other axes, y_s and z_s.

(ii) α_{x_s} is the influence coefficient for the bending deflexion along the x_s-axis, under the action of the force at the point B. Similarly, α_{y_s} is defined in terms of the y_s-axis.

(iii) α_{z_s} is the influence coefficient for extension along the z_s--axis under the action of the force at B.

(iv) β_{x_s} is the influence coefficient for bending deflexion along the x_s-axis due to the moment acting at B. Likewise for the y_s-axis.

All the coefficients α and β are for linear deviation, i.e., deflexion.

The eqs. (4.5.1) can be combined:

$$\tilde{\vec{u}}^{e\ell} = \tilde{\alpha}(\tilde{\vec{F}} - \mu^B \tilde{\vec{w}}^B) + \tilde{\beta}\tilde{\vec{M}}, \tag{4.5.2}$$

where

$$\tilde{\alpha} = \begin{bmatrix} \alpha_{x_s} & 0 & 0 \\ 0 & \alpha_{y_s} & 0 \\ 0 & 0 & \alpha_{z_s} \end{bmatrix}, \qquad \tilde{\beta} = \begin{bmatrix} 0 & \beta_{x_s} & 0 \\ -\beta_{y_s} & 0 & 0 \\ 0 & 0 & 0 \end{bmatrix} \tag{4.5.3}$$

and the tilde over the vector means that the vector is expressed by three projections onto the axes of the cane coordinate system $O_s x_s y_s z_s$.

The vectors without tilde will denote three projections onto the axes of the absolute (external) system.

Let us introduce the transition matrix. Let A be the transition matrix from the proper to the external coordinate system. Then

$$\vec{F} = A\tilde{\vec{F}}, \qquad \tilde{\vec{F}} = A^{-1}\vec{F} \tag{4.5.4}$$

and likewise for other vectors. Now eq. (4.5.2) can be written in terms of the external system:

$$\vec{u}^{e\ell} = \alpha(\vec{F}-\mu^B\vec{w}^B) + \beta\vec{M}, \tag{4.5.5}$$

where

$$\alpha = A\tilde{\alpha}A^{-1}, \qquad \beta = A\tilde{\beta}A^{-1} \tag{4.5.6}$$

and the vectors are expressed in terms of the external system.

By a procedure like that for linear displacement, the equation for angular displacement $\vec{\varphi}^{e\ell}$ is found to be

$$\begin{aligned}
\varphi^{e\ell}_{x_s} &= -\gamma_{x_s}(F_{y_s} - \mu^B w^B_{y_s}) + \delta_{x_s} M_{x_s}, \\
\varphi^{e\ell}_{y_s} &= \gamma_{y_s}(F_{x_s} - \mu^B w^B_{x_s}) + \delta_{y_s} M_{y_s}, \\
\varphi^{e\ell}_{z_s} &= 0\cdot(F_{z_s} - \mu^B w^B_{z_s}) + \delta_{z_s} M_{z_s},
\end{aligned} \tag{4.5.7}$$

where

(i) γ_{x_s} is the influence coefficient for the bending angle around the x_s-axis due to the force acting at B. Likewise for the y_s-axis.

(ii) δ_{x_s} is the influence coefficinet for the bending angle around the x_s-axis due to the moment acting at B. Likewise for the y_s-axis.

(iii) δ_{z_s} is the influence coefficient for the torsion around the z_s-axis due to the moment acting at B.

Let us combine the relations (4.5.7):

$$\tilde{\vec{\rho}}^{e\ell} = \tilde{\gamma}(\tilde{\vec{F}} - \mu^B \tilde{\vec{w}}^B) + \tilde{\delta}\tilde{\vec{M}}, \tag{4.5.8}$$

where

$$\tilde{\gamma} = \begin{bmatrix} 0 & -\gamma_{x_s} & 0 \\ \gamma_{y_s} & 0 & 0 \\ 0 & 0 & 0 \end{bmatrix}, \qquad \tilde{\delta} = \begin{bmatrix} \delta_{x_s} & 0 & 0 \\ 0 & \delta_{y_s} & 0 \\ 0 & 0 & \delta_{z_s} \end{bmatrix} \tag{4.5.9}$$

Introducing the transition matrix, (4.5.8) becomes

$$\vec{\rho}^{e\ell} = \gamma(\vec{F} - \mu^B \vec{w}^B) + \delta\vec{M}, \tag{4.5.10}$$

where

$$\gamma = A\tilde{\gamma}A^{-1}, \qquad \delta = A\tilde{\delta}A^{-1}. \tag{4.5.11}$$

Thus we have derived the procedure for the transformation of the influence coefficients. The coefficients α_{x_s}, α_{y_s}, α_{z_s}, β_{x_s}, β_{y_s}, γ_{x_s}, γ_{y_s}, δ_{x_s}, δ_{y_s}, δ_{z_s} hold for the cane coordinate system and can be found from tables, for a given form of cross section. By using the procedure described, the elastic equations are written and used in the external coordinate system. This is necessary for the algorithm described in (4.4). Note that eqs. (4.5.5) and (4.5.10) also hold when the fixed end point A of the cane is moving, in which case, the cane coordinate system is moving too. In this case, $\vec{w}^B$ is the absolute acceleration of the point B with respect to the external, immobile system, $\vec{u}_i^{e\ell}$ and $\vec{\rho}_i^{e\ell}$ represent the cane elastic deflexion and tilt, and the transition matrix A and the matrix coefficients α, β, γ, δ are functions of time. Let us now apply these considerations to the articulated mechanism - the manipulator. As we said earlier, the segment "i" can be regarded as having a fixed-end at a moving point S_i (Fig. 4.9, 4.12). Taking into account the assumption that displacements from nominal motion are small, the matrix influence coefficients will be calculated for nominal motion. The b. - f. system of the i-th segment introduced

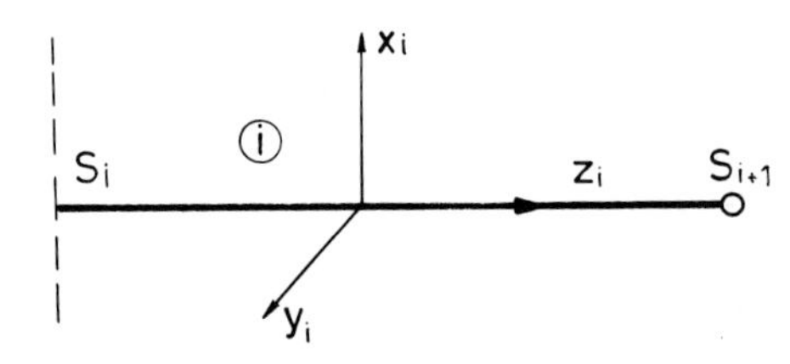

Fig. 4.12. Segment as a fixed-end cane

earlier i.e. $O_i x_i y_i z_i$ represents the cane coordinate system. Thus, the influence coefficients $\alpha_{x_s}, \ldots, \delta_{z_s}$ are taken from tables for the adopted cross-section and they represent input data. The algorithm produces the matrices $\tilde{\alpha}_i$, $\tilde{\beta}_i$, $\tilde{\gamma}_i$, $\tilde{\delta}_i$ according to (4.5.3), (4.5.9) and then the matrix coefficients α_i, β_i, γ_i, δ_i according to

$$\begin{aligned} \alpha_i &= A_i \tilde{\alpha}_i A_i^{-1}, \qquad \beta_i = A_i \tilde{\beta}_i A_i^{-1} \\ \gamma_i &= A_i \tilde{\gamma}_i A_i^{-1}, \qquad \delta_i = A_i \tilde{\delta}_i A_i^{-1}. \end{aligned} \tag{4.5.12}$$

The transition matrix A_i was calculated in the block of nominal dynamics.

In practical manipulator realizations, the most common segment forms are cylindrical or rectangular tubes (Figs. 4.13. and 4.14); so it is such cross-sections which determine from the tables the expressions for the influence coefficients. Let us first consider a cane in the form of a cylindrical tube with one fixed end (Fig. 4.13).

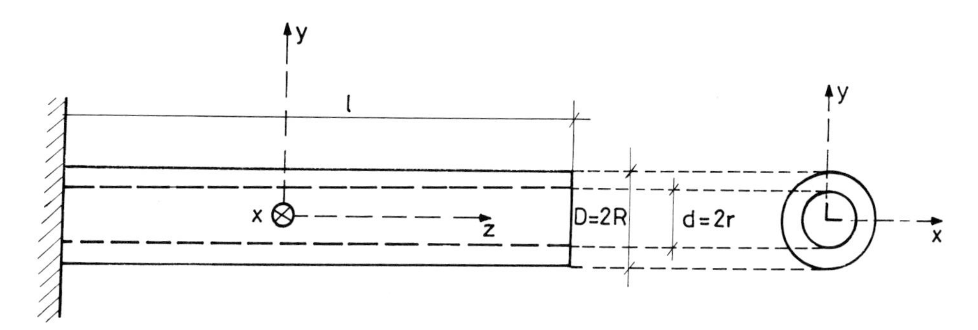

Fig. 4.13. Cylindrical tube segment

The moments of inertia for such a cross-section are

$$I_x = I_y = \frac{R^4 \pi}{4}(1-\psi^4) = \frac{D^4 \pi}{64}(1-\psi^4) \tag{4.5.13}$$

where R is the external radius, D = 2R the outer diameter, $\psi = r/R$ the ratio of the internal and external radii.

If at the free end of the cane a force is acting, the influence coefficients α_x, α_y of the deflexion due to bending and the in fluence coefficient α_z due to extension are

$$\alpha_x = \frac{\ell^3}{3EI_y}, \qquad \alpha_y = \frac{\ell^3}{3EI_x}, \qquad \alpha_z = \frac{\ell}{EA}, \tag{4.5.14}$$

where E is Young´s modulus for the adopted material and A is the area of cross-section:

$$A = R^2\pi(1-\psi^2) = \frac{D^2\pi}{4}(1-\psi^2)$$

The influence coefficients γ_x, γ_y of the tilt due to bending are

$$\gamma_x = \frac{\ell^2}{2EI_x}, \qquad \gamma_y = \frac{\ell^2}{2EI_y}, \qquad \gamma_z = 0. \tag{4.5.15}$$

If at the end of the cane a moment is acting, the influence coefficients of the deflection due to bending are

$$\beta_x = \frac{\ell^2}{2EI_y}, \qquad \beta_y = \frac{\ell^2}{2EI_x}, \qquad \beta_z = 0 \tag{4.5.16}$$

and the influence coefficients δ_x, δ_y of the tilt due to bending and δ_z due to torsion are

$$\delta_x = \frac{\ell}{EI_x}, \qquad \delta_y = \frac{\ell}{EI_y}, \qquad \delta_z = \frac{32\ell}{(1-\psi^4)GD^4\pi}, \tag{4.5.17}$$

where G is the torsion modulus:

$$G = \frac{E}{2(1+\nu)}, \tag{4.5.18}$$

where ν is Poisson´s coefficient. $\nu = 0.29$ for aluminium.

Let us now consider a cane in the form of a rectangular tube (Fig. 4.14).

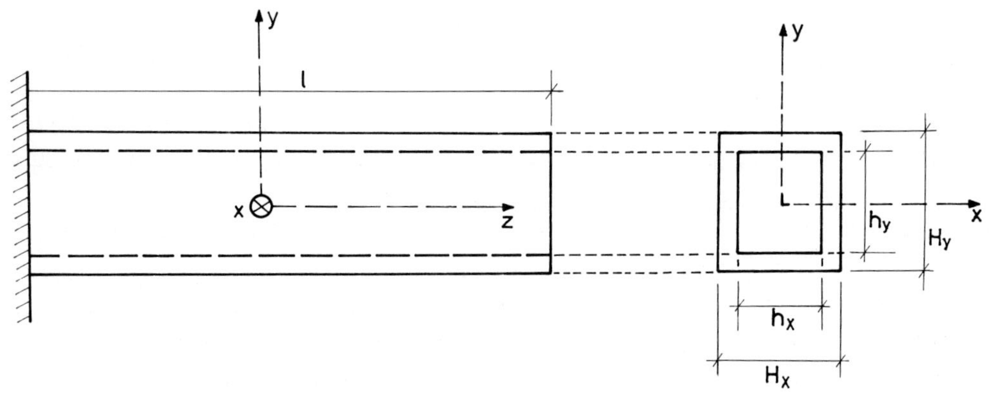

Fig. 4.14. Segment in the form of a rectangular tube

The moments of inertia of this cross-section are

$$I_x = \frac{H_x H_y^3 - h_x h_y^3}{12}, \qquad I_y = \frac{H_x^3 H_y - h_x^3 h_y}{12}. \tag{4.5.19}$$

For the moments of inertia so calculated, the expressions for the influence coefficients (4.5.14) - (4.5.17) are valid. The difference is that for the influence coefficients α_z due to extension in expression (4.5.14) the area A is calculated as $A = H_x H_y - h_x h_y$. There is also a difference in the case of torsion (4.5.17), where

$$\delta_z = \frac{\ell}{GJ}, \qquad J = \frac{1}{12}(H_x^3 H_y + H_x H_y^3 - h_x^3 h_y - h_x h_y^3).$$

4.6. Results of the Method and Their Application

As already explained, the micro-dynamic analysis yields time functions of the deflexion (the block-vector u), of the tilt (the block-vector φ), and the micro-reactions (F_R^m, M_R^m). In other words, one obtains $\vec{u}_i$, $\vec{\varphi}_i$, $\vec{F}_{R_i}^m$, $\vec{M}_{R_i}^m$, $i=1,\ldots,n$ for a series of time instants.

Let us now consider the micro-reactions in joint S_i and let us introduce the value

$$s_i \vec{F}_{R_i}^m + (1-s_i)\vec{M}_{R_i}^m . \tag{4.6.1}$$

By projecting this value onto the axis $\vec{e}_i$ (the rotational or linear joint axis), the supplementary component of motor load due to elastic deformations is

$$\Delta P_i = \vec{e}_i \{ s_i \vec{F}_{R_i}^m + (1-s_i)\vec{M}_{R_i}^m \}, \tag{4.6.2}$$

The algorithm desribed in this chapter represents a simplified method for the calculation of elastic oscillations. The computation which has to be performed in each time instant is easy enough and not time-consuming. But, the frequency of elastic oscillations will couse small subintervals for integration, if precise calculation is desired. So, if we consider some longer manipulation task, the whole simulation procedure may become time-consuming. There are several ways of surmaunting this disadvantage. For instance, it is convenient to separate the integration of nominal and of micro dynamics and to use different time increments. It would accelerate the algorithm to some extent. Another approach may also be applied. From the standpoint of micro-dynamics, we may consider the system (4.2.2) as a stationary one over the nominal time subinterval Δt and use the analytical solutions.

If one considers a manipulator with six segments and six degrees of freedom, it is most interesting to know the deflection of the tip, $\vec{u}_6$, and the tilt on the tip, $\vec{\varphi}_6$. One can thus determine the magnitude of the positioning and orientation error (with respect to nominal motion), as a consequence of segments elasticity. This is particularly profitable in the so-called "dynamic method for evaluation and choice of industrial manipulators", i.e. computer-aided design [19-22], and in that constraints are introduced for the maximum elastic displacement. In that case, we are interested mainly in the maximal value of elastic deviation. If computer time saving is necessary, we will not compute the whole micro dynamics, but make some estimations. If a manipulator is not too fast and it moves with no sudden accelerations or decelerations, then, we may neglect the amplitude of oscillations and use only the quasi static deviation (4.4.37). For faster manipulators with sudden changes in acceleration, we have to estimate the amplitude of oscillations and find some upper bound of the whole deviation. After some numerical investigations, we suggest the introduction of an amplitude limit value a^m which is added to the quasistatic deviation in order to estimate the maximal deviation:

$$u^m = u^{qs} + a^m \qquad (4.6.3)$$

The value a_m is changed after each sudden change in acceleration, according to

$$a^{m(new)} = |u^{qs(new)} - u^{qs(old)}| + a^{m(old)} \qquad (4.6.4)$$

So, during the simulation, the deviation u^m is computed for each time instant and compared with the constraint.

Let us also mention that in the case of computing only the quasi static deviation, it is not necessary to discretize the segment mass, but it may be considered as continual which leads to more precise calculation.

4.7. Example

We consider a cylindrical manipulator, illustrated in Fig. 4.15. The segments are taken in the form of cylindrical tubes of outer radius R = 0.022 meters and inner radius r = 0.016 meters. Construction material is a light aluminium alloy AℓMg 3. The joint with three degrees of freedom, by which the gripper is connected to the minimal configuration, is articulated into three series-connected simple rotational joints according to the real axes in the complex joint.

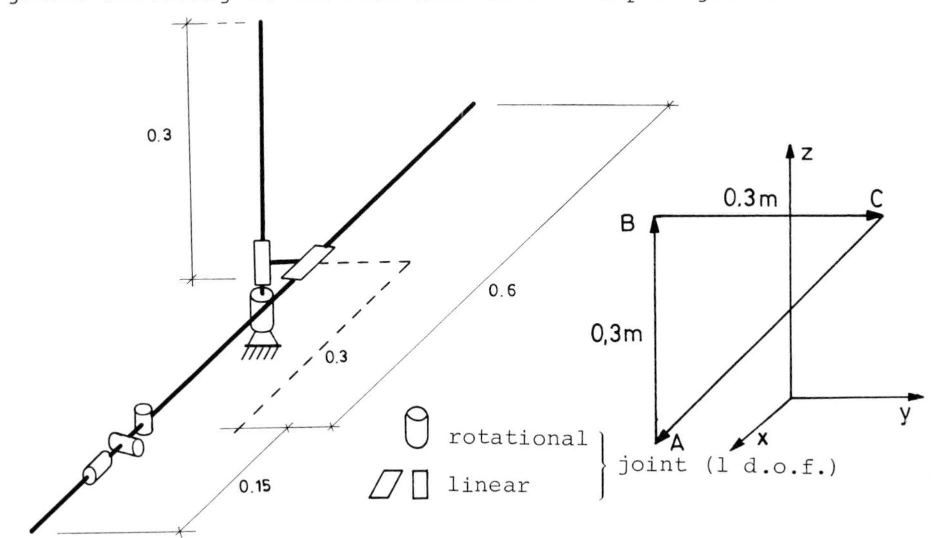

Fig. 4.15. Cylindrical manipulator with 6 d.o.f.

Fig. 4.16. Trajectory of object transfer

The manipulation task consists in transfering a working object of 5 kilograms mass along trajectory ABCA (Fig. 4.16), maintaining a constant orientation in space of the gripper. On the rectilinear portions of the trajectory, the velocity profile is traiangular. Total time of the task is T = 3 s and each of the three rectilinear portions is passed in the same time $T_1 = T_2 = T_3 = T/3 = 1$ s. The initial position of the manipulator is given in Fig. 4.15.

The results of the nominal dynamics were presented in Chapter III. Fig. 4.17. illustrates the simulation results of micro-motion. The quasi static displacement (determined by $\ddot{u} = 0$ in (4.2.2)) is given by (4.4.37). This quasi static displacement is a consequence of static forces and nominal inertial forces. Fig. 4.17. illustrate the time history of such quasi static displacement at the manipulator tip, i.e.

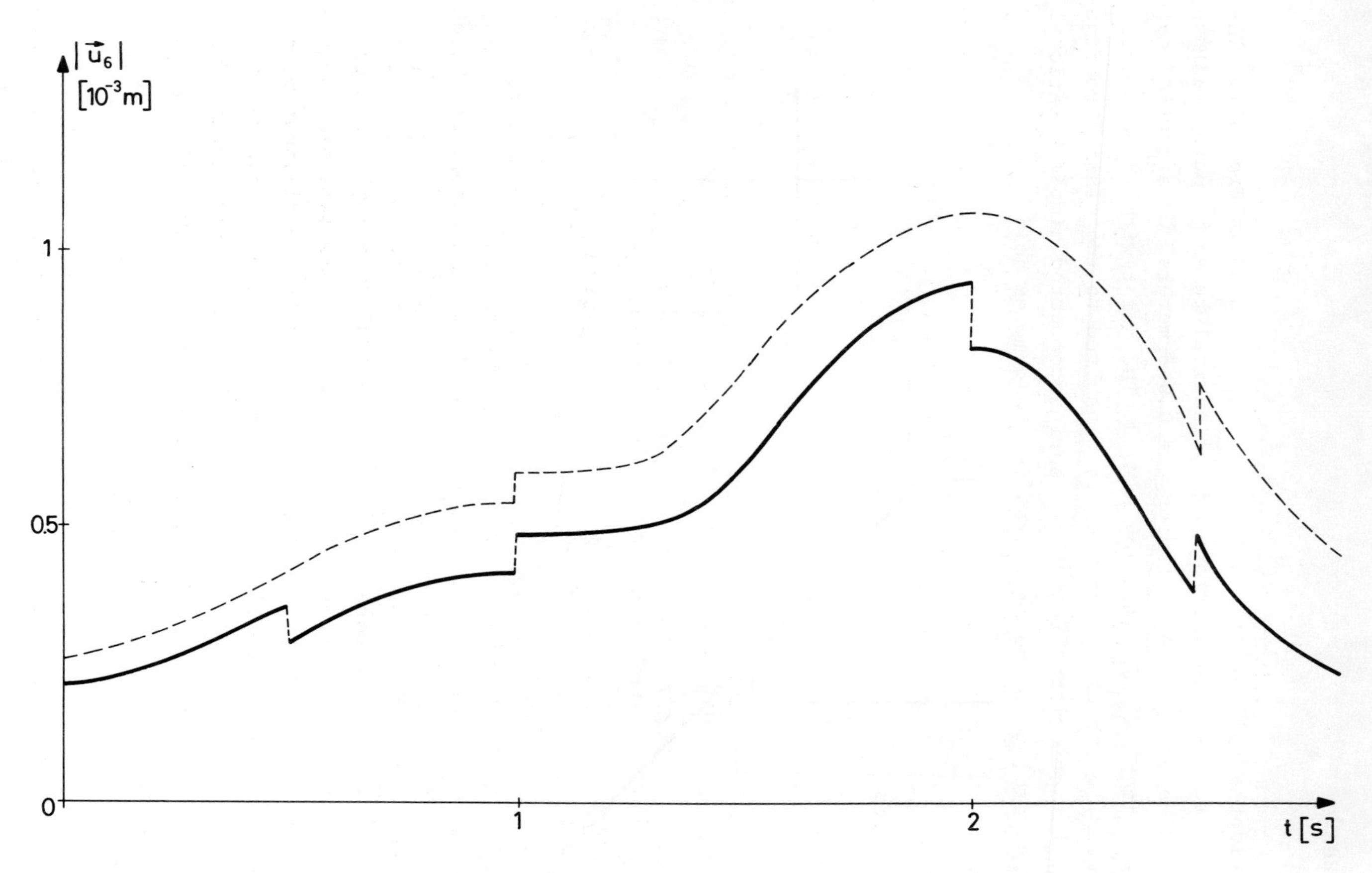

Fig. 4.17. Module of quasistatic elastic deflection of manipulator tip

of u_6^{qs}. The elastic displacement is a vector and to the vertical coordinate axis were applied the vector intensities, i.e., the absolute value of the manipulator tip displacement from its nominal motion. The dotted curve represents $u^m = u^{qs} + a^m$.

The deflexion, as a movement, must be a continual time function. Hence, the discontinuities in Fig. 4.16. deserve explanation. As we said the figure illustrates the quasi static deflexion due to statical and nominal inertial forces. As the nominal acceleration changes abruptly at particular points of the trajectory, the inertial forces also change abruptly, and so too does the quasi static deflexion, i.e., discontinuities appear.

References

[1] Stepanenko Yu., "Method for Analysis of Spatial Lever Mechanisms", (in Russian), Mekhanika mashin, Vol. 23, Moscow, 1970.

[2] Juričić D., Vukobratović M., "Mathematical Modelling of a Bipedal Walking System", ASME Publ. 27-WA/BHF-13.

[3] Vukobratović M., Stepanenko Yu., "Mathematical Models of General Anthropomorphic Systems", Math. Biosciences, Vol. 17, 1973.

[4] Stepanenko Yu., Vukobratović M., "Dynamics of Arbiculated Open-Chain Active Mechanisms", Math. Biosciences, Vol. 28, No 1/2, 1976.

[5] Vukobratović M., "Computer Method for Mathematical Modelling of Active Kinematic Chains via Generalized Coordinates", Journal of Mechanisms and Machine Theory, Vol. 13, No 1, 1978.

[6] Vukobratović M., Legged Locomotion Robots and Anthropomorphic Mechanisms, Monograph, Institute "Mihailo Pupin", Beograd, Yugoslavia, 1975.

[7] Orin D., Vukobratović M., R.B. Mc Ghee., G.Hartoch, "Kinematic and Kinetic Analysis of Open-Chain Linkoges Utilizing Newton-Euler Methods", Math. Biosciences, Vol. 42, 1978.

[8] Vukobratović M., "Synthesis of Artificial Motion", Journal of Mechanisms and Machine Theory, Vol. 13, No 1, 1979.

[9] Vukobratović M., Potkonjak V., "Contribution to the Forming of Computer Methods for Automatic Modelling of Active Spatial Mechanisms Motion", PART1: "Method of Basic Theoremes of Mechanics", Journal of Mechanisms and Machine Theory, Vol. 14, No 3, 1979.

[10] Vukobratović M., Potkonjak V., "Contribution to the Computer Methods for Generation of Active Mechanisms via Basic Theorems of Mechanics", Teknitcheskaya kibernetika ANUSSR, No 2, 1979.

[11] Paul R., Luh J., Bender J., Brown R., Remington M., Walker M., Advanced Industrial Robot Control Systems, First Report, School of Electrical Engineering, Purdue University 1978, West Lafayette, Indiana 47907.

[12] Vukobratović M., "Computer Method for Mathematical Modelling of Active Kinematic Chains via Euler´s Angles", Journal of Mechanisms and Machine Theory, Vol. 13, No 1, 1978.

[13] Vukobratović M., "General Mathematical Models of Active Spatial Mechanisms and Their Application in Robotics", invited paper, IX Internat. Conf. on Nonlinear Oscillations 1981, Kiev, USSR.

[14] Vukobratović M., Potkonjak V., "Contribution to Automatic Forming of Active Chain Models via Lagrangian Form", Journal of Applied Mechanics, No 1, 1979.

[15] Potkonjak V., Vukobratović M., "Two Methods for Computer Forming of Dynamic Equations of Active Mechanisms", Journal of Mechanism and Machine Theory, Vol. 14, No 2, 1979.

[16] Vukobratović M., Nikolić I., "Further Development of Methods for Automatical Forming of Dynamic Models of Industrial Robots", Proc. of International Conf. on Systems Engineering, Coventry, Sept., 1980.

[17] Popov E.P., Vereščagin A.F., Ivkin A.M., Leskov A.G., Medvedov V.S., "Design of Robot Control Using Dynamic Models of Manipulator Devices", Proc. of VI IFAC Symp. on Automatic Control in Space, Erevan, SSSR, 1974.

[18] Popov E.P., Vereschagin A.F., Zenkevich A.F., Zenkevich S.A., Manipulation Robots: Dynamics and Algorithms, (in Russian), "Nauka", Moscow, 1978.

[19] Vukobratović M., Potkonjak V., Hristić D., "Dynamic Method for the Evaluation and Choice of Industrial Manipulators", Proc. 9th International Symposium on Industrial Robots, Washington, 1979.

[20] Vukobratović M., Potkonjak V., "Transformation Blocks in the Dynamic Simulation of Typical Manipulator Configurations", Journal of Mechanism and Machine Theory, Vol. 17, No 3, 1982.

[21] Vukobratović M., Stokić D., "Synthesis of Programmed Functional Motion and Nominal Dynamics of Industrial Nonredundant Manipulation Robots", (in Russian), Teknitcheskaya kibernetika ANUSSR, No 3, 1981.

[22] Dubovski S.,"A Basis for High Quality Dynamic Analysis with an Application to the Dynamics and Control of Flexible Manipulators", Proc. of the National Science Foundation, Workshop on New Directions for Kinematics Research, Stanford University, 1976.

[23] Truckenbrodt A., "Dynamics and Control Methods for Moving Flexible Structures and Their Application to Industrial Robots", Proc. of 5th World Congres on Theory of Machines and Mechanisms, publ. ASME, 1979.

[24] Truckenbrodt A., "Modelling and Control of Flexible Manipulator Structures", Proc. of 4th IFToMM Symp. on Theory and Practice of Robots and Manipulators, Poland, 1981.

[25] Lakhota N.A., Rahmanov E.V., Shvedov V.N., "Control of Clasic Manipulators Along Trajectory", (in Russian), Teknicheskaya Kibernetika, No 2, 1980.

[26] Akulenko L.D., Mihailov S.A., Tchernousko F.L., "Modelling of Dynamics with Elastic Segments", (in Russian), Journal of Rigid Body Mechanics, No 3, 1981.

[27] Vukobratović M., Potkonjak V., "Computer Aided Design of Industrial Manipulators via Their Dynamic Properties", Journal of Mechanisms and Machine Theory, Vol. 17, No 2, 1982.

[28] Vukobratović M., Potkonjak V., "Computer Method for Dynamic Modelling of a Manipulators with Elastic Properties", Journal of Mechanisms and Machine Theory, Vol. 17, No 2, 1982.

Chapter 5
Dynamical Method for the Evaluation and Choice of Industrial Manipulators

5.1. Introduction

One of the basic reasons for studying the dynamics of active mechanisms applicable to robotics is that it is desirable to be thoroughly acquainted with the dynamical properties of robots-manipulators during their design.

Computer-oriented methods for the construction of mathematical models of active mechanisms [1 - 20] and appropriate algorithms for the simulation of manipulator dynamics have made possible wider applications of computers to the design of industrial manipulators.

It was realized at the beginning that for successful manipulator design it is necessary to first analyze and examine the dynamical characteristics of various configurations in order to depict the one best suited to the particular application. Hence investigations were directed towards the development of an algorithm for the simulation of the complete manipulator dynamics. Such an algorithm, which would work for an arbitrary manipulator configuration and arbitrary manipulation tasks, would permit calculation of all dynamical values, characteristic of manipulator operation in task execution and thus permit fast analysis of a great number of various configurations. In Chapter II of this book we described the computer-aided methods for the formation and solution of the dynamical model, and in Chapters III and IV we described the simulation algorithm [18, 19]. The input data for the algorithm are manipulator configuration, initial state and the manipulation task. As output one obtains time functions of the drives, reactions in the joints, stresses in segments, forces and moments in the gripper-workpiece interface, diagrams of torque - r.p.m. of each driving motor, total energy consumption and the value of elastic deviations from nominal manipulator motion.

Then we proceeded to the definition and elaboration of the dynamical criteria for the evaluation and comparison of manipulators and the development of the procedures for determining their optimal configuration for some particular application [18, 20]. We repeat that by manipulator configuration we mean its structure (kinematic scheme) and its parameters (dimensions, inertial properties, i.e. masses and tensors of inertia, etc.).

The first aim of such criteria is to assist the choice of the optimal parameters during the design of a manipulator, and to facilitate automation of the design process. The dynamical criteria are also suitable for evaluation and comparison of the manipulators offered on the market.

Test tasks: Since each manipulator is being designed for particular tasks, but not for one job only, a set of test tasks is defined on which the manipulator will be tested, i.e. on which simulation of the various manipulation configurations will be performed in order to chose the best one. Of course, the set of test tasks must correspond to the intended manipulator application.

The choice of test tasks is complex and in principle is performed by the user of the method, i.e., the designer, according to the intended manipulator application.

We will not enter into more detailed consideration of the problem of test task choice.

5.2. Defining the Dynamical Criteria

From the standpoint of the efficiency of industrial manipulators in practical use, two aspects can be distinguished: operation speed and energy consumption. Hence three criteria are defined for the evaluation and optimization of manipulators:

(a) Velocity criterion (or time criterion) - criterion of work speed.

(b) Energy criterion - criterion of least energy consumption.

(c) Combined criterion.

These criteria serve both for the comparison of manipulators and of functional movements and tasks, because each task in practice can be performed by means of mutually different movements, different velocity profiles, in different time periods. It should be mentioned that it is also possible to evaluate or choose the manipulation systems according to supplementary criteria associated with the realization of control algorithms. However, this is not within the scope of this monograph so we shall discuss only some dominant influences in the choice of manipulators based on "mechanical" criteria.

We shall elaborate these criteria in more detail.

(a) Velocity (time) criterion. Let T denote the time of the set manipulation task. Optimization with respect to the velocity criterion is reduced to determining that manipulator configuration which permits the greatest work speed, i.e. the least time T. Thus, minimization of time T is the criterion.

(b) Energy criterion. Let E denote the total energy the manipulator uses in performing the test task. Optimization means the minimimization of E, i.e. finding the configuration which ensures minimal consumption of energy.

(c) Combined criterion. It is useful to define a combined criterion, taking into account both the work velocity and energy consumption. In particular, a criterion would be useful whereby the connection between the speed and energy could be weakened or strengthened at wish and in which a greater or smaller significance could be assigned to each of these criteria. We shall call such a criterion "efficiency" and denote it by G. Efficiency is represented by a real number G from the interval $[0, 1]$. It demonstrates to what extent the manipulator considered satisfies the requirements. Value $G = 0$ shows that the requirements are completely unsatisfied and $G = 1$ shows that they are fully satisfied. Between these extremes are various grades of satisfaction. Thus, validation or evaluation of the manipulator is carried out aimed at greater efficiency G.

For the sake of defining the efficiency G as a combined criterion let us first define the velocity (time) efficiency G_T and energy efficiency G_E as follows:

Let T_o be the greatest time allowed, i.e., the longest permissible and sensible operation time, and T_1 be the shortest time already satisfying all our demands, so that some shorter time is not needed. Then the time T_o has the corresponding efficiency $G_T(T_o) = 0$, and the time T_1 the efficiency $G_T(T_1) = 1$.

Likewise, let E_o be the greatest energy, and E_1 the smallest. Then $G_E(E_o) = 0$, $G_E(E_1) = 1$.

Let us introduce

$$t(T) = T_o - T, \qquad e(E) = E_o - E. \tag{5.2.1}$$

Then $t_o = 0$, $t_1 = T_o - T_1$, $e_o = 0$, $e_1 = E_o - E_1$ as well $G_t(t_o) = 0$, $G_t(t_1) = 1$, $G_e(e_o) = 0$, $G_e(e_1) = 1$. As the efficiencies were determined in the end point of intervals $[t_o, t_1]$, $[e_o, e_1]$ only, let us suppose a linear change in between and illustrate efficiency $G_t(t)$, $G_e(t)$ graphically (Fig. 5.1).

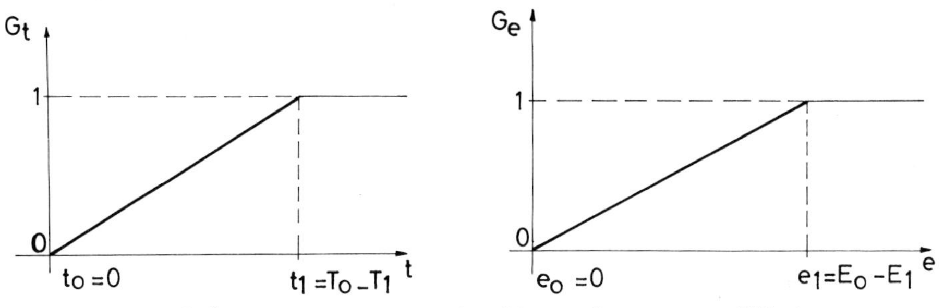

Fig. 5.1. Diagrams of velocity and energy efficiency

Total validation of manipulator, i.e. its "efficiency" is now obtained by averaging G_t and G_e. As we said, this averaging must satisfy the following demands:

I The possibility must exist of strengthening or weakening at wish the connection between G_t and G_e because sometimes is very important to achieve both high velocity and energy efficiency. Sometimes this is not so important, and sometimes it is sufficient for one of the efficiences G_t or G_e to be high.

II The possibility must exist of assigning at wish a greater or smaller importance to one efficiency relative to the other.

For averaging, which satisfies the set demands I, II, the function of "weighing disjunctive-conjunctive media" can be taken

$$G = G(G_t, G_e, r, w_t, w_e) = (w_t G_t^r + w_e G_e^r)^{1/r}, \tag{5.2.2}$$

$$w_t + w_e = 1, \quad \text{and} \quad w_t,\ w_e \varepsilon [0,\ 1],\ r\varepsilon(-\infty,\ +\infty) \tag{5.2.3}$$

Before describing the main features of such an averaging function, a few words should be said about its appearance and application. The function was developed based on continual logic [21, 22]. On account of its features it is partucularly suited to the definition of complex criteria and has already been used as such, for instance in [23] for the evaluation of complex systems. We should mention that the problematics of [23] does not relate in any way to the problems described in this monograph.

Let us consider the function (5.2.2). W_t and W_e represent the weighing factors, for which (5.2.3) holds, by means of which each of the variables G_t and G_e can be assigned the desired importance relative to the other. Hence, demand II has been fulfilled.

The real number r is called the averaging factor. By varying the factor r, the connection between the variables G_t and G_e can be weakened or strengthened, thus, fulfilling demand I. Let us consider this in more detail. For $r = -\infty$ the function (5.2.2) represents a true conjunction. By increasing r, the connection between G_t and G_e is weakened but still has a conjunctive character (i.e. $G \neq 0 \iff G_t \neq 0 \wedge G_e \neq 0$). For $r = 0$ the function represents the geometrical mean and for $r = 1$, the arithmetic mean, i.e. disjunctive-conjunctive unassignment is obtained. For $r > 1$, the function disjunctive characteristics, changing to a true disjunction when $r = +\infty$.

This discussion will not be proven exactly. More details can be found in |23|. We shall illustrate it here by means of a qualitative demonstration of level-curves of the function G (5.2.2) in the plane G_t, G_e (Fig. 5.2). Arrows in the sketches show the sense of the growth of the function G.

However, some basic details should be explained. A more precise definition of the averaging function G will be given with particular attention paid to the boundary values in the cases $r \to -\infty$, $r \to 0$, $r \to +\infty$. Thus, for two non-negative variables a_1 and a_2 ($a_1 < 1$, $a_2 < 1$),

$$G(a_1,\ a_2,\ r,\ w_1,\ w_2) = \begin{cases} \min(a_1,\ a_2), & \text{for } r = -\infty \quad (5.2.4a) \\ (w_1 a_1^r + w_2 a_2^r)^{1/r}, & \text{for } -\infty < r < 0 \quad (5.2.4b) \end{cases}$$

$$G(a_1, a_2, r, w_1, w_2) = \begin{cases} a_1^{w_1} \cdot a_2^{w_2}, & \text{for } r = 0 \quad (5.2.4c) \\ (w_1 a_1^r + w_2 a_2^r)^{1/r}, & \text{for } 0<r<+\infty \quad (5.2.4d) \\ \max(a_1, a_2), & \text{for } r=+\infty \quad (5.2.4e) \end{cases}$$

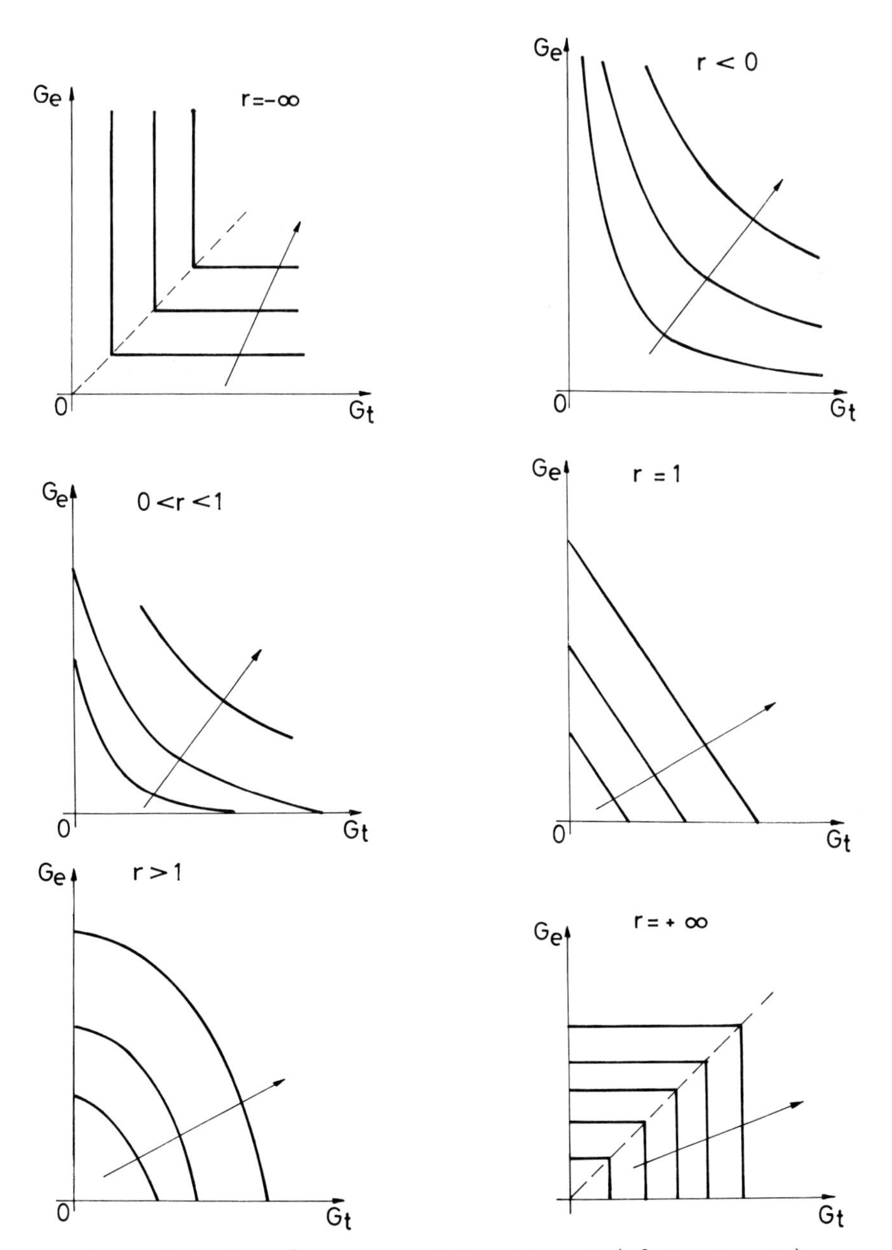

Fig. 5.2. Level-curves of function G (plane G_t, G_e)

In accordance with the continual logic concept (5.2.4a) is considered a true conjunction (Fig. 5.2a), and (5.2.4e) a true disjunction (Fig. 5.2f). Between these two extreme is a region of varying strengths of the conjuctive and disjunctive properties. For $r<1$, conjuctive characteristics prevail so this region is denoted quasi-conjunction. For $r=1$ there is an equilibrium of conjunctive and disjunctive characteristics, i.e., a case of conjunctive-disjunctive non-assignement. For $r>1$, the disjunctive character (quasi-disjunction) prevails. In accordance with this the following values were defined and denominated: c - degree of conjunction, d - degree of disjunction. This can be illustrated in the following way:

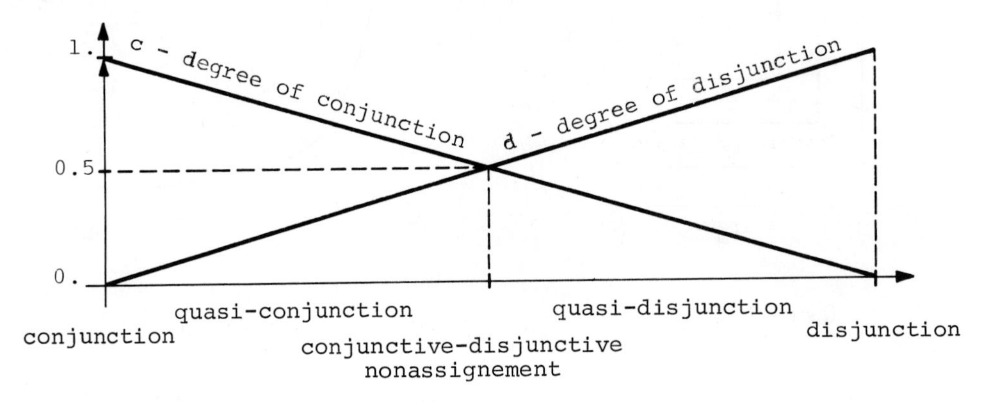

Thus, $c = 1$, $d = 0$ for true conjunction
$c = 0,5$, $d = 0,5$ for conjunctive-disjunctive nonassignement
$c = 0$, $d = 1$ for true disjunction

From this one can understand the character and purpose of the values introduced.

A more precise definition of the degrees c and d is

$$c = \frac{\bar{G}(+\infty) - \bar{G}(r)}{\bar{G}(+\infty) - \bar{G}(-\infty)} = 2 - 3\bar{G}(r),$$

$$d = \frac{\bar{G}(r) - \bar{G}(-\infty)}{\bar{G}(+\infty) - \bar{G}(-\infty)} = 3\bar{G}(r) - 1,$$

where $\bar{G}(r)$ is the average value of medium G defined by (5.2.4) over the observed interval.

However, the choice of the degrees c and d is based on the desired degree of conjunction according to the diagram above. In [23] the functions r = r(d) and r = r(c) were calculated and illustrated graphically. This we have shown in Fig. 5.3.

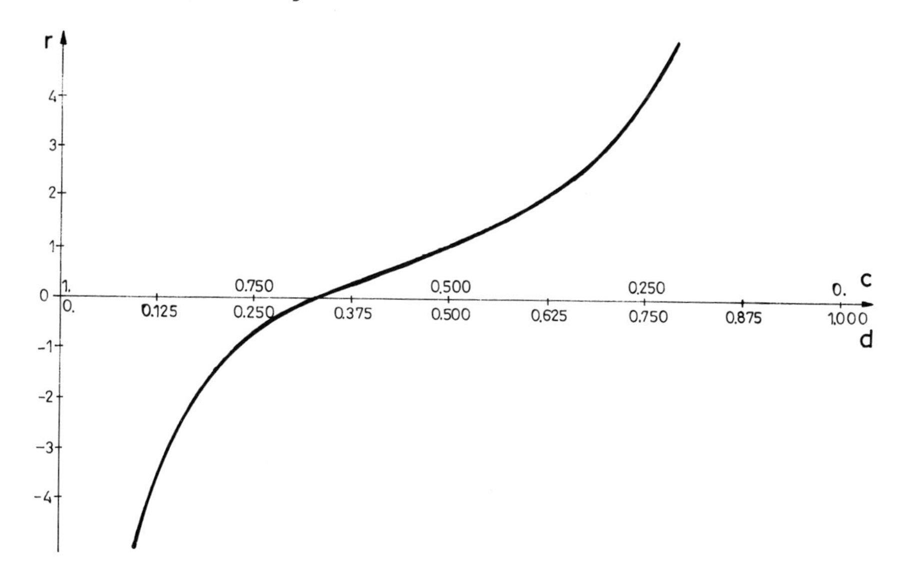

Fig. 5.3. Dependence of factor r on degrees c and d

The table with factor r for a series of equidistant values of the conjunction degree c, i.e., disjunction degree d (Fig. 5.4) have also been transfered from [23]. From the Table (Fig. 5.4) the corresponding factor r is chosen from the desired degree c, i.e. the desired intensity of the connection between G_t and G_e.

Further, Fig. 5.5. illustrates qualitatively the level-curves of the function G in the (t, e)-plane. The level-curves are illustrated for r<0 only because this is the interesting case (conjuctive function character).

Let us discuss the meaning of the conjunctive character of the function, i.e. quasi-conjunction. As we said, $G \neq 0 \iff G_t \neq 0 \wedge G_e \neq 0$. Further, the stronger the conjunction, the more the value and growth of function G is influenced by the smaller of the values being averaged. Then, efficient growth of G is achieved only by simultaneously augmentating G_t and G_e.

Thus for defining the combined criterion, it is necessary to prescribe

Degree of conjunction c	Degree of disjunction d	Averaging factor r
1.	0.	- infinitive
0.9375	0.0625	-9.0600
0.8750	0.1250	-3.5100
0.8125	0.1875	-1.6548
0.7500	0.2500	-0.7203
0.6875	0.3125	-0.1478
0.6250	0.3750	0.2612
0.5625	0.4375	0.6194
0.5	0.5	1.
0.4375	0.5625	1.4490
0.3750	0.6250	2.0185
0.3125	0.6875	2.7917
0.2500	0.7500	3.9293
0.1875	0.8125	5.8023
0.1250	0.8750	9.5207
0.0625	0.9375	20.6303
0.	1.	- infinitive

Fig. 5.4. Factor of averaging for various values of the degrees of conjunction and disjunction

T_o, T_1, E_o, E_1, w_t, w_e, r. These values are prescribed by the user of the method according to the manipulator purpose and its practical application. The discussion of this combined criterion was not intended to be very precise because we think that this much is sufficient to understanding its idea as well as its practical application. The user (designer) chooses the degree c depending on how important it is, that both the velocity and energy efficiency be high. For a chosen degree c, from Table Fig. 5.4, or diagram Fig. 5.3., the factor r can be read. The weighing factors w_t, w_e are chosen by the user depending on how important one efficiency is relative to the other, bearing in mind (5.2.3).

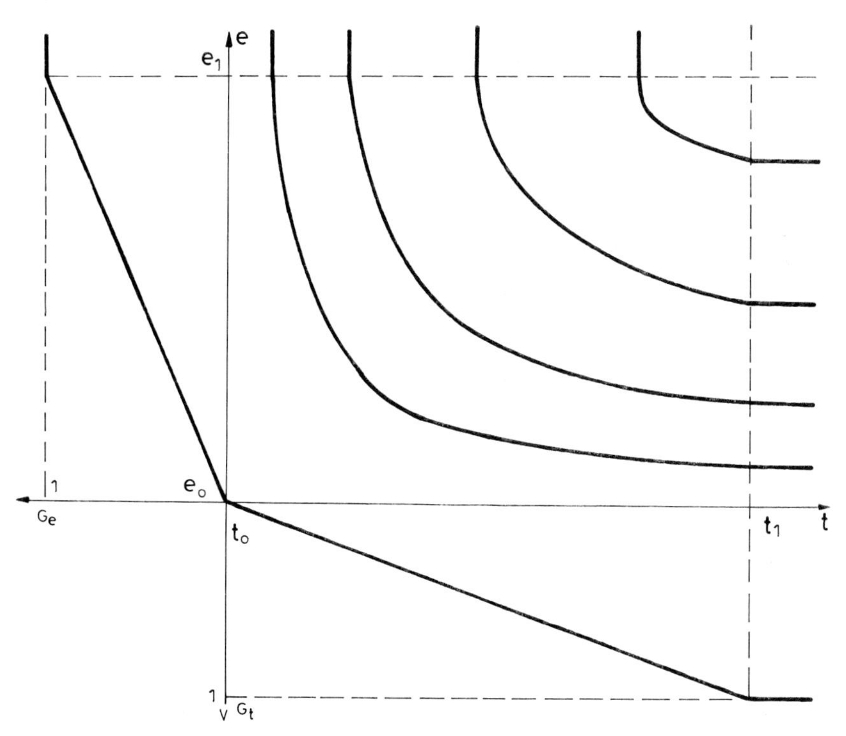

Fig. 5.5. Level-curves of function G in plane t, e; r<0.

Use of the dynamical criteria. After certain numerical examinations, it was concluded, that the combined criterium is particularly suitable for the evaluation and validation of various manipulators considering their intended practical applications. Criteria (a) and (b) i.e. the velocity and energy criteria are suitable as criteria for minimization in the procedure of the optimal structure and manipulator parameters choice during the design stage. The use of the criteria will be ilustrated by examples in Para. 5.4. Some of the examples are rather theoretical in order to show the optimization methodology or some properties of the criteria. On the other hand, the resting examples follow directly from the practical application of the method considered.

5.3. Definition of Limitations

In order to define the optimization problem, quite apart from the criteria, limitations have to be introduced. Thus, optimization consists of the minimization of the chosen criterion and the satisfaction of certain limitations. From a practical standpoint, the following limitations can be defined:

(i) Limitation of reachability. This is the requirement that the manipulator can make the required test-movement (geometrically).

(ii) Limitation of stress. This is the requirement that the stresses in the manipulator segments do not exceed certain permitted values.

(iii) Limitation of drives. This is the requirement that the driving motors in the manipulator joints can produce the forces and torques necessary for the task.

(iv) Limitation of elasticity. This is the requirement that the deviations due to elastic deformations of the segments do not exceed certain permissable values.

It should be mentioned that limitation (i) is of a formal nature and that it is not essentially considered here. This limitation should be understood more as a certain equalization of the manipulators from the standpoint of their reachability.

The optimization procedure, i.e., the choice of the optimal parameters, will be explained in the following paragraph.

5.4. Examples

Example 1.

Four of the most common kinematic schemes of a manipulator with 6 degrees of freedom were observed. These schemes, together with adopted segment lengths (in meters), are presented in Fig. 5.6.

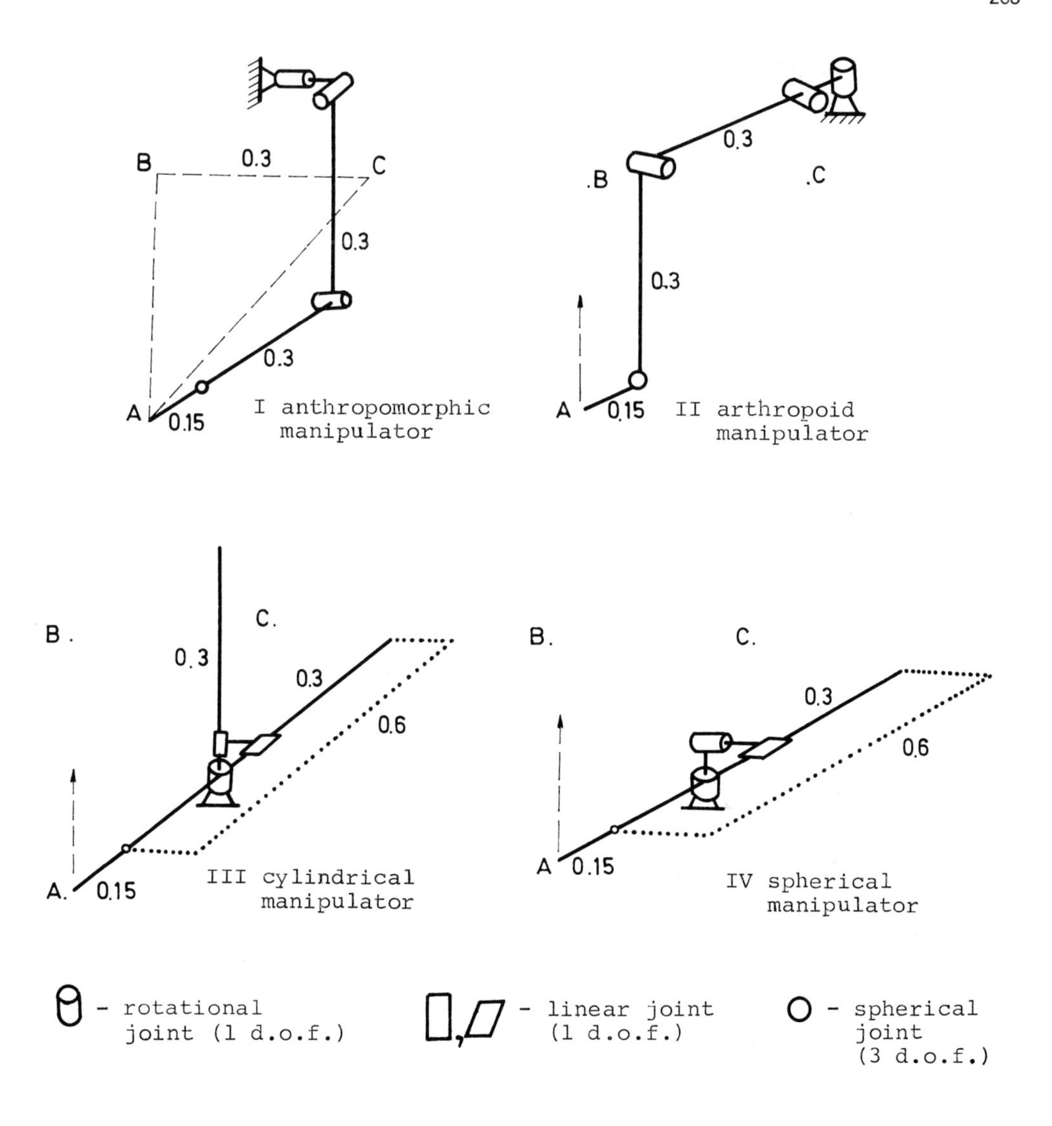

Fig. 5.6. Most common manipulators schemes

Test-task. As a test-task the manipulator tip traverses the trajectory ABCA in Figs. 5.6. and 5.7. starting from its initial position with the last manipulator segment maintaining its initial orientation in space. The initial position is given in Fig. 5.6. The last manipulator segment carries a payload of 5 kilograms. A triangular velocity profile of the tip velocity was chosen. The tip starts from the rest position and on the part AB of the trajectory accelerates constantly up to the mid-point and then constantly deccelerates so that at point B the velocity is zero. This is repeated in the trajectory parts BC and CA (Fig. 5.8).

Let T denote the task execution time, which can be varied.

Manipulator segments were chosen in the form of circular tubes of inner radius r and outer radius R, with constant ratio r/R = 0.75 (Fig. 5.9). Such a cross-section (circular ring) was chosen because it is the most common in practice. A rectangular cross-section tube (also common) would be treated in an exactly analogous way.

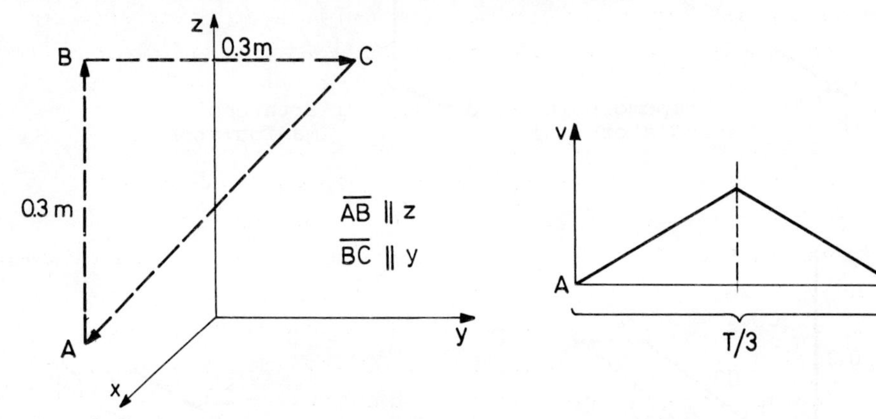

Fig. 5.7. Trajectory of object transfer

Fig. 5.8. Triangular velocity profile

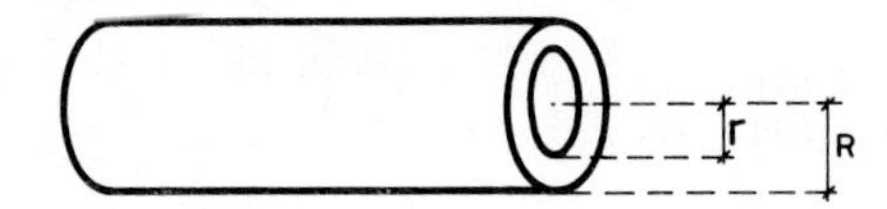

Fig. 5.9. Segment in the form of cylindrical tubes

As material for segment production the light alloy AℓMg3 was adopted. Data about the density (ρ), permitted bending stress (σ_d) and torsion stress (τ_d), Young´s modulus (E_y) are given in Table (Fig. 5.10)

ρ	$2.7\ g/cm^2 = 2700\ kg/m^3$
σ_d	$\frac{1}{k}\ 30\ kp/mm^2 = 2.943 \cdot 10^8\ N/m^2 \cdot \frac{1}{k}$
τ_d	$\frac{1}{k}\ 25\ kp/mm^2 = 2.4525 \cdot 10^8 \cdot N/m^2 \cdot \frac{1}{k}$
E_y	$7.848 \cdot 10^{10} \cdot N/m^2$

k = 5, safety coefficient

Fig. 5.10. Material data

As driving motors for the manipulator joints D.C. permanent magnet INDOX motors were chosen, Frame 23, type 2315-P20-0, produced by INDIANA GENERAL. The reduction ration is 100 in rotational and 500 in linear joints.

Optimization by the velocity criterion. As several kinematical schemes are in question, optimization is performed for each structure (kinematical scheme) separately and the optimal parameters for that structure obtained. The results for the various structures are then compared and the best one chosen.

In order to subject all the examined structures (kinematical schemes) I-IV to same conditions, let us suppose that the maximal extensions are the same. In doing so, the segment lengths were determined and they are given in Fig. 5.6. The lengths will not be changed during optimization. This is in order to avoid an analysis of reachability, which is a separate and complex question. In this simplified example the cross-section of all the segments will be the same and the motor masses will be neglected. Thus, due to a constant ratio $\psi = \frac{r}{R} = 0.75$, the only optimization parameter is the outer tube radius R. Thus R should be chosen so as to minimize the task execution time T and enable the limitations (i) - (iv) to be satisfied.

The procedure is the following: for a selected value of R a series of simulations is performed successively reducing the task execution time T. One proceeds in this way until the drive limitation (iii) is violated, i.e., until the driving motors cannot produce the manipulator working speed. Then R is reduced and the procedure repeated. Thus, the curve $T_{min}(R)$ i.e. minimal execution time depending on R is obtained. Let us consider first the limitations (i) - (iii) only. The procedure of reducing R is repeated until limitation (ii) is violated, i.e., until the stresses in segments exceed the permissable values. If further reduction of R is required, T has to be increased. Consequently, minimum time (maximum velocity) appears in both limitations (ii) and (iii). Fig. 5.11. illustrates these results (for the anthropomorphic manipulator), i.e., the curve $T_{min}(R)$ and limitations (ii) and (iii). The minimum appears at the point M_1 designated by a circle. Let us now introduce the limitation of elasticity (iv). We impose the condition that the manipulator tip linear deviation due to segment elasticity be less than 0.001 m. In this case we consider only the quasi-static deflexion due to the nominal dynamics. By introducing limitation (iv) the permissable domain is narrowed and the minimum point moves into position M_2, designated by a square in Fig. 5.11. The coordinates $(R^T_{opt}, T^{abs}_{min})$ correspond to the point M_2. The dotted line in Fig. 5.11. represents the corresponding energy consumptions. The abrupt decline of energy consumption to the left of M_1 is due to the abrupt increase of working time T, i.e., to velocity decline.

Fig. 5.12. illustrates the curves $T_{min}(R)$ for the various structures I-IV, only to the right of the points M_1. Points M_1 (circles) represent the minimum without limitation (iv), points M_2 (squares) represent the real minimum when limitation (iv) is taken into account. The significantly best characteristic is that of the spherical manipulator. This is due to good compensation of masses around the second joint axis (counted from the base) Fig. 5.6. However, in some other test-task this might not be the case. Generally one can conclude that compensation must be considered.

Analysis of energy consumption was first performed in the case of the anthropomorphic structure I. Let E be the total energy the manipulator uses in performing the set test-task. Fig. 5.13. illustrates the resulting family of curves representing the dependence of energy consumption E on the radius R. Each of these curves corresponds to a certain time interval T of task execution (work speed), as indicated in the figure. Each curve has an upper bound defined by the drive limitation (iii), and a lower bound defined by the stress (ii) or elasticity limitation (iv). These bounds also determine the range of possible values of the radius R for a certain execution time T. The optimum-minimum is, for every T, the lower limit of the corresponding curve.

Fig. 5.14. illustrates the family of curves representing the dependence of energy consumption E on the work speed, i.e., time T, for the anthropomorphic structure I. Each curve corresponds to a certain radius R. It should be noted that the curve is steeper in the region of greater work speeds because the inertial forces are more influential there. In Fig. 5.15. the same family of curves is illustrated for the cylindrical structure III. Evidently, such a structure demonstrates greater sensitivity to work speed, i.e., the curves are steeper.

We proceeded further to a comparison of energy consumption for the various structures I-IV. Fig. 5.16. shows the table in which the energy consumptions in the various structures I-IV are compared. Clearly, from the point of view of minimum energy consumption, the order of the structures examined is: cylindrical, spherical, arthropoid, anthro-

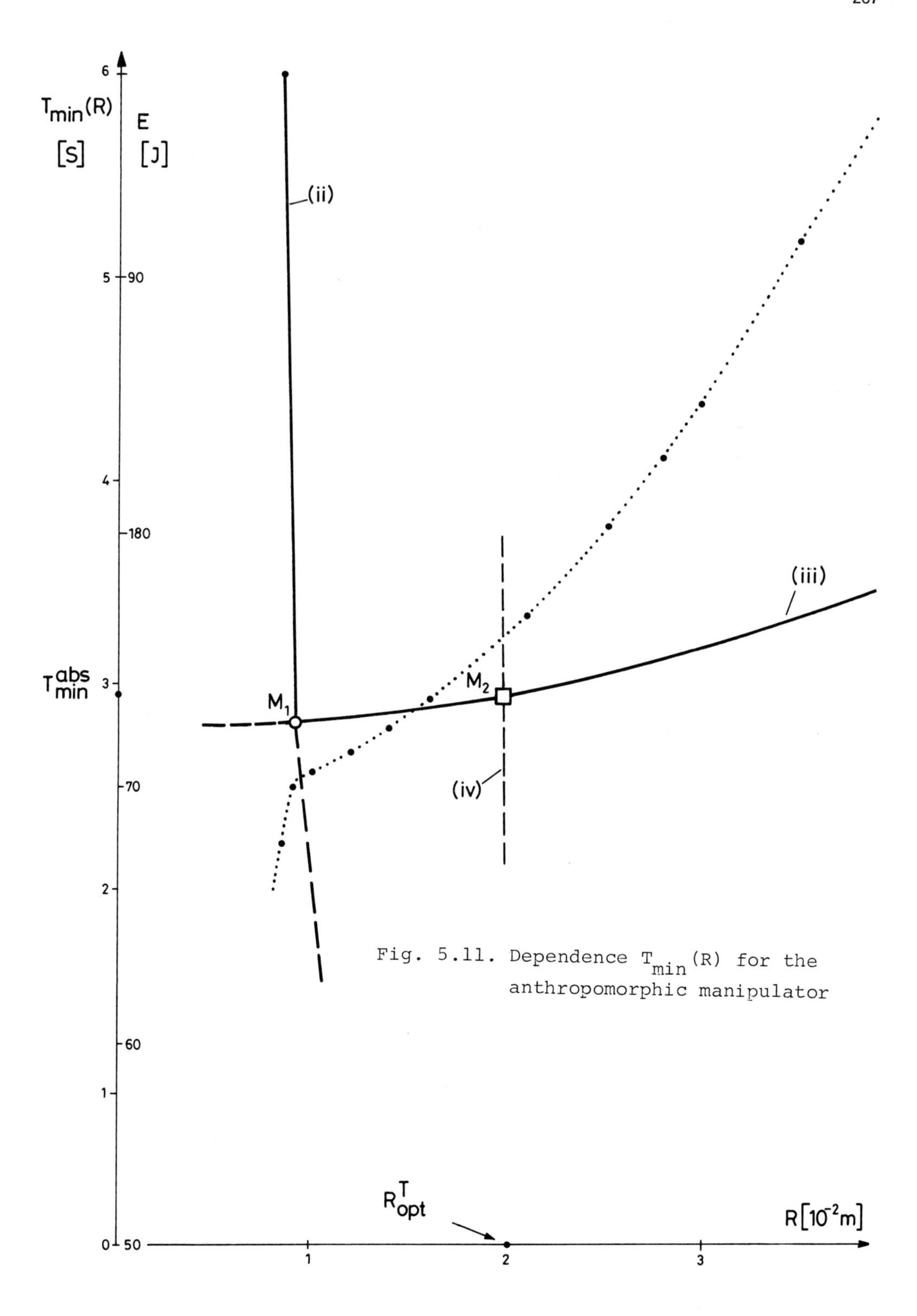

Fig. 5.11. Dependence $T_{min}(R)$ for the anthropomorphic manipulator

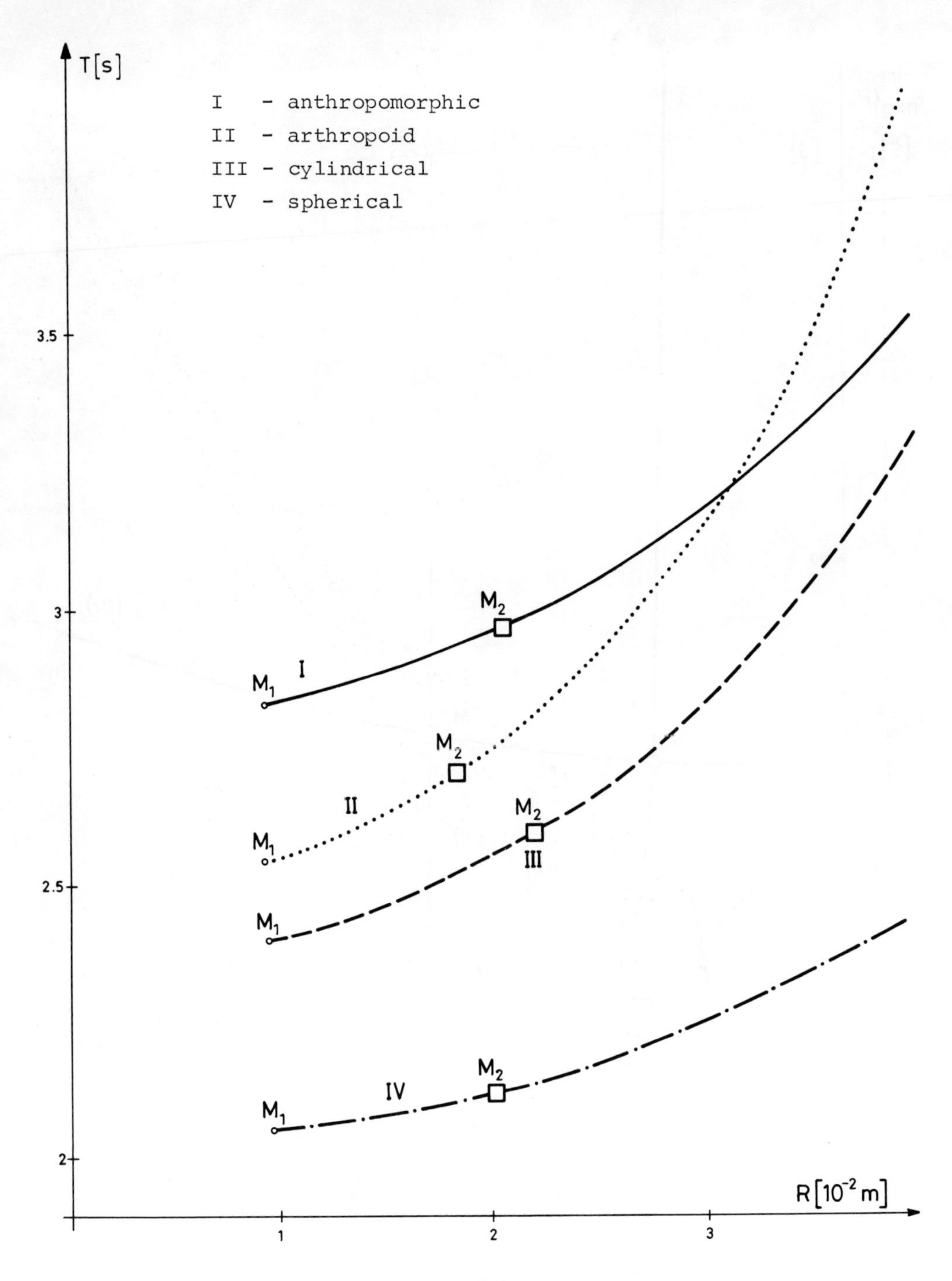

Fig. 5.12. Dependence $T_{min}(R)$ for various structures

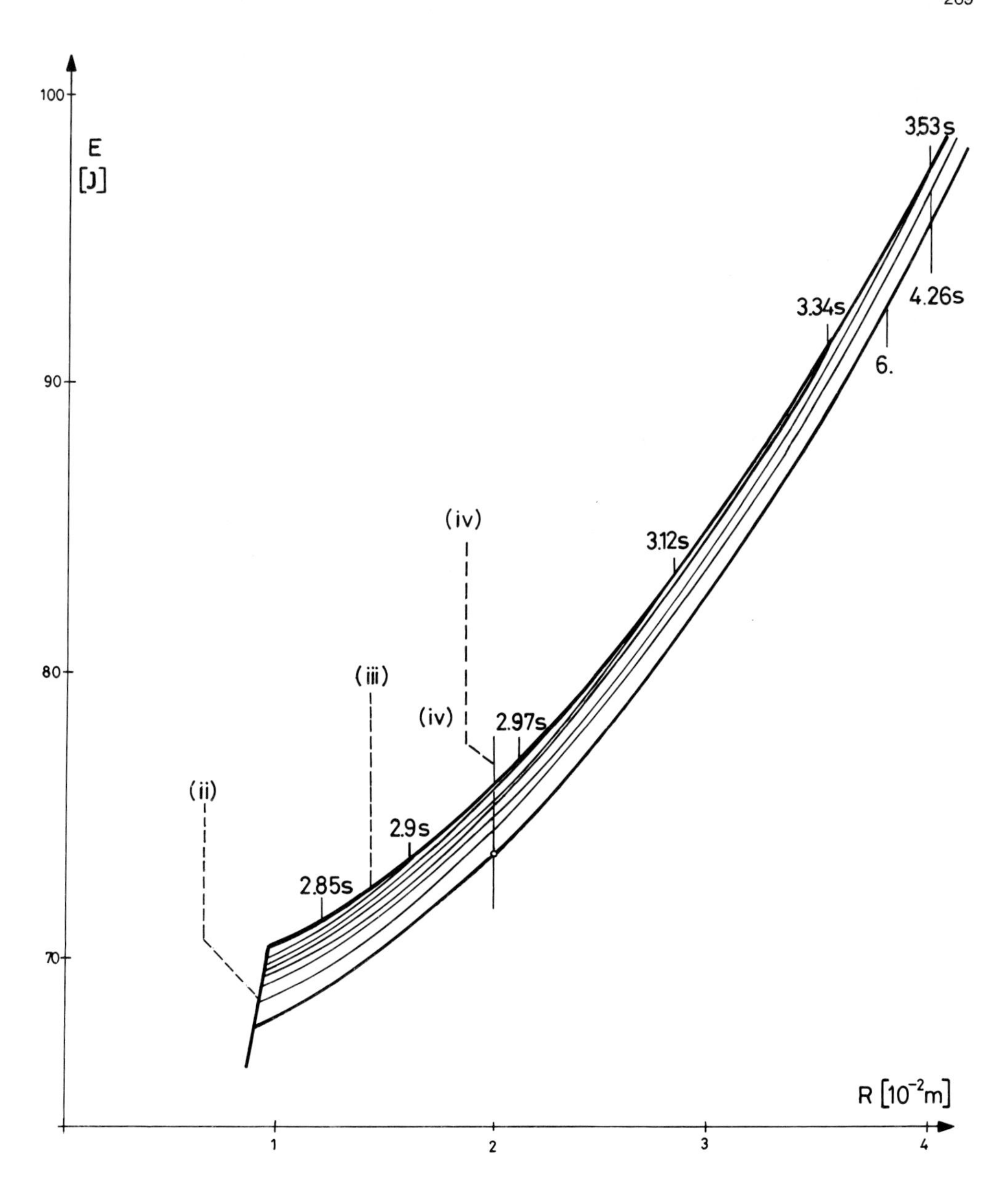

Fig. 5.13. Change in energy consumption depending on radius R for the anthropomorphic structure I

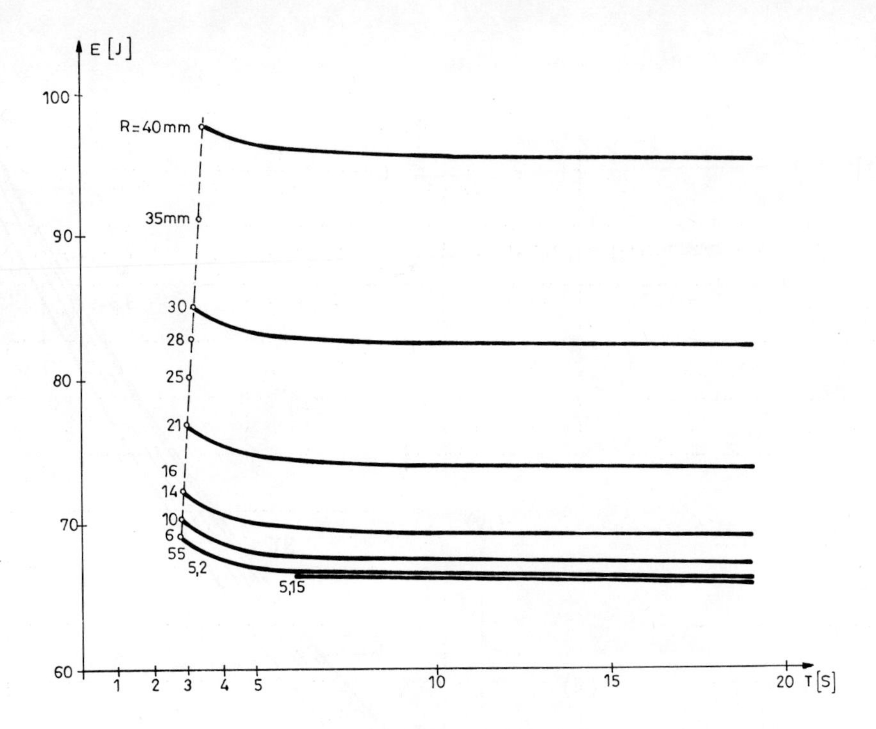

Fig. 5.14. Dependence of energy consumption on time T, for the anthropomorphic structure I

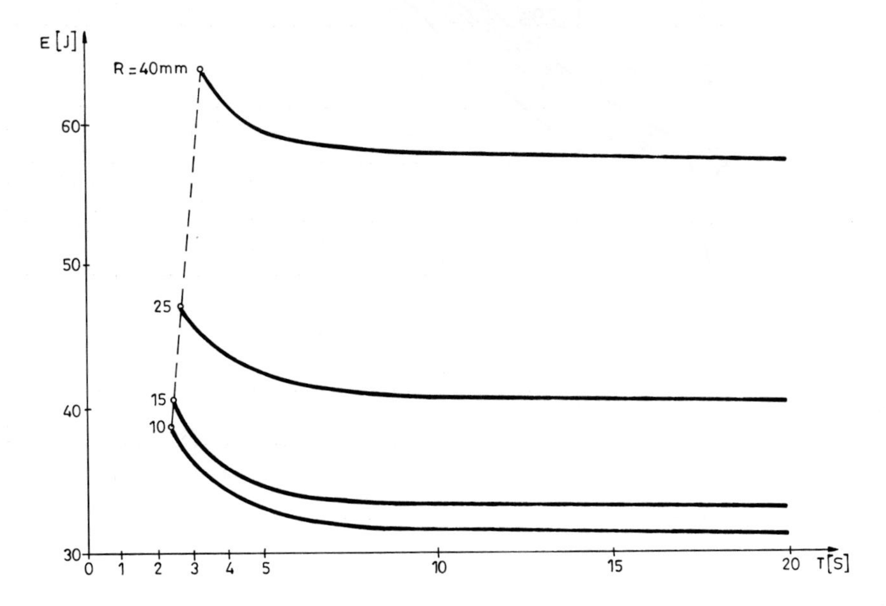

Fig. 5.15. Dependence of energy consumption on time t, for the cylindrical structure III

pomorphic.

	R = 0.04 m	r = 0.03 m	
	T = 10s	T = 6.2s	T = 4.26s
I anthropomorphic	95.6 J	96.1 J	97.1 J
II arthropoid	90.1 J	90.6 J	92 J
III cylindrical	58 J	59 J	60.9 J
IV spherical	68.3 J	69 J	70.3 J

	R = 0.025	r = 0.01875	
	T = 10	T = 6.2	T = 3.4
I	77.5	77.9	79.6
II	73.9	74.5	76.6
III	40.9	41.7	44.9
IV	56.9	57.6	60

	R = 0.014	r = 0.0105	
	T = 10	T = 6.2	T = 3.4
I	69.4	69.9	71.4
II	66.8	67.3	69.2
III	33.3	34.1	37
IV	51.9	52.5	54.8

Fig. 5.16. Comparison of energy consumption for the various structures I-IV

Other results. It is also interesting to study the dependance of the stresses in segments on the radius R. Let σ_{max} (R, T) be the maximal stress appearing in the manipulator segments during the task execution

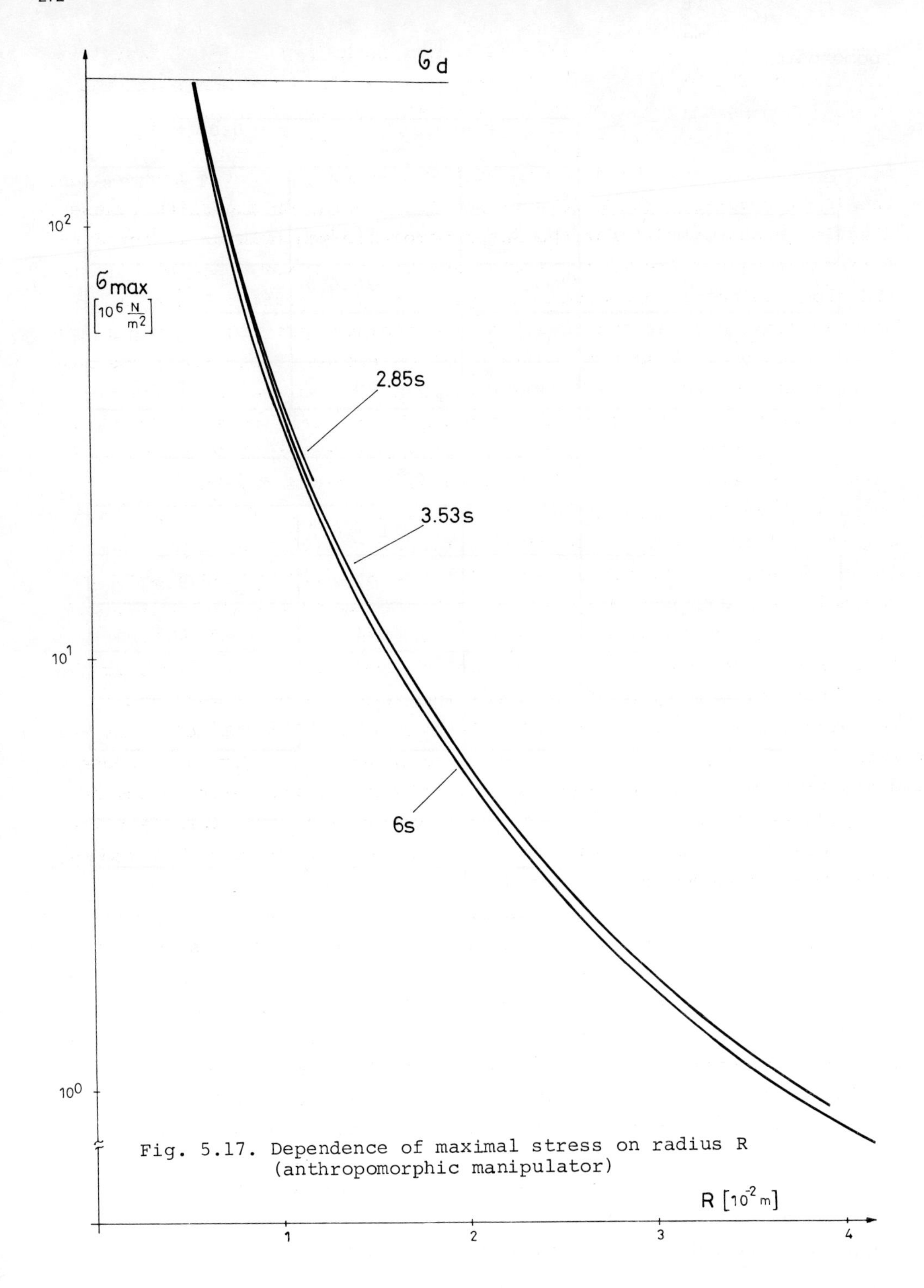

Fig. 5.17. Dependence of maximal stress on radius R (anthropomorphic manipulator)

time T when the segment's external radius is R. Fig. 5.17. illustrates the family of curves σ_{max} (R, T) for three different times T and the violation of limitation (ii), i.e., of the permitted stress σ_d.

Let us now consider the diagrams of torque - r.p.m. of the motors, i.e., the (P-n) diagrams, which result from the simulation algorithm. These diagrams were recorded for the anthropomorphic manipulator I, for different values of time T. (P-n) diagrams for the motor in joint 2 (counted from the base) were given in Fig. 5.18. The straight lines on the diagram represent the characteristics (obtained from catalogues) of the selected motors. These are the (P_{max}-n) characteristics, i.e., the maximal torque for particular values of r.p.m. Such characteristics restrict the domain within which the real P-n diagram, obtained by simulation, is defined. Let, us explain these characteristics in more detail. The straight line (*) connecting stall torque (P_M) and no-load speed (n_M) follows from the constraint on maximal input voltage. This characteristic does not take care of motor dynamics since rotor inertia is neglected. The horizontal line (**) follows from the constraint on maximal rotor current. The complete motor dynamics will not be discussed here, but in Volume 6 of this series. Thus, the catalogue characteristics represent the drive limitation (iii). It can be seen from the illustration, that by augmenting the work speed, i.e. by reducing T, the diagrams spread out. Thus, for T=3.5 s, 3s and 2.5 s the diagrams are within the permissible domain, which means that the motor can produce manipulator work at these speeds. For T=2 s the diagram extends beyond the permissible domain, i.e., the drive constraint is violated and the algorithm indicates that movement at this speed is impossible to achieve using the chosen motor.

P-n diagrams for joint 3 are given in Fig. 5.19. As may be seen, for T=3 s and 2.2 s the constraint is not violated in joint 3. For T=2.2 s, the diagram is discontinued before the end since the constraint is violated in joint 2 at that moment and the operation of the simulation algorithm is interrupted. For shorter times (1.65 s - 1.1 s in the Fig.), the constraint is violated in joint 3 as well; as T is shorter, i.e., the speed is greater, the constraint is violated earlier.

Obviously, such diagrams are very useful for choosing the right driving motors. The diagrams obtained by the simulation algorithm are compared against catalog maximal characteristics.

It should be pointed out, however, that for practical reasons we did not consider the possibility of allowing intermittent motor overload. Since the capabilities of the actuators were evaluated in these calculations on the basis of the exact "dynamical" needs of the manipulation robot, neglecting these motor capabilities results in a certain power

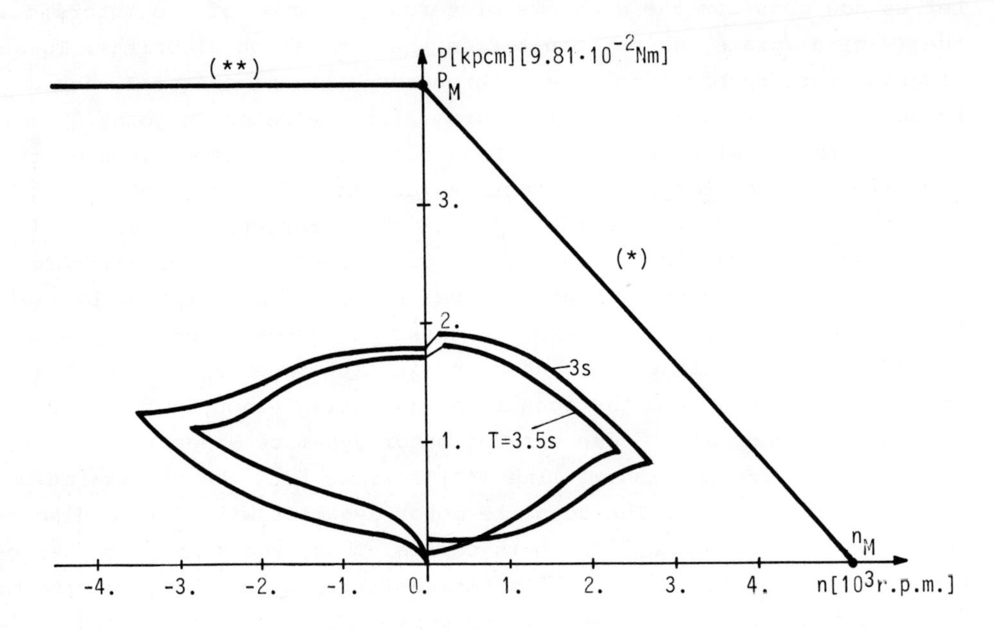

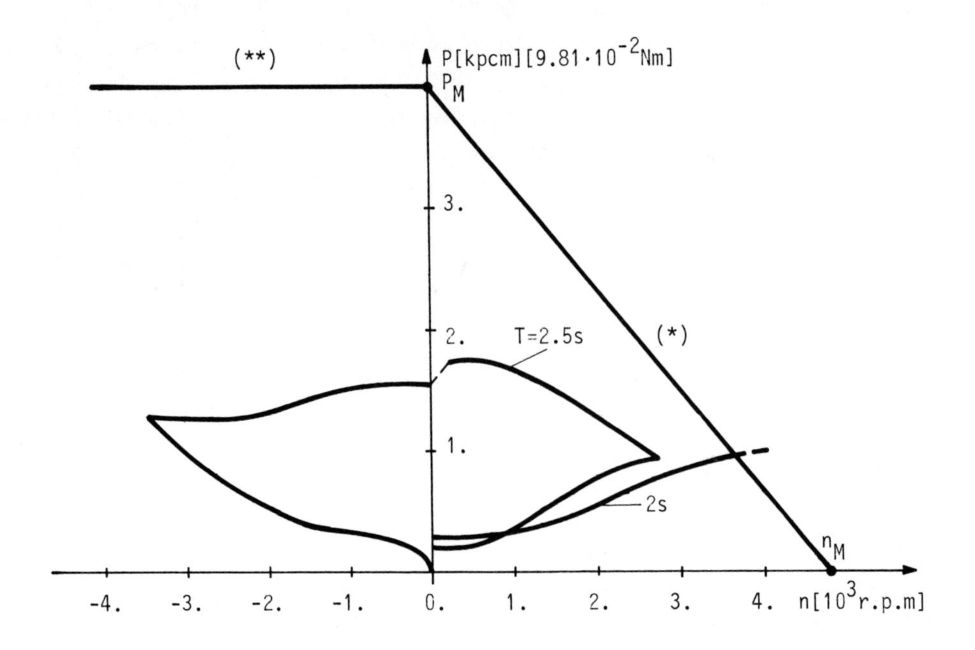

Fig. 5.18. P-n diagrams for joint 2

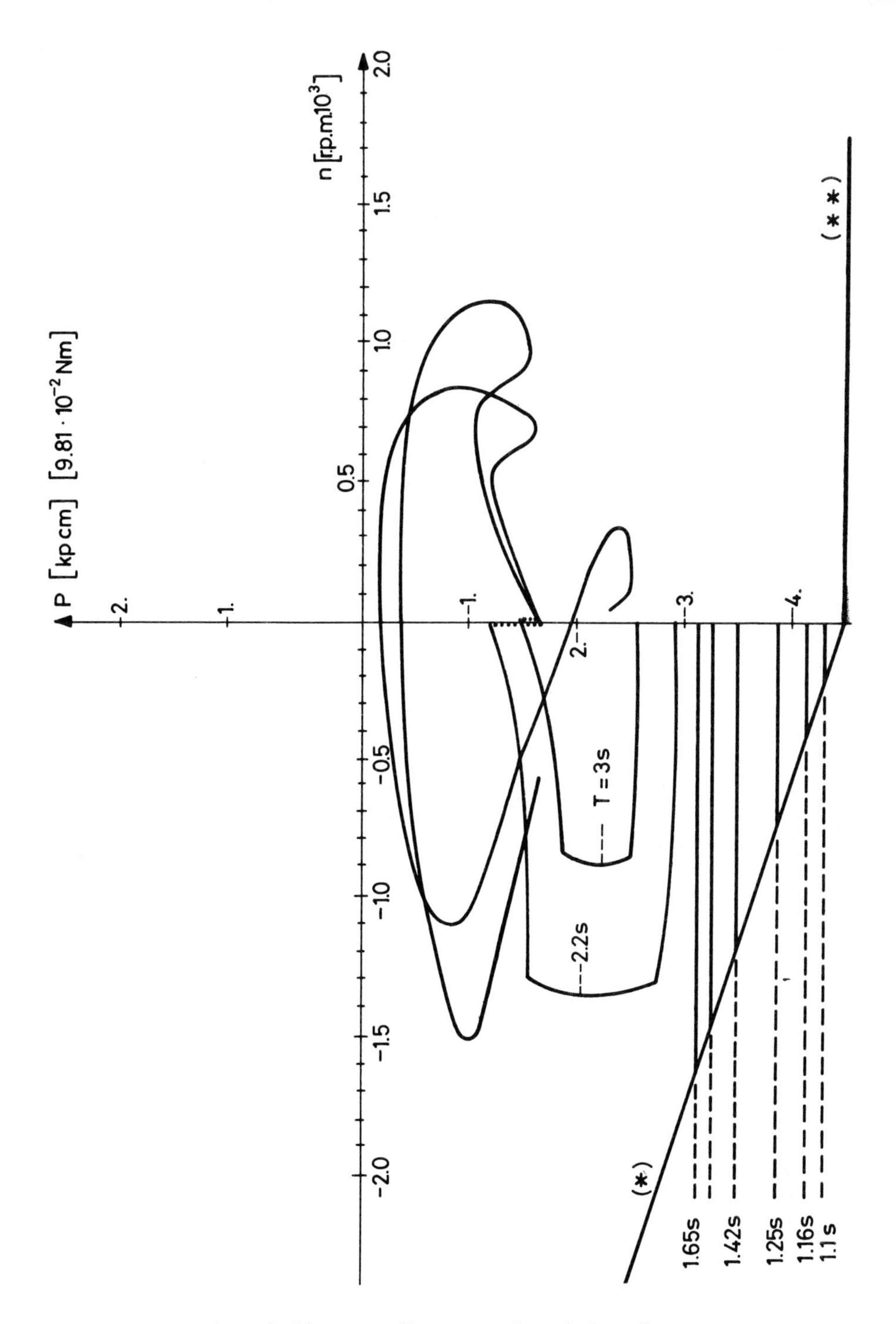

Fig. 5.19. P-n diagrams for joint 3

reserve.

Example 2

In this example we consider the structural design of a manipulator for a practical industrial application. The manipulator has been chosen for spray-painting. However, greater positioning precision is required if this manipulation device is to be used for other industrial tasks.

The choice of parameters is often determined by the design of the manipulator itself. However, some dimensions can be varied considerably. They are then chosen mainly on the basis of experience or approximate calculations. Such design results in over dimensioned manipulator segments, with augmented masses. This in turn produces an increase in energy consumption, speed reduction and the need for more powerful driving motors. On the other hand, it is possible to err in the other direction and choose dimensions which are too small. This can lead to insufficient mechanical rigidity of the device and the occurrence of greater elastic deviations. In this example we demonstrate the procecure for the systematic choice of the optimal dimensions of the manipulator.

The kinematical scheme of the manipulators is spherical. It possesses 6 d.o.f. (Fig. 5.20). The segments are enumerated by the numbers 1 to 6 and shown in the drawing. Such a structure has resulted in a choice of 5 rotational and 1 linear generalized coordinate.

Manipulator parameters. The dimensions and other parameters of segments 1, 2, 4, 5, 6 were conditioned by the design solutions themselves and by the actuators chosen. The dimensions of these segments are given in Fig. 5.20. and the masses and tensors of inertia are given in the table in Fig. 5.21.

The third segment is connected by means of a linear (sliding) joint to the second segment and it is intended that it consist of two cylindrical quides and a hydraulic piston and piston rod.

Piston/piston rod mass is 3.07 kg. The maximal extension of the guides out of the second segment must be 0.8 meters, which, for design reasons, results in the guides being 1.14 meters long. Thus the cross-section dimensions of the guides remain to be chosen. This will strongly influence the manipualtor dynamics, as we will demonstrate.

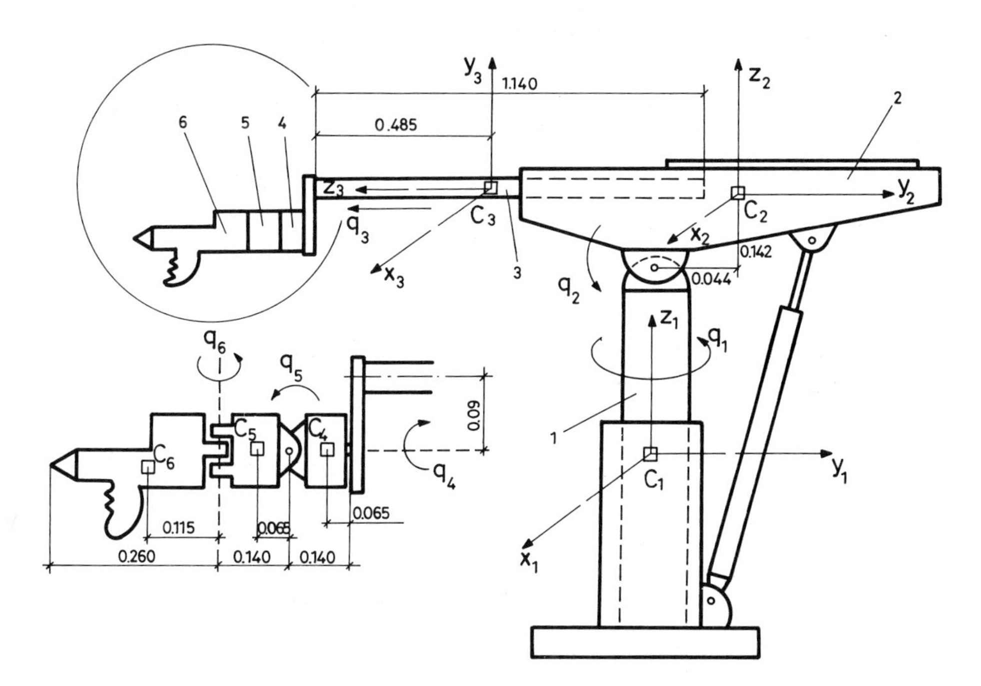

Fig. 5.20. Spherical industrial manipulator

No of segment i	Mass m_i [kg]	Tensor of inertia		
		J_{x_i} [kgm^2]	J_{y_i}	J_{z_i}
1	61.46			0.322
2	27.8	2.98		3.701
3	2.33	0.004	0.004	0.004
4	2.33	0.004	0.004	0.004
5	2.33	0.007	0.009	0.009

Fig. 5.21. Inertial properties of segments (empty spaces denote insignificant data)

The test-task was chosen in accordance with the manipulator application. The task is to paint-spray a panel shown in Fig. 3.22. The initial manipulator position is given in Fig. 5.22. and the spray-painting is done from a constant distance. Thus the manipulation task reduces to the tip moving along the path B C C_1 B_1 B_2 C_2 ... (Fig. 5.23). First, however, it is necessary to bring the tip from the initial position A to the starting position B. So the complete trajectory is A B C C_1 B_1 ... (Fig. 5.23) maintaining the spray-gun´s initial orientation in space.

Since motion along C_1 B_1, B_2 C_2,... means periodically traversing path BC, the simulation and evaluation will be performed on the trajectory ABC only.

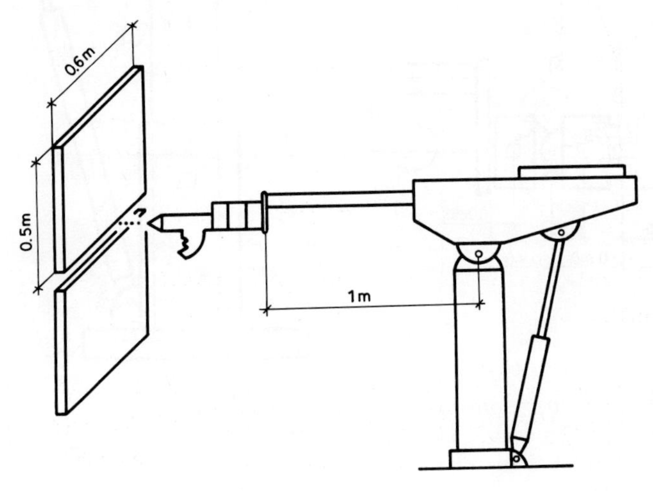

Fig. 5.22. Spray-painting (initial position)

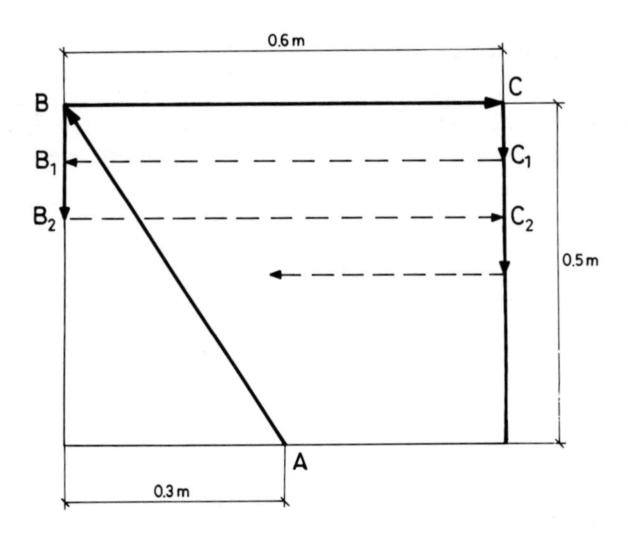

Fig. 5.23. Spraying trajectory

Bringing the tip into the starting position for spray-painting (part AB) is performed in the course of $T_1 = 1$ s, with a triangular velocity profile. The painting itself should be carried out with constant velocity of $v = 0.3$ m/s. Hence, over section BC a trapezoidal velocity profile was adopted (Fig. 5.24).

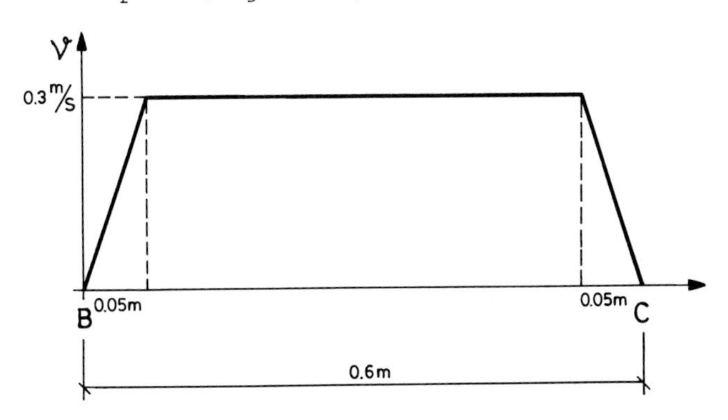

Fig. 5.24. Trapezoidal velocity profile

Material for the guides is steel of specific density $\sigma = 7.85 \cdot 10^3$ kg/m^3 and Young's modulus of elasticity $E = 2.1 \cdot 10^{11}$ N/m^2.

The choice of the dimensions of the cross-sections has been made according to the criterion of minimal energy consumption. Due to the desired positioning accuracy, a limitation was introduced, that the deviation (deflexion) of the manipulator tip due to elastic deformation of the segment be not greater than $\varepsilon = 0.001$ m.

First, a full circular cross-section was chosen and an initial radius of the guides $R = 0.02$ meters was selected. Simulation was then carried out and it was found that the elastic deflexion does exceed the permitted value. The mass of the third segment, for the chosen R, is $m_3 = 25.5$ kg.

We then proceeded to a systematic choice. A series of simulations was performed, increasing the radius R successively as well as the energy consumption. Thus, the value of $R = 0.024$ meters was obtained for which the limitation is satisfied. This represents the optimum. Energy consumption was as follows: over AB, $E_{AB} = 132.69$ J; over BC, $E_{BC} = 17.28$ J. The mass of the third segment was $m_3 = 35.46$ kg.

Since this radius value results in too large a segment mass m_3 and great energy consumption, the cross-section was modified and adopted

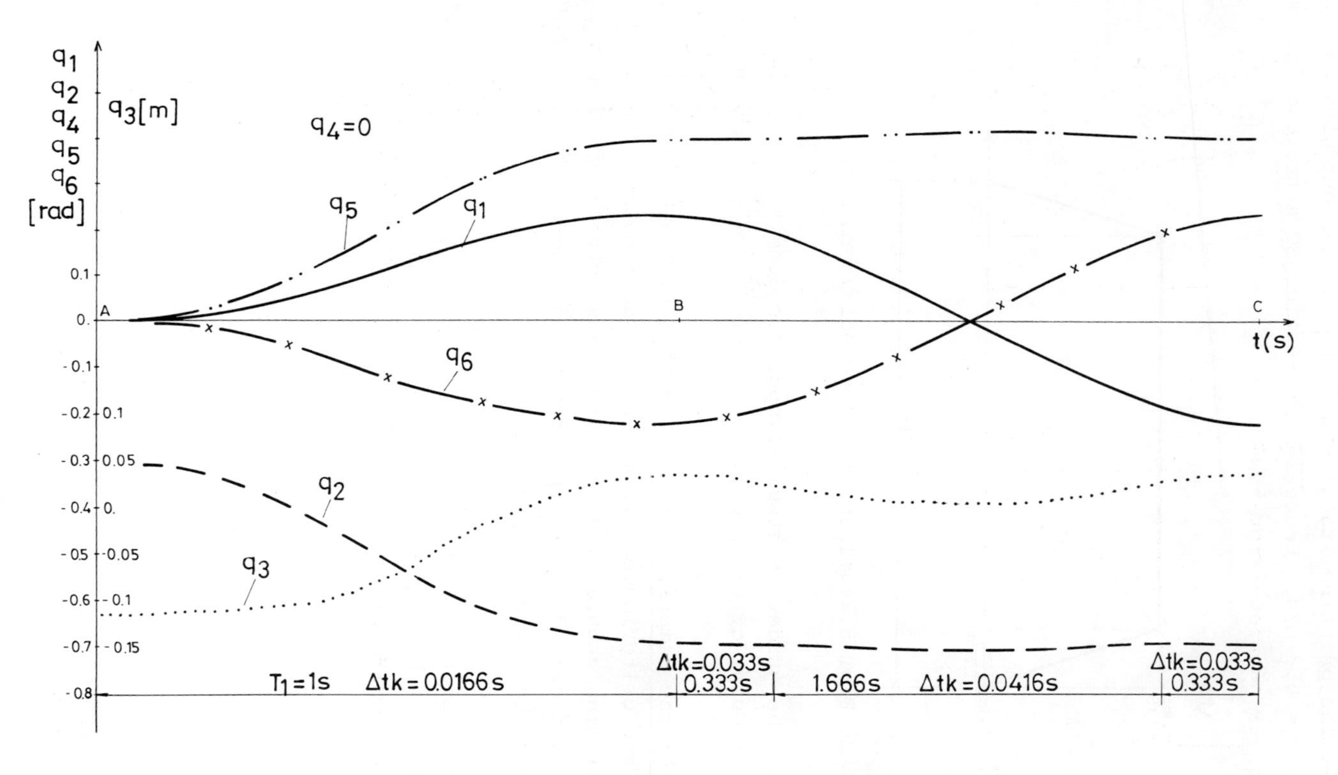

Fig. 5.25. Time history of the generalized coordinates

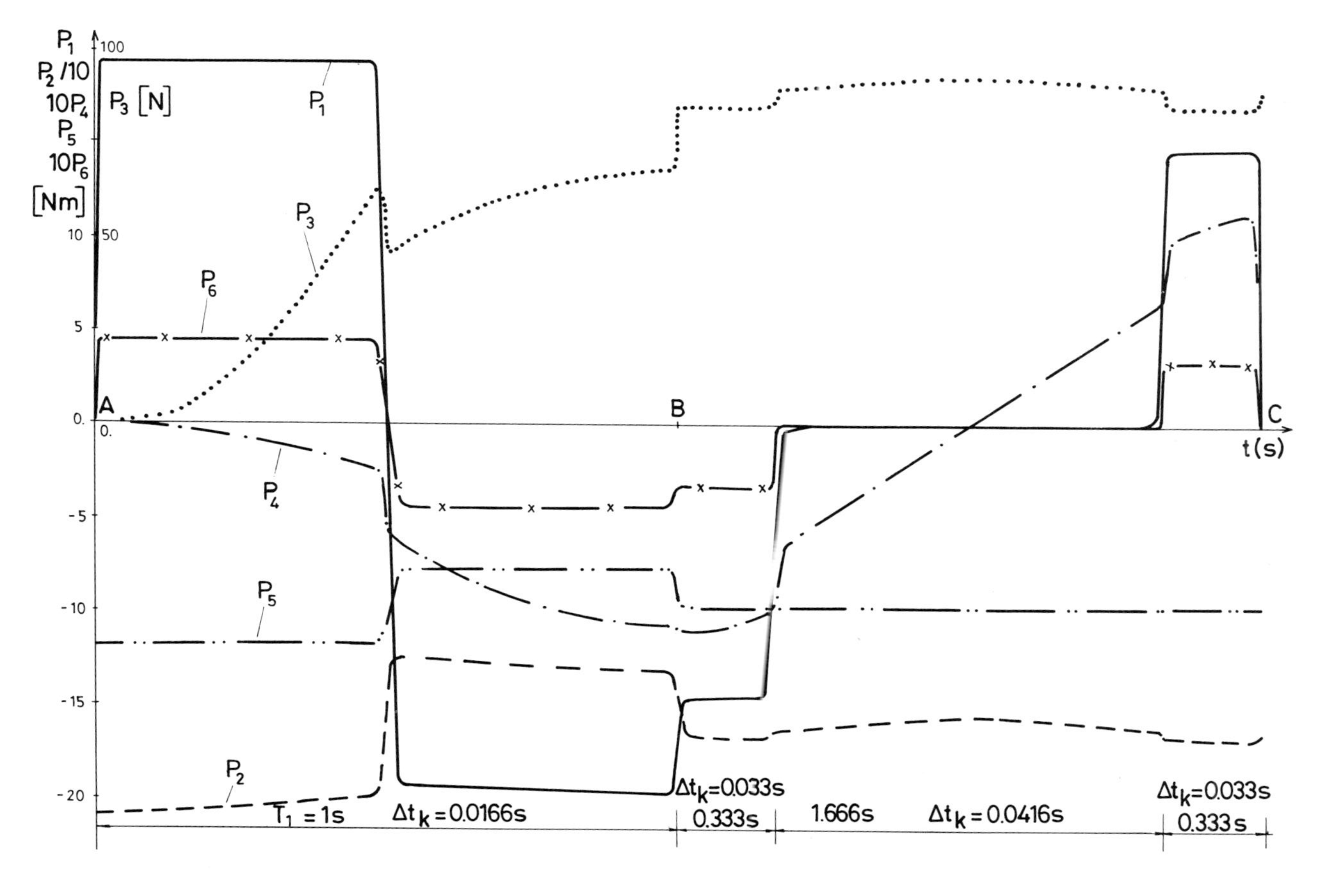

Fig. 5.26. Drives in the joints

in the form of a circular ring (tubular guide) with the ratio of internal to external radius $\psi = \frac{r}{R} = 0.8$.

For such a cross-section, by means of a series of simulations, the optimal values of $R = 0.025$ m, $r = 0.02$ m were obtained. The third segment mass is then $m_3 = 15.72$ kg and energy consumption is $E_{AB} = 82.46$ J and $E_{BC} = 10.10$ J, which evidently represents a significant improvement.

We note that, in considering the deflection, only the quasi static deflection due to statical and nominal inertial forces was taken into account.

It should be mentioned that in the choice of the cross-section and of its dimensions, care must be taken about the available standard forms and dimensions. For this reason, after obtaining the "optimal" dimensions (demonstrated in the last example), comparison is made with the existing standards and the closest greater values are adopted. On account of this and other design limitations we cannot speak about some unconditionally optimal parameters of the manipulation mechanism, calculated as above. On the other hand, these limitations present no obstacle to the design of manipulation robots on the basis of their exact dynamical models.

Fig. 5.25. illustrates the time history of the generalized coordinates and Fig. 5.26. illustrates the drives in the manipulator joints for such an optimal case. Note that the time axis scale is not constant because the value of time step Δt was changed for the sake of precision.

Example 3

In this example we consider a spot-welding manipulator of arthropoid kinematical scheme (Fig. 5.27). The design goal is the dimensioning of the tubular segments (2. and 3. in Fig. 5.28) from the standpoint of the work speed criterion.

Manipulator configuration. The kinematical scheme and the segment lengths are given in Fig. 5.28. Masses and moments of inertia are given in Table Fig. 5.29.

Segments No. 2 and 3 are in the form of circular tubes with constant internal to external radius ratio $\frac{r}{R} = 0.9$. Thus optimal dimensioning reduces to the choice of the external radius R, permitting maximal work speed, i.e. minimal test-task execution time.

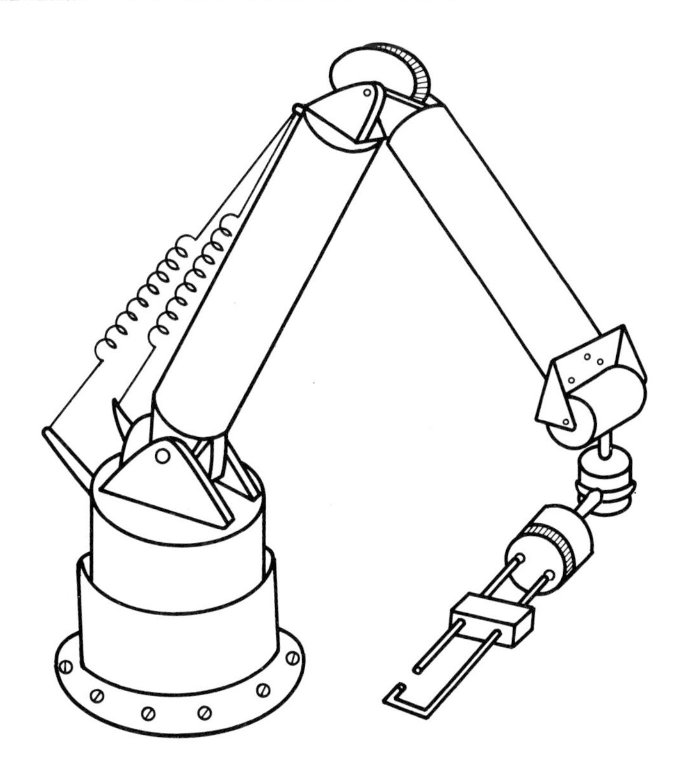

Fig. 5.27. Arthropoid spot-welding manipulator

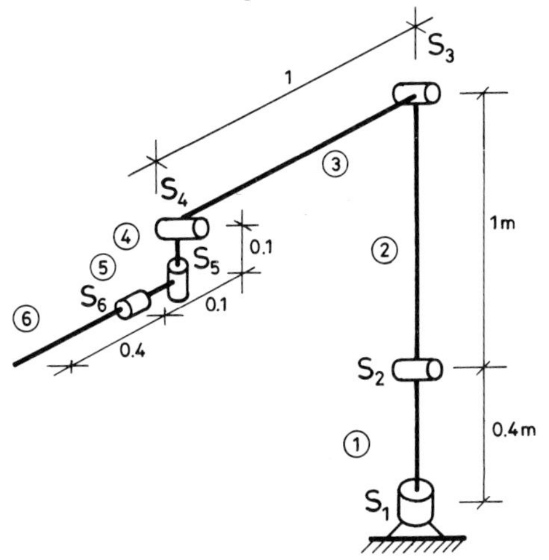

Fig. 5.28. Manipulator kinematical scheme

Segment	m_i [kg]	$I_{i_{xi}}$ [kgm^2]	$I_{i_{yi}}$	$I_{i_{zi}}$
1				0.68
4	4	0.02	0.02	0.02
5	4	0.02	0.02	0.02
6	12	0.16	0.16	0.05

Fig. 5.29. Masses and moments of inertia (empty spaces denote insignificant data)

Material for the segments 2 and 3 is the light alloy AℓMg3. Data about this material are given in the table in Fig. 5.10.

The drives of the joints S_2 and S_3 are in the form of so-called disc-type D.C. servomotors with reducers. Maximal output torque is 620 Nm, maximal output r.p.m. is 36 r.p.m. The mass of one motor is 7 kg. The manipulator is equiped with spring-loaded weight compensators, as shown in Fig. 5.27.

Test-task. The manipulator was required to move along the path ABCA illustrated in Fig. 5.30. The initial manipulator position is given in Fig. 5.28. For reasons of clarity, Fig. 5.30 illustrates the gripper and its orientation in positions A, B, C only. Total task execution time is denoted by T. Let the movement AB be performed in the time interval $T_1 = T/4$, movement BC in $T_2 = T/2$, and CA in $T_3 = T/4$. All changes of position and orientation are executed with a triangular velocity profile.

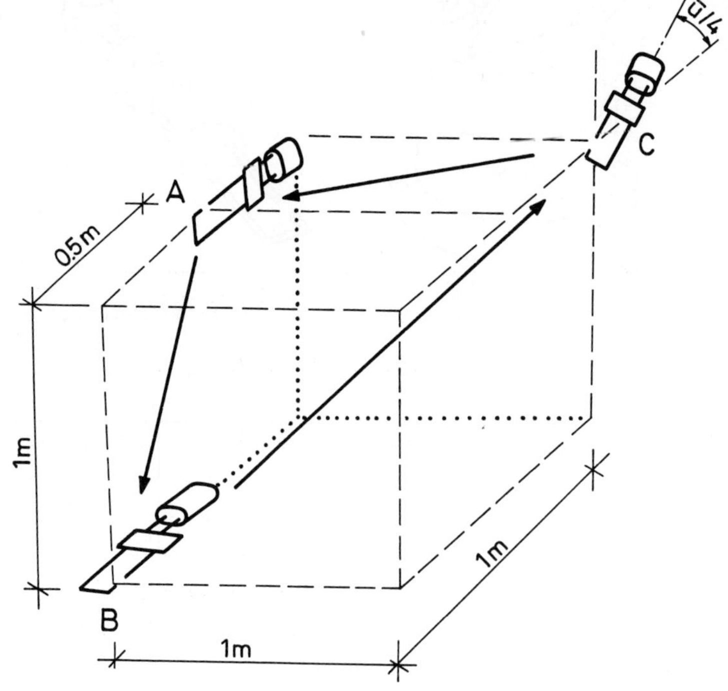

Fig. 5.30. Manipulation test-task

Dimensioning of the cross-sections of segments 2 and 3 was based on the velocity (i.e. time) criterion as in the procedure in Example 1 of this chapter. Fig. 5.31 illustrates the dependence of $T_{min}(R)$, i.e. the minimal task execution time, on the radius R. Note the drive limitation (iii) and elastic deformation limitation (iv). The intersection point of these limitations yields optimum-minimum with coordinates (R^T_{opt}, T^{abs}_{min}). R^T_{opt} is the optimal value of the radius (from the viewpoint of

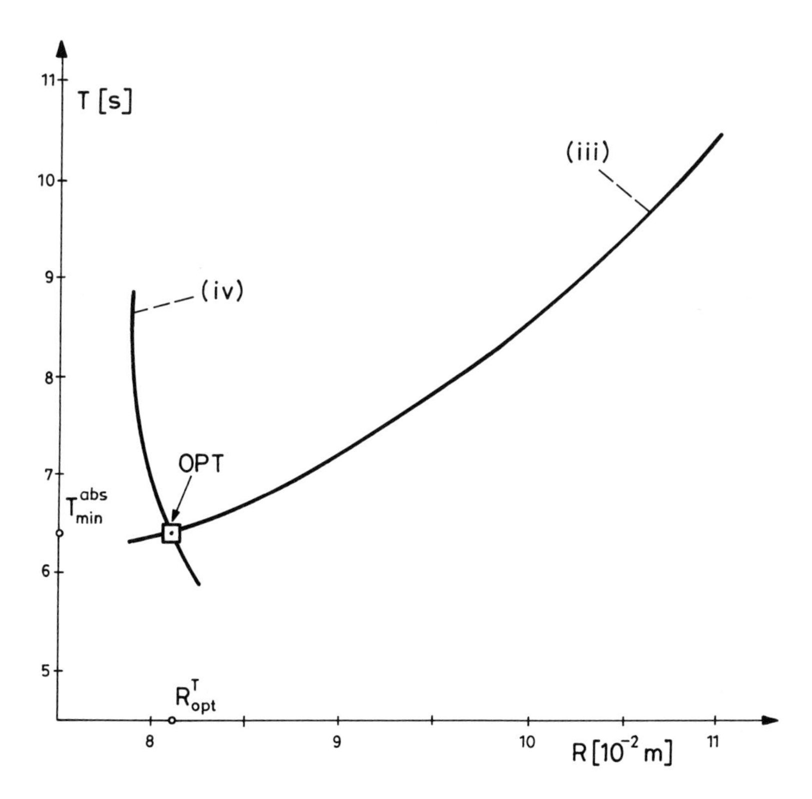

Fig. 5.31. Dependence of $T_{min}(R)$

the time criterion) and T^{abs}_{min} is the corresponding absolute time minimum for the execution of the task. It was found that R^T_{opt} = 0.081 m and T^{abs}_{min} = 6.4 s.

Figs 5.32 and 5.33 illustrate the simulation diagrams for the case of the optimal cross-section and minimal time. Fig. 5.32 illustrates the time function of the drives in the mechanism joints and Fig. 3.33 illustrates likewise the generalized coordinates. The time increment Δt is constant for the whole movement.

Fig. 5.34 shows the P-r.p.m. (P is the torque) for the motor in joint 2. Straight lines represent the motor characteristics.

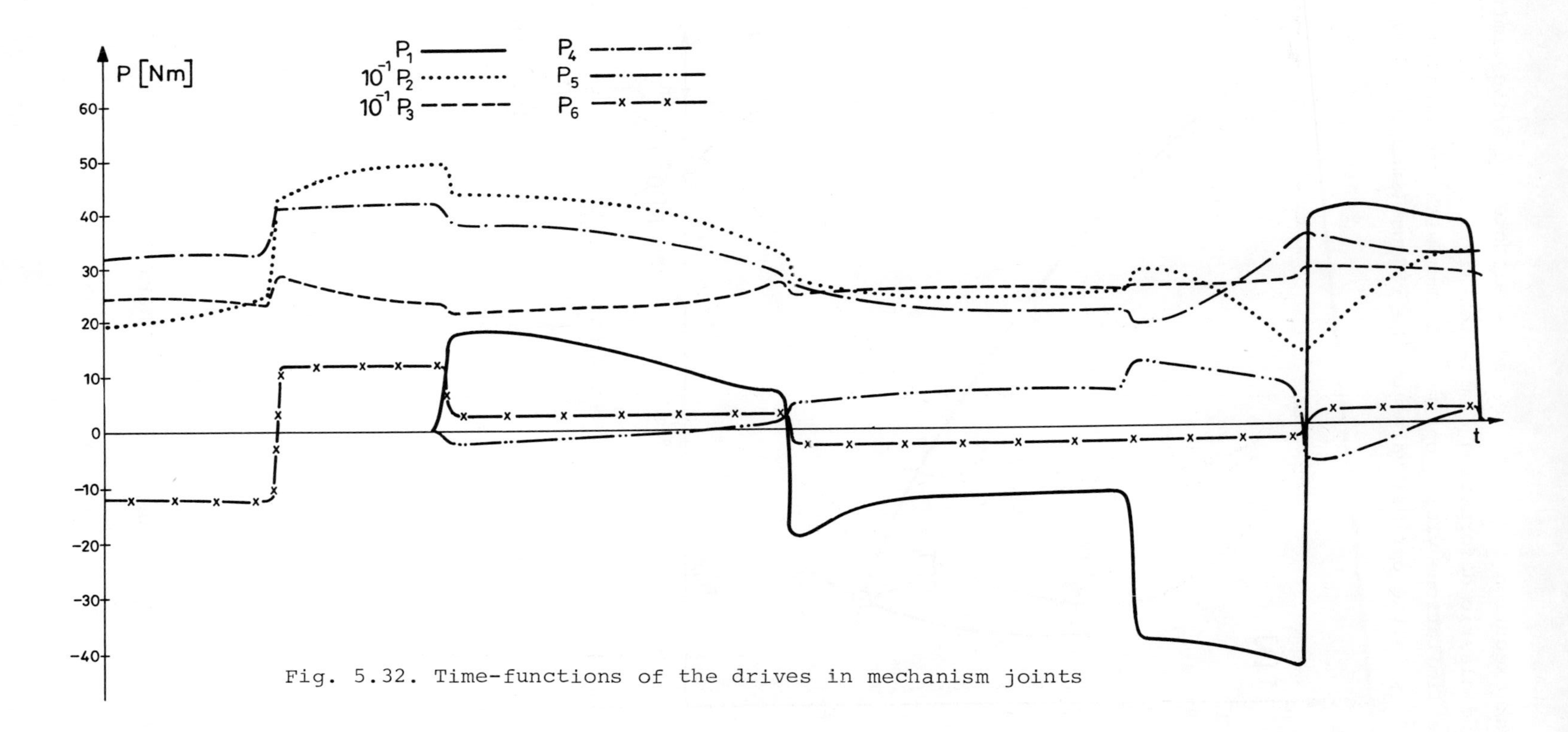

Fig. 5.32. Time-functions of the drives in mechanism joints

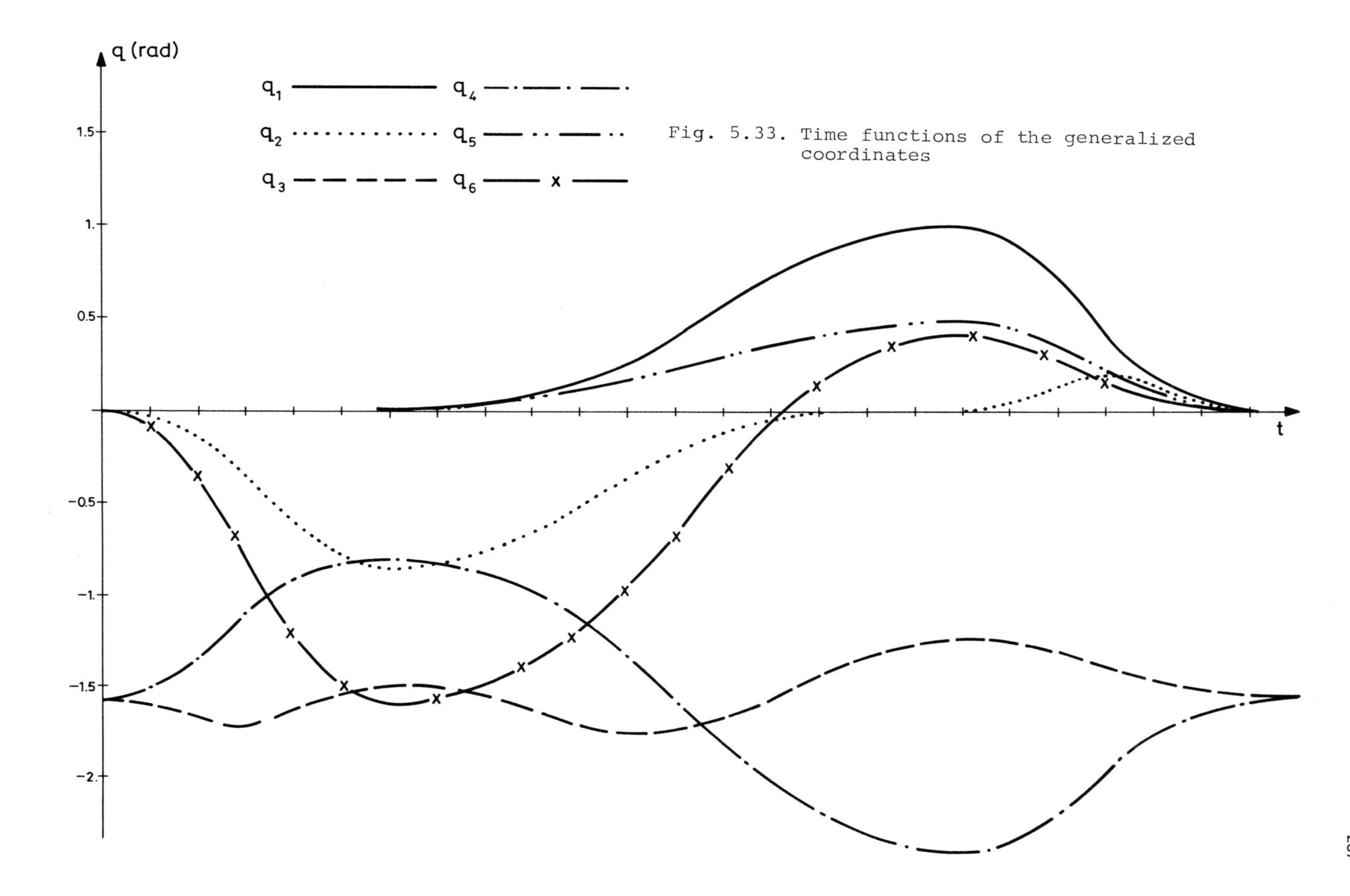

Fig. 5.33. Time functions of the generalized coordinates

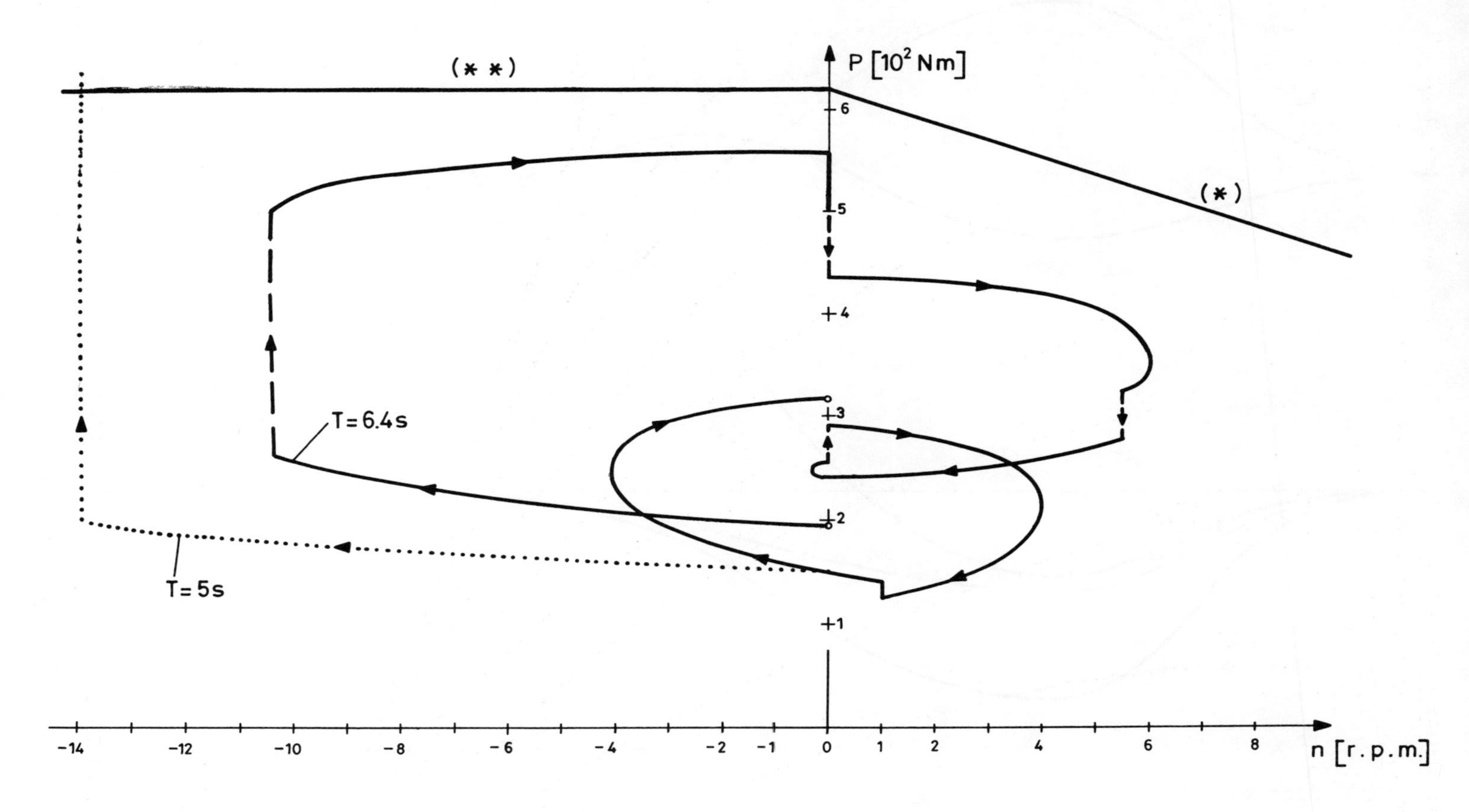

Fig. 5.34. Dependence torque - r.p.m. for joint 2

<u>Selecting the best reduction ratio.</u> In this example we have used the disc-type D.C. servomotors with reducers. The reduction ratio of the

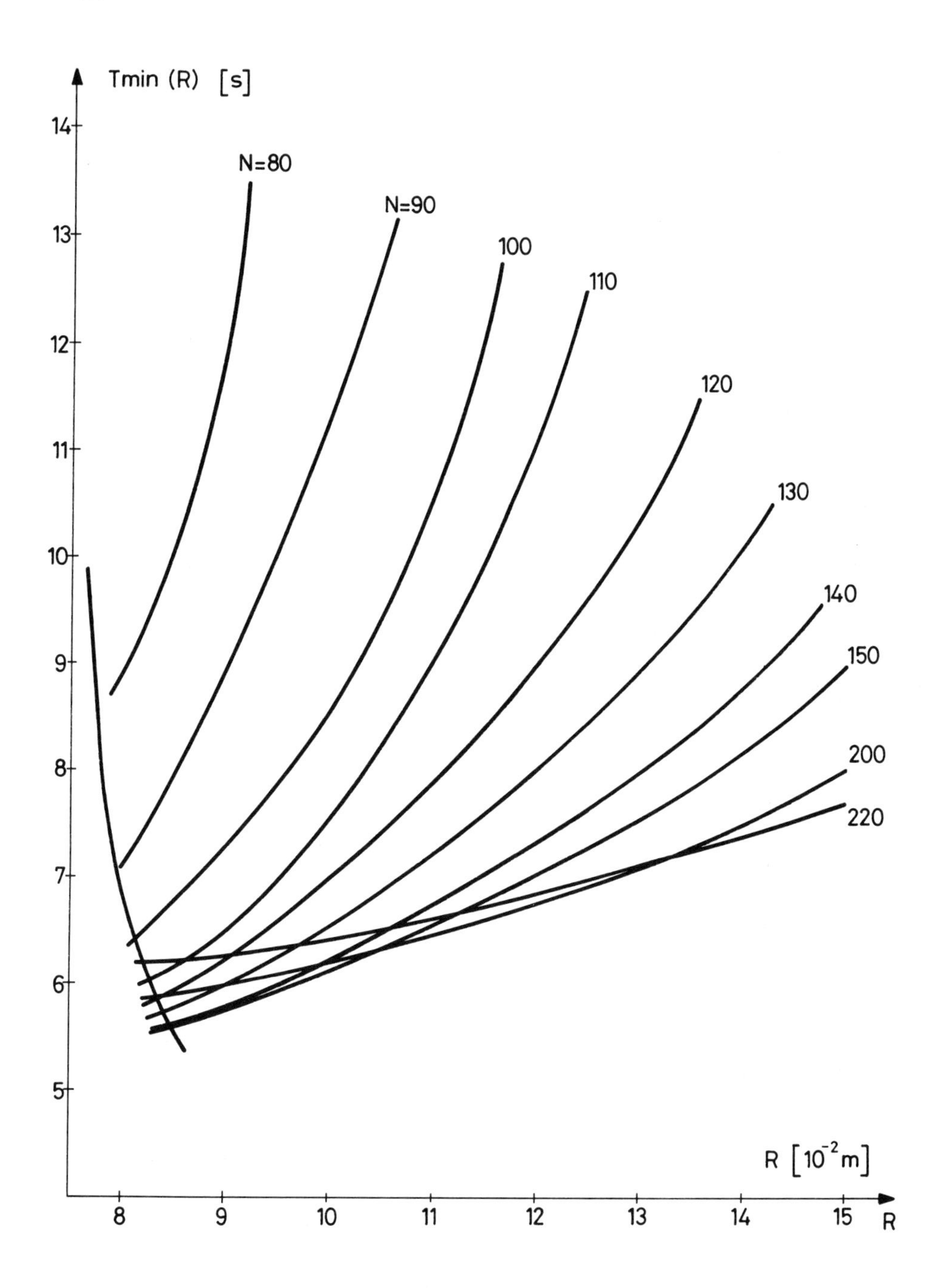

Fig. 5.35. Dependence of $T_{min}(R)$ for different values of N

reducer was N = 100. So, the maximal motor torque is ca 10Nm and the maximal r.p.m. is 3600. Now, it will be shown that this value of reduction ratio is not the best one. The influence of the reduction ratio will be discussed and the optimal value determined which allows the shortest execution time i.e. the greatest operation speed. In Fig. 5.35. there are presented the curves $T_{min}(R)$ for different values of reduction ratio N. It is evident that the optimal value of N depends on the chosen value of R. Fig. 5.36. represents the minimal execution time depending on the ratio N. The curves are recorded for two values or radius R. From Figs. 5.35. and 5.36. one can conclude that with increase of the radius R (heavier segments) the minimum of time with respect to ratio N moves towards greater values of N.

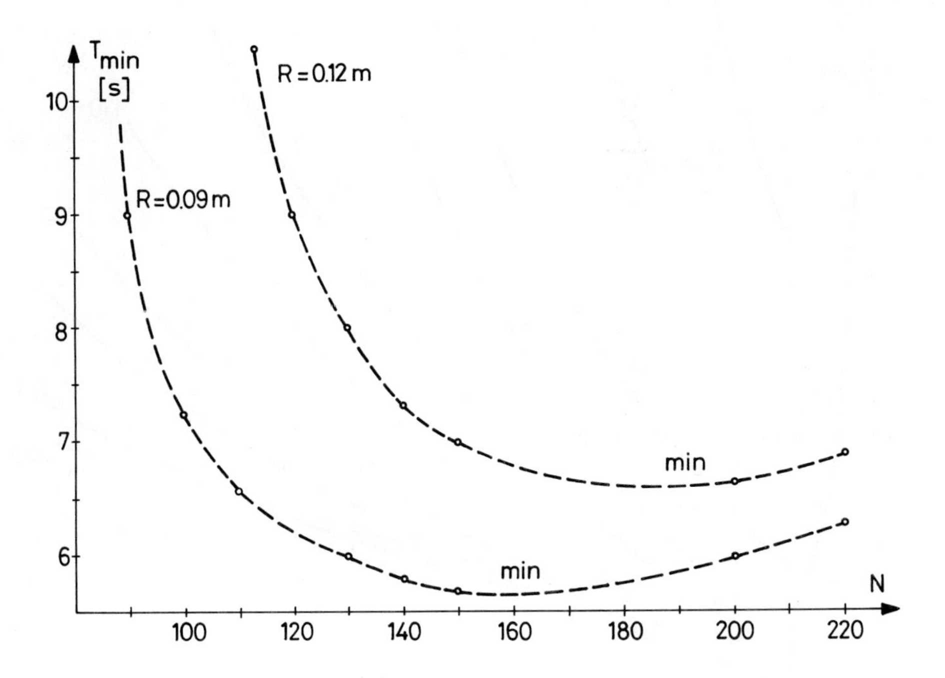

Fig. 5.36. Dependence of $T_{min}(N)$ for two values of R

For the optimal radius R^{T}_{opt} = 0.084m the best reduction ratio is about 160 and the corresponding minimum of execution time is T^{abs}_{min} = 5.5s which is considerably better then the one previously determined. This illustrative example shows a systematic approach to the choice of the reduction ratio from the aspect of maximal operation speed. Such a systematic choice allows an optimal utilization of actuator.

Two-parameter optimization. In the previous discussion it was assumed that the cross-sections of segments 2 and 3 were equal. Thus, we dealt with one-parameter optimization problem. This assumption was introduced in order to simplify the problem. It should be said that with some configurations such an assumption is not necessary because one-parameter optimization follows directly from the configuration itself. This is the case with cylindrical and spherical manipulator configurations because there is only one main segment which forms the manipulator arm. The introduction of the reduction ratio as another variable parameter did not change essentially the optimization procedure which became a series of one-dimensional optimizations.

Now, we will present real two-parameter optimization. The assumption that the cross-sections of segments 2 and 3 are equal is not necessary and surely does not lead to true optimal dimensions. So, we now consider segments 2 and 3 as having different and independent cross-sections with outer radii R_2 and R_3. Thus, in determining the optimal dimensions, there are two variable parameters R_2 and R_3 and two-parameter optimization is necessary.

The optimization is performed on the basis of energy criterion. We adopt the reduction ratio $N = 150$ and the execution time $T = 6s$. This problem is very complicated for optimization because there is no explicit function of the criterion and no explicit expressions for the limitations. Both the value of the criterion and the answer as to whether a point is feasible or not follow from the simulation algorithm. This fact restricts to a great extent the possibility of selecting the optimization method among those given in literature. Here we use one variant of the feasible directions method [25]. This method is almost the only one that suits the problem considered. The method is illustrated in Fig. 5.37.

Let us describe the optimization method. The procedure starts at a feasible point A (Fig. 5.37). Four probes are made around the point A. We locate the improved feasible point B and accept it. We proceed in the same manner until we reach point D. At D, no probe produces an improved feasible point. We then choose a new point E by interpolating between the best feasible point H and the best nonfeasible point G. Such a procedure leads us towards the optimum (quadratic point in Fig. 5.37).

In the example considered the starting point was $R_2=R_3=0.082m$. By applying the method, the optimal values of R_2 and R_3 were obtained: R_2= 0.086m; R_3 = 0.064m.

It should be stressed that there are no problems in expanding this procedure and using it for three-dimensional or multidimensional optimization.

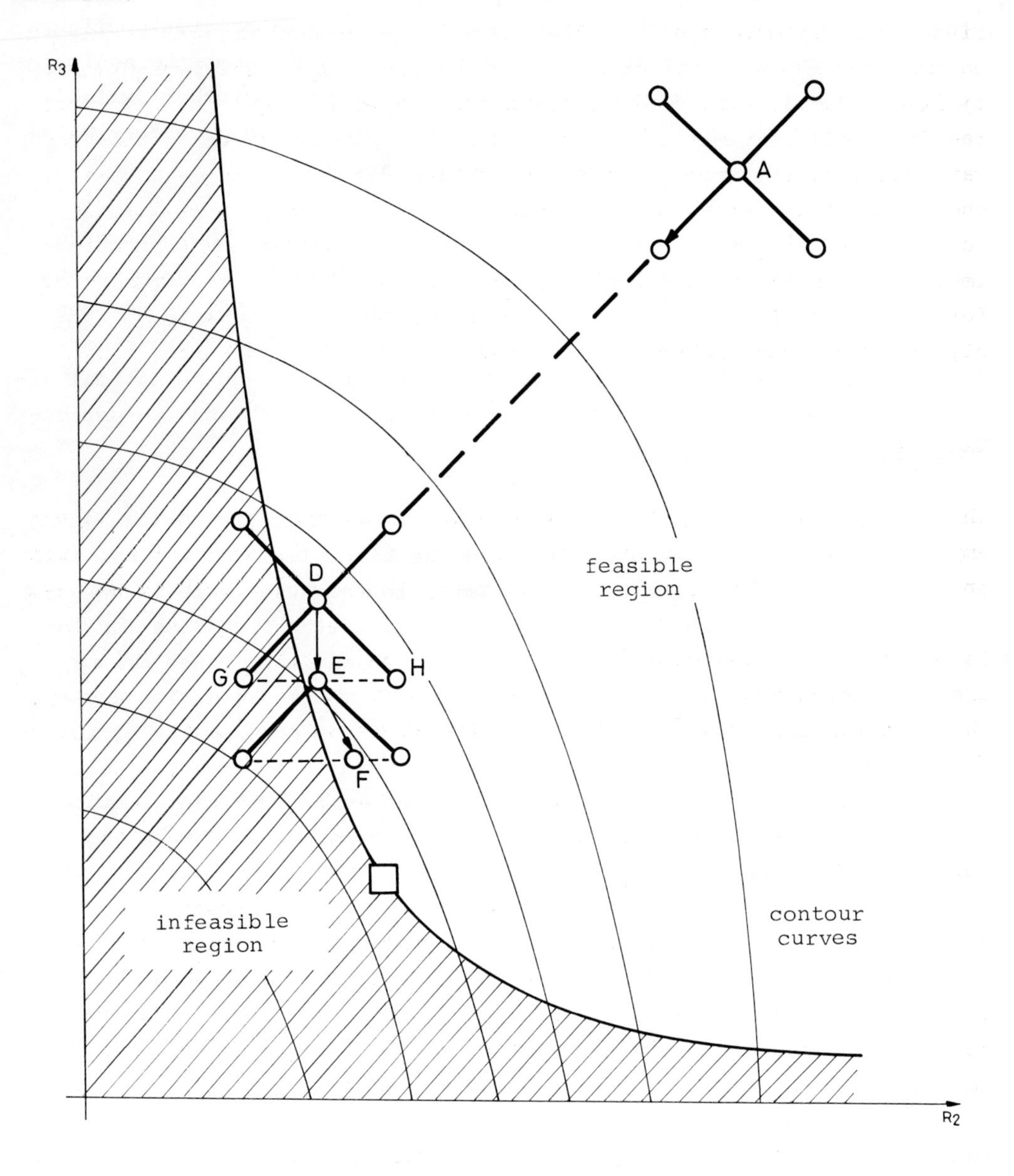

Fig. 5.37. Method of feasible directions

It should be mentioned that the actuators in contemporary manipulator designs because of the important dynamic influence of their elevated weight, particularly with larger manipulator, are most frequently removed from the joints and placed near to or in the central columns i.e. the manipulator base.

Finally we should mention that the problem of the elaboration of the criterion for the choice and evaluation of manipulation robots still needs investigating. Widely applicable results are needed to enable designers to resolve their dilemmas in the choice of the correct systems. This remark particularly concerns the use of the combined criteria in the choice and evaluation, because this represents a compromise solution for a number of cases in industrial practice. However, what is most important, is the methodology of a general simulation algorithm which, for the first time, incorporates modern concepts in the design of manipulation robots and uses their exact mathematical dynamical models.

Example 4.

This example deals with the combined criterion. The purpose of the example is only to demonstrate some properties of the combined criterion and to contribute to better understanding of the criterion and its use.

As already said, the combined criterion is especially convenient for the evaluation and comparison of manipulators in the case of choosing the most suitable device from those offered by manufacturers.

In order to use the combined criterion, one has to prescribe some characteristic values depending on manipulation tasks which should be performed by the device in practical operation. These characteristic values are: T_o, T_1, E_o, E_1, w_t, w_e, r, and they are defined in paragraph 5.2. Let us consider a manipulation task for which it holds: $T_o = 10$ s, $T_1 = 5$ s, $E_o = 100$ J, $E_1 = 50$ J. As this is a theoretical example given to illustrate some properties of the criterion, we shall not restrict our discussion to some prescribed values of w_t, w_e, and r, but we shall discuss the results for various values of w_t, w_e, r.

Let us assume that six manipulators are offered, and we have to choose the most suitable one according to the combined criterion. For each manipulator, the maximal velocity (minimal time) and energy consumption are defined by the values T and E (paragraph 5.2). These values refer to the considered manipulation task or a set of test-tasks. For each value T and E, the corresponding elementary efficiencies G_t and G_e can be computed (Fig. 5.1).

The values of T, E, G_t, G_e as well as the difference $\Delta G = G_t - G_e$, for each considered manipulator, are shown in the table Fig. 5.38.

Manipulator	T	E	G_t	G_e	$\Delta G=G_t-G_e$
1	5.5	95.	0.9	0.1	0.8
2	6.	90.	0.8	0.2	0.6
3	7.	80.	0.6	0.4	0.2
4	7.5	75.	0.5	0.5	0.
5	8.5	65.	0.3	0.7	-0.4
6	9.9	51.	0.02	0.98	-0.96

Fig. 5.38. Velocities (times) and energy consumptions

Now, the total efficiency G for each manipulator is computed via (5.2.2) or (5.2.4). This final result is computed for various values of weighting factors w_t, w_e, and for various values of averaging factor r.

Some of the results and analyses are presented in Fig. 5.39. These curves represent the total efficiency depending on the conjunction degree c (i.e. depending on the factor r). The curves are computed for all six manipulators and they hold for $w_t = w_e = 0.5$. It is clear that the total efficiency decreases for stronger conjunction, i.e., when c increases. It can also be concluded that this decrease is more expressed for the manipulators having a larger difference $|\Delta G|$ between the elementary efficiencies. If these two efficiencies are equal, then the total efficiency G does not depend on c. This conclusion quite agrees with the earlier statement (paragraph 5.2) that for stronger conjunction the smaller of the two elementary efficiencies becomes predominant.

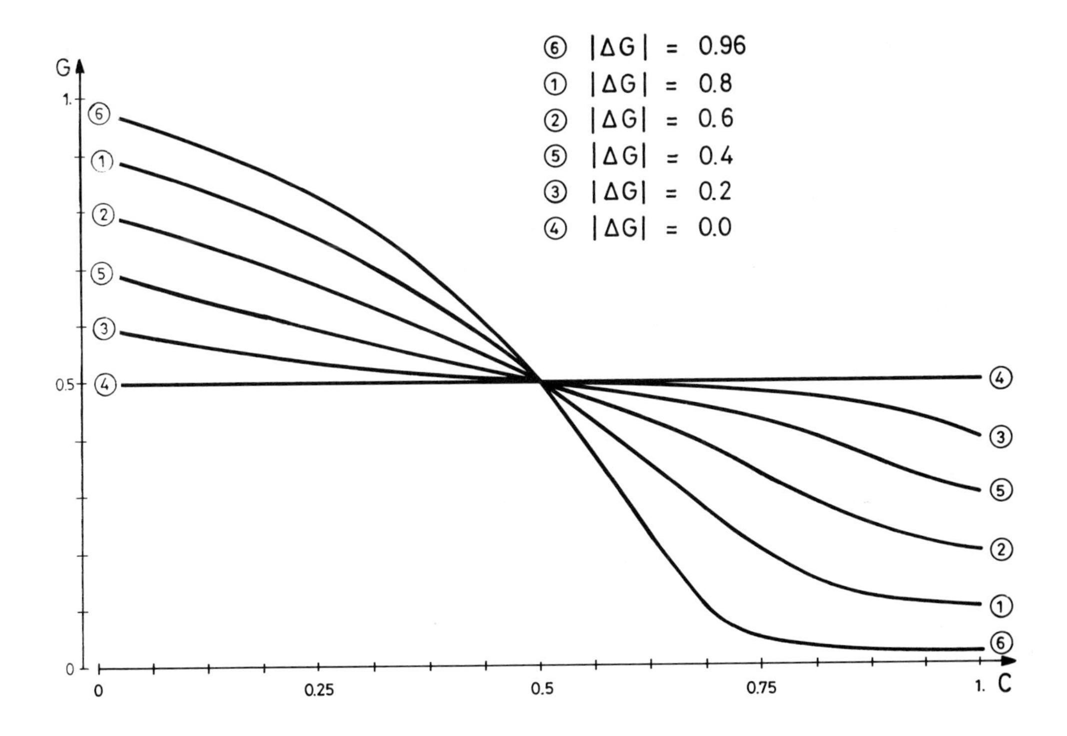

Fig. 5.39. Total efficiency depending on conjunction degree c

Let us now discuss the influence of weighting factors w_t and w_e. For manipulator 1., the characteristics G - c are shown in Fig. 5.40. They are presented for various values of factors w_t and w_e. Manipulator 1 has a high time efficiency $G_t = 0.9$ and a low energy efficiency $G_e=0.1$. So, if $w_t > w_e$ then the total efficiency G keeps a high value even for a strong conjunction. But, if $w_t < w_e$ then G decreases fast.

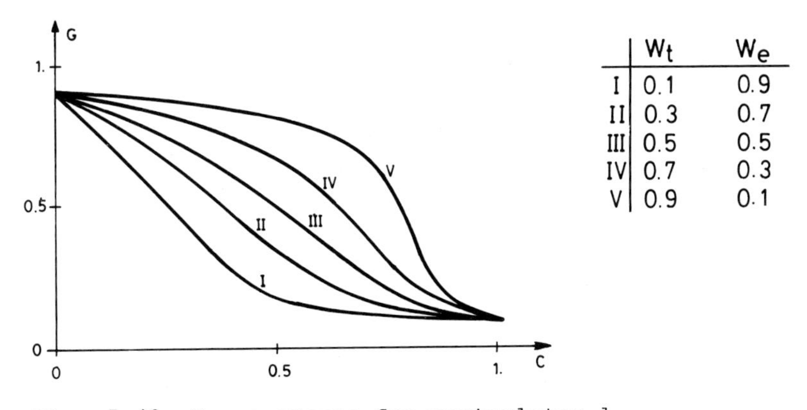

	w_t	w_e
I	0.1	0.9
II	0.3	0.7
III	0.5	0.5
IV	0.7	0.3
V	0.9	0.1

Fig. 5.40. G - c curves for manipulator 1

It should now be emphasized that the computation of the averaging function (5.2.2) is not a simple task. If we just write this formula in some programming lanquage, it will result in an unpermissible error. So, a special subroutine has been developed for precise computation of the averaging function.

In this paragraph examples of optimization have been presented.One may note that in all the examples only one parameter optimization has been performed. It should be emphasized that this fact does not make the examples theoretical, but they still remain real problems from manipulator design. Let us explain it. Numerous parameters have to be determined during the design of a manipulator mechanism. The values for most of them follow directly from the adopted design solutions. Among the remaining ones, the lengths of segments are determined by the reachability conditions and so they should not be optimized. Hence, the only parameters to be optimized are usually some cross-section parameters. In previous examples the form of the cross-section was adopted as a circular tube with constant ratio ψ ($\psi = r/R$), and so, only optimization of the external radius was performed. But, if we adopt the rectangular cross-section, or assume that the cross-sections of the two segments may be different, then the multiparameter optimization problem arises. There are also some other parameters which may be optimized; for instance the reduction ratio in the manipulator joints. If a complete dynamical model (including the actuators) is considered then some variable parameters of actuators should be discussed too. As the whole simulation procedure has to be carried out in each point of parameter space, it is desirable to reduce the number of variable parameters and find the most essential ones. Usually this number reduces to two or three. We also mention the possibility of taking the cross section moments of inertia as variable parameters. Such an approach allows us not to adopt the cross-section form arbitrarily, but on the basis of optimization results.

CONCLUSION. This monograph was aimed at a detailed treatment of the dynamics of active mechanisms, particularly of open configuration mechanisms, applied by manipulation robots. Firstly, all methods of the formation of mathematical models of the open active spatial mechanisms are computer-oriented, so one can speak about automatic formation of the mathematical models of such mechanisms. The methods presented in this book possess broad possibilities concerning the modifications of the manipulation mechanism configurations, their parameters, and the types of the kinematic chain active joints. All these possibilities which involve simple variation of all parameters and are relevant to the various types of manipulation mechanism and the automatic processing of their models on digital computers, have reduced the very complex problem of the mathematical modelling of system dynamics to the level of simulation routine. Such a modelling basis also has a positive feedback into the synthesis of functional movements and the synthesis of control in general. In particular, by adopting the concept of two-stage control synthesis, the role of the centralized model of the manipulation mechanism acquires much more dynamical significance because by its use at the basis of the algorithm for the synthesizing program kinematics one obtains the required control signals of the nominal (programmed) manipulator working state. Thus, such dynamical models have been able to be used as a source of information about the system dynamics in the synthesis of the dynamical control of manipulators and robots in general. A second very important role for complete dynamical models consists, of course, in their use for advanced design of the manipulation systems themselves. For example, the design of actuator energy requirements (which depend on the set manipulation task), dimensioning without unnecessary reserves while taking into account the exact dynamical loads and the limitations concerning the required rigidity of the manipulation system mechanism.

Finally, dynamical models of the manipulation mechanisms have made possible the correct choice and evaluation of manipulators according to the type of task and type of the working states; the choice being made on the basis of "dynamical" criteria containing at once the various possibilities (limitations) of the chosen actuators and the limitations based on the permissable stress and elasticity of the structure itself. It should be mentioned that the results achieved in this domain can be extended by further application of the combined criterion as well by means of a multi-parameter optimization of the manipulation mechanisms. However, we arrived at a method which uses the dynamical models of spatial mechanisms for modern computer-aided design of manipulators, for the synthesis of functional movements (nominal dynamcis) and finally for the synthesis of control. This will be the subject of the book, dedicated to manipulator control [24].

References

[1] Stepanenko Yu., "Method of Analysis of Spatial Lever Mechanisms", (in Russian), Mekhanika mashin, Vol. 23, Moscow, 1970.

[2] Juričić D., Vukobratović M., "Mathematical Modelling of a Bipedal Walking System", ASME Publ. 72-WA/BHF-13.

[3] Vukobratović M., Stepanenko Yu., "Mathematical Models of General Anthropomorphic Systems", Math. Biosciences, Vol. 17, 1973.

[4] Stepanenko Yu., Vukobratović M., "Dynamics of Articulated Open--Chain Active Mechanisms", Math. Biosciences, Vol. 28, No 1/2, 1976.

[5] Vukobratović M., "Computer Method for Mathematical Modelling of Active Kinematic Chains via Generalized Coordinates", IFToMM Journal of Mechanisms and Machine Theory, Vol. 13, No 1, 1978.

[6] Vukobratović M., Legged Locomotion Robots and Anthropomorphic Mechanisms, Monograph, 1975, Institute "M.Pupin", Beograd, Yugoslavia.

[7] Orin D., Vukobratović M., R.B. Mc Ghee., G.Hartoch, "Kinematic and Kinetic Analysis of Open-Chain Linkages Utilizing Newton--Euler Methods", Math. Biosciences, Vol. 42, 1978.

[8] Vukobratović M., "Synthesis of Artificial Motion", Journal of Mechanism and Machine Theory, Vol. 13, No 1, 1978.

[9] Vukobratović M., Potkonjak V., "Contribution to the Forming of Computer Methods for Automatic Modelling of Active Spatial Mechanisms Motion" PART1: "Method of Basic Theoremes of Mechanics", Journal of Mechanisms and Machine Theory, Vol. 14, No 3, 1979.

[10] Vukobratović M., Potkonjak V., "Contribution to the Computer Methods for Generation of Active Mechanisms via Basic Theoremes of Mechanics" (in Russian), Tehnitcheskaya kibernetika ANUSSR, No 2, 1979.

[11] Luh, J.Y.S., Walker, M.W., Paul R.P.C., "On-Line Computational Scheme for Mechanical Manipulators", Trans. of ASME Journal of Dynamic Systems, Measurement and Control, June, 1980, Vol 102.

[12] Vukobratović M., "Computer Method for Mathematical Modelling of Active Kinematic Chains via Euler´s Angles", IFToMM Journal of Mechanisms and Machine Theory, Vol. 13, No 1, 1978.

[13] Vukobratović M., Potkonjak V., "Contribution to Automatic Forming of Active Chain Models via Lagrangian Form", Journal of Applied Mechanics No 1, 1979.

[14] Hollerbach, J.M., "A Recursive Formulation of Lagrangian Manipulator Dynamics", Proc. of JACC, Aug. 13-15, 1980, San Francisco.

[15] Potkonjak V., Vukobratović M., "Two Methods for Computer Forming of Dynamic Equations of Active Mechanisms", Journal of Mechanism and Machine Theory, Vol. 14, No 3, 1979.

[16] Popov E.P., Vereschagin A.F., Ivkin A.M., Leskov A.G., Medvedov V.S., "Design of Robot Control Using Dynamic Models of Manipulator Devices", (in Russian) Proc. of VI IFAC Symp. on Automatic Control in Space, Erevan, USSR, 1974.

[17] Popov E.P., Vereschagin A.F., Zenkevich S.A., Manipulation Robots: Dynamics and Algorithms, (in Russian), "Nauka", Moscow, 1978.

[18] Vukobratović M., Potkonjak V., Hristić D., "Dynamic Method for the Evaluation and Choice of Industrial Manipulators", Proc 9th International Symposium on Industrial Robots, Washington, 1979.

[19] Potkonjak V., Vukobratović M., "Transformation Blocks in Dynamic Simulation of Typical Manipulator Configurations", Journal of Mechanisms and Machine Theory, Vol. 17 No 3, 1982.

[20] Vukobratović M., Potkonjak V., "Contribution to Computer-Aided Design of Industrial Manipulators Using Their Dynamic Properties", Journal of Mechanisms and Machine Theory, Vol. 17, No 2, 1982.

[21] Ginzburg S.A., Mathematical Continual Logic and Function Presentation, (in Russian), "Energiya" Publ., Moscow, 1968.

[22] Losonczi L., "About a New Class of Medium Values", (in German), Acta Sci. Math., No. 32, Szeged, 1971.

[23] Dujmović J.J., "Weight Conjuctive and Disjunctive Means and Their Application in System Evaluation", Publications of Electrical Engn. Faculty, Series: Mathematics-Physics, 1974, Beograd, Yugoslavia.

[24] Vukobratović M., Stokić D., Synthesis and Control Algorithms of Manipulation Robots, Monograph, Springer-Verlag, 1982.

[25] Gottfried B.S., Weisman J., Introduction to Optimization Theory, Prentice-Hall, 1973.

Subject Index

Lecture Notes in Control and Information Sciences

Editors: A. V. Balakrishnan, M. Thoma

Volume 28

Analysis and Optimization of Systems

Proceedings of the Fourth International Conference on Analysis and Optimization of Systems, Versailles, December 16–19, 1980
Editors: **A. Bensoussan, J. L. Lions**
1980. 167 figures, 23 tables. XIV, 999 pages (206 pages in French). ISBN 3-540-10472-0

Volume 29
M. Vidyasagar

Input-Output Analysis of Large-Scale Interconnected Systems

Decomposition, Well-Posedness and Stability
1981. VI, 221 pages. ISBN 3-540-10501-8

Volume 30

Optimization and Optimal Control

Proceedings of a Conference Held at Oberwolfach, March 16–22, 1980
Editors: **A. Auslender, W. Oettli, J. Stoer**
1981. VIII, 254 pages. ISBN 3-540-10627-8

Volume 31
B. Rustem

Projection Methods in Constrained Optimisation and Applications to Optimal Policy Decisions

1981. XV, 315 pages. ISBN 3-540-10646-4

Volume 32
T. Matsuo

Realization Theory of Continuous-Time Dynamical Systems

1981. VI, 329 pages. ISBN 3-540-10682-0

Volume 33
P. Dransfield

Hydraulic Control Systems – Design and Analysis of Their Dynamics

1981. VII, 227 pages. ISBN 3-540-10890-4

Volume 34
H. W. Knobloch

Higher Order Necessary Conditions in Optimal Control Theory

1981. V, 173 pages. ISBN 3-540-10985-4

Volume 35

Global Modelling

Proceedings of the IFIP-WG 7/1 Working Conference, Dubrovnik, Yugoslavia, September 1–5, 1980
Editor: **S. Krčevinac**
1981. VIII, 232 pages. ISBN 3-540-11037-2

Volume 36

Stochastic Differential Systems

Proceedings of the 3rd IFIP-WG 7/1 Working Conference, Visegrád, Hungary, September 15–20, 1980
Editors: **M. Arató, D. Vermes, A. V. Balakrishnan**
1981. VI, 250 pages. ISBN 3-540-11038-0

Volume 37
R. Schmidt

Advances in Nonlinear Parameter Optimization

1982. VI, 159 pages. ISBN 3-540-11396-7

Volume 38

System Modeling and Optimization

Proceedings of the 10th IFIP Conference, New York City, USA, August 31 to September 4, 1981
Editors: **R. F. Drenick, F. Kozin**
1982. Approx. 910 pages. ISBN 3-540-11691-5

Springer-Verlag
Berlin
Heidelberg
New York